Graph Theory: Researches and Applications

Edited by **Jen Blackwood**

CLANRYE INTERNATIONAL

New Jersey

Published by Clanrye International,
55 Van Reypen Street,
Jersey City, NJ 07306, USA
www.clanryeinternational.com

Graph Theory: Researches and Applications
Edited by Jen Blackwood

© 2015 Clanrye International

International Standard Book Number: 978-1-63240-252-3 (Hardback)

Printed in the United States of America.

Contents

Preface

Graph theory is considered to be a crucial analysis tool in computer science and mathematics. Due to the built-in simplicity of graph theory, it can be employed to model several distinct physical and abstract systems like transportation and communication networks, models for business administration, psychology, and political science and so on. The aim of this book is not just to describe the present state and development tendencies of this theory, but to educate the reader enough to enable him/her to embark on the research complications of their own. Taking into consideration the huge amount of knowledge regarding graph theory and practice are elucidated in this book, it focuses on primary topics of theoretical researches. This book intends to serve as a valuable source of reference for students associated with various fields like system sciences, engineering, social sciences, mathematics, computer sciences, etc. as well as for practitioners and software professionals.

This book is a result of research of several months to collate the most relevant data in the field.

When I was approached with the idea of this book and the proposal to edit it, I was overwhelmed. It gave me an opportunity to reach out to all those who share a common interest with me in this field. I had 3 main parameters for editing this text:

1. Accuracy – The data and information provided in this book should be up-to-date and valuable to the readers.

2. Structure – The data must be presented in a structured format for easy understanding and better grasping of the readers.

3. Universal Approach – This book not only targets students but also experts and innovators in the field, thus my aim was to present topics which are of use to all.

Thus, it took me a couple of months to finish the editing of this book.

I would like to make a special mention of my publisher who considered me worthy of this opportunity and also supported me throughout the editing process. I would also like to thank the editing team at the back-end who extended their help whenever required.

Editor

A Graph Theoretic Approach for Certain Properties of Spectral Null Codes

Khmaies Ouahada and Hendrik C. Ferreira

Department of Electrical and Electronic Engineering Science,
University of Johannesburg, Auckland Park, 2006
South Africa

1. Introduction

In this chapter, we look at the spectral null codes from another angle, using graph theory, where we present a few properties that have been published. The graph theory will help us to understand the structure of spectral null codes and analyze their properties differently.

Graph theory [1]–[2] is becoming increasingly important as it plays a growing role in electrical engineering for example in communication networks and coding theory, and also in the design, analysis and testing of computer programs.

Spectral null codes [3] are codes with nulls in the power spectral density function and they have great importance in certain applications such as transmission systems employing pilot tones for synchronization and track-following servos in digital recording [4]–[5].

Yeh and Parhami [6] introduced the concept of the index-permutation graph model, which is an extension of the Cayley graph model and applied it to the systematic development of communication-efficient interconnection networks. Inspiring the concept of building a relationship between an index and a permutation symbol, we make use in this chapter of the spectral null equations variables in each grouping by representing only their corresponding indices in a permutation sequence form. In another way, these indices will be presented by a permutation sequence, where the symbols refer to the position of the corresponding variables in the spectral null equation.

Presenting a symmetric-permutation codebook graphically, Swart *et al.* [7] allocated states to all symbols of a permutation sequence and presented all possible transpositions between these symbols by links as depicted for a few examples in Fig. 1 [7].

The Chapter is organized as follows: Section II introduces definitions and notations to be used for spectral null codes. Section III presents few graph theory definitions. Section IV presents the index-graphic presentation of spectral null codes. Section V makes an approach between graph theory and spectral null codes where we focus on the relationship between the cardinalities of the spectral null codebooks and the concepts of distances in graph theory and also we elaborate the concept of subgraph and its corresponding to the structure of the spectral null codebooks. We conclude with some final remarks in Section VI.

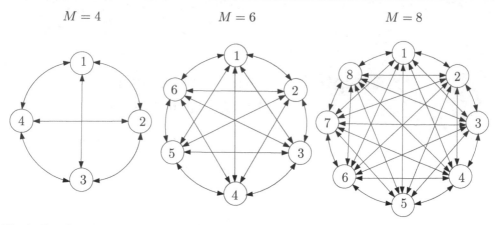

Fig. 1. Graph representation for permutation sequences

2. Spectral null codes

The technique of designing codes to have a spectrum with nulls occurring at certain frequencies, i.e. having the power spectral density (PSD) function equal to zero at these frequencies, started with Gorog [8], when he considered the vector $\mathbf{X} = (x_1, x_2, ..., x_M)$, $x_i \in \{-1, +1\}$ with $1 \leq i \leq M$, to be an element of a set S, which is called a codebook of codewords with elements in $\{-1, +1\}$. We investigate codewords of length, M, as an integer multiple of N, thus let

$$M = Nz,$$

where N represents the number of groupings in the spectral null equation and z represents the number of elements in each grouping. The values of $f = r/N$ are frequencies at spectral nulls (SN) at the rational submultiples r/N [9]. To ensure the presence of these nulls in the continuous component at the spectrum, it is sufficient to satisfy the following spectral null equation [10],

$$A_1 = A_2 = \cdots = A_N, \tag{1}$$

where

$$A_i = \sum_{\lambda=0}^{z-1} x_{i+\lambda N}, \quad i = 1, 2, \ldots, N, \tag{2}$$

which can also be presented differently as,

$$
\begin{aligned}
&\overbrace{\phantom{A_1 = x_1 + x_{1+N} + x_{1+2N} + x_{1+3N} + \cdots + x_{1+(z-1)N}}}^{z} \\
A_1 &= x_1 + x_{1+N} + x_{1+2N} + x_{1+3N} + \cdots + x_{1+(z-1)N} \\
A_2 &= x_2 + x_{2+N} + x_{2+2N} + x_{2+3N} + \cdots + x_{2+(z-1)N} \\
A_3 &= x_3 + x_{3+N} + x_{3+2N} + x_{3+3N} + \cdots + x_{3+(z-1)N} \\
&\;\vdots \qquad\;\; \vdots \qquad\;\;\; \vdots \qquad\;\;\; \vdots \qquad\qquad \vdots \\
A_N &= x_N + x_{2N} + x_{3N} + x_{4N} + \cdots + x_{zN}.
\end{aligned}
\tag{3}
$$

If all the codewords in a codebook satisfy these equations, the codebook will exhibit nulls at the required frequencies. henceforth we present the channel symbol -1 with binary symbol 0.

Definition 2.1. *A spectral null binary block code of length M is a subset $C_b(M, N) \subseteq \{0,1\}^M$ of all binary M-tuples of length M which have spectral nulls at the rational submultiples of the symbol frequency $1/N$.*

Definition 2.2. *The spectral null binary codebook $C_b(M, N)$ is a subset of the M dimensional vector space $(\mathbb{F}_2)^M$ of all binary M-tuples, where $\mathbb{F}_2$ is the finite field with two elements, whose arithmetic rules are those of mod-2 arithmetic.*

For codewords of length M consisting of N interleaved subwords of length z, the cardinality of the codebook $C_b(M, N)$ for the case where N is a prime number is presented by the following formula [10],

$$|C_b(M, N)| = \sum_{i=0}^{M/N} \binom{M/N}{i}^N, \tag{4}$$

where $\binom{M/N}{i}$ denotes the combinatorial coefficient $\frac{(M/N)!}{i!(M/N-i)!}$.

Example 2.3. *If we consider the case of $M = 6$, we can predict two types of spectral with different nulls since N can take the value of $N = 2$ or $N = 3$. Their corresponding spectral null equations are presented respectively as follows:*

$$x_1 + x_3 + x_5 = x_2 + x_4 + x_6 \tag{5}$$

$$x_1 + x_4 = x_2 + x_5 = x_3 + x_6 \tag{6}$$

The corresponding codebooks for (5) and (6) are respectively as follows:

$$C_b(6,2) = \begin{Bmatrix} 0\,0\,0\,0\,0\,0 \\ 0\,0\,0\,0\,1\,1 \\ 0\,0\,0\,1\,1\,0 \\ 0\,0\,1\,0\,0\,1 \\ 0\,0\,1\,1\,0\,0 \\ 0\,0\,1\,1\,1\,1 \\ 0\,1\,0\,0\,1\,0 \\ 0\,1\,1\,0\,0\,0 \\ 0\,1\,1\,0\,1\,1 \\ 0\,1\,1\,1\,1\,0 \\ 1\,0\,0\,0\,0\,1 \\ 1\,0\,0\,1\,0\,0 \\ 1\,0\,0\,1\,1\,1 \\ 1\,0\,1\,1\,0\,1 \\ 1\,1\,0\,0\,0\,0 \\ 1\,1\,0\,0\,1\,1 \\ 1\,1\,0\,1\,1\,0 \\ 1\,1\,1\,0\,0\,1 \\ 1\,1\,1\,1\,0\,0 \\ 1\,1\,1\,1\,1\,1 \end{Bmatrix},$$

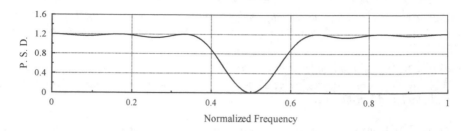

Fig. 2. Power spectral density of codebook $N = 2, M = 6$.

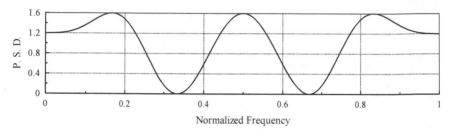

Fig. 3. Power spectral density of codebook $N = 3, M = 6$.

and

$$\mathcal{C}_b(6,3) = \begin{Bmatrix} 0\,0\,0\,0\,0\,0 \\ 0\,0\,0\,1\,1\,1 \\ 0\,0\,1\,1\,1\,0 \\ 0\,1\,0\,1\,0\,1 \\ 0\,1\,1\,1\,0\,0 \\ 1\,0\,0\,0\,1\,1 \\ 1\,0\,1\,0\,1\,0 \\ 1\,1\,0\,0\,0\,1 \\ 1\,1\,1\,0\,0\,0 \\ 1\,1\,1\,1\,1\,1 \end{Bmatrix}.$$

The cardinalities of $\mathcal{C}_b(6,2)$ and $\mathcal{C}_b(6,3)$ are respectively equal to 20 and 10. This also can be easily verified from (4).

We can see clearly the power spectral density $\mathcal{C}_b(6,2)$ and $\mathcal{C}_b(6,3)$ respectively presented in Figures 2 and 3 where the nulls appear to be multiple of $1/N$ as presented in Definition 2.1.

3. Graph theory: Preliminary

We present a brief overview of related definitions for certain graph theory fundamentals which will be used in the following sections.

Definition 3.1. *[1]–[2]*

(a) *A graph G = (V, E) is a mathematical structure consisting of two finite sets V and E. The elements of V are called vertices, and the elements of E are called edges. Each edge has a set of one or two vertices associated with it.*

(b) *A graph $G' = (V', E')$ is a subgraph of another graph $G = (V, E)$ iff $V' \subseteq V$ and $E' \subseteq E$.*

Definition 3.2. [1]–[2] *The graph distance denoted by $G_d(u,v)$ between two vertices u and v of a finite graph is the minimum length of the paths connecting them.*

Definition 3.3. [1]–[2] *The adjacency matrix of a graph is an $M \times M$ matrix $\mathcal{A}_d = [a_{i,j}]$ in which the entry $a_{i,j} = 1$ if there is an edge from vertex i to vertex j and is 0 if there is no edge from vertex i to vertex j.*

4. Index-graphic presentation of spectral null codes

The idea of the index-graphic presentation of the spectral null codes is actually based on the presentation of the indices of the variables in each grouping of the spectral null equation (1).

Definition 4.1. *We denote by $I_p(i, \lambda)$ the permutation symbol of the corresponding index of the variable $x_{i+\lambda N}$ in (2).*

$$I_p(i,\lambda) = i + \lambda N \quad where \quad \begin{cases} i = 1, 2, \ldots, N, \\ \lambda = 0, 1, \ldots, z - 1. \end{cases} \tag{7}$$

Definition 4.2. *We denote by $\mathcal{P}_{I_p}(M, N)$ the index-permutation sequence from a spectral null equation for variables of length $M = Nz$ as presented.*

$$\mathcal{P}_{I_p}(M, N) = \prod_{i=1}^{N} \prod_{\lambda=0}^{z-1} I_p(i, \lambda). \tag{8}$$

The product sign in (8) is not used in its traditional way, but just to give an idea about the sequence and the order of the permutation symbols.

Example 4.3. *To explain the relationship between the spectral nulls equation, the index-permutation sequences and their graph presentation, we take the case of $M = 4$ where we have only two groupings since $N = 2$.*

$$A_1 = A_2 \rightarrow x_1 + x_3 = x_2 + x_4 \tag{9}$$

We can see from (9), that the indices of the variables x_i, using (8), are represented by the symbols $I_p(1, 0) = 1$, $I_p(1, 1) = 3$, $I_p(2, 0) = 2$ and $I_p(2, 1) = 4$. The index-permutation sequence is then $\mathcal{P}_{I_p}(4, 2) = (13)(24)$.

An index-permutation symbol is presented graphically by just being lying on a circle, which it is called a state. The state design follow the order of appearance of the indices in (9). The symbols are connected in respect of the addition property of their corresponding variables in (9) as depicted in Fig. 4.

Spectral null codebooks have the all-zeros and all-ones codewords [10], where all the variables y_i are equal. We call the corresponding spectral null equation, which is $x_1 = x_2 = x_3 = x_4$ as the all-zeros spectral null equation, which still satisfying (9) since it is a special case of it. If we substitute the variables in (9) by using the all-zeros spectral null equation, we obtain the following relationships:

$$\begin{cases} x_1 + x_3 = x_2 + x_4, \\ x_1 = x_2 = x_3 = x_4, \end{cases} \Rightarrow \begin{cases} x_2 + x_3 = x_1 + x_4, \\ x_1 + x_2 = x_3 + x_4. \end{cases} \tag{10}$$

Equation (10) shows the resultant equations derived from (9) and the all-zeros spectral null equation. Fig. 5 shows that the same graph G_1 in Fig. 4 is actually a special case of the graph G_2 when we take into consideration the all-zeros spectral null equation.

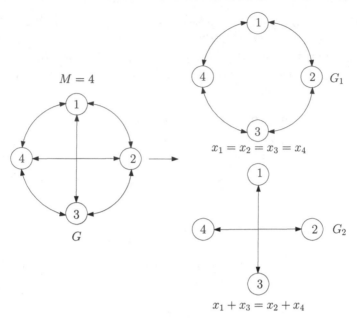

$$x_1 = x_2 = x_3 = x_4$$

$$x_1 + x_3 = x_2 + x_4$$

Fig. 4. Equation representation for Graph $M = 4$

Since the obtained relationship between the variables $x_1 = x_2 = x_3 = x_4$ is a special case of the equation representing the graph G_2 in Fig. 4, we limit our studies to (1) and to its corresponding graph to study the cardinality and other properties of the code.

Fig. 4 shows that the graph G, which is the general form of all possible permutations is the combinations or the union, $G = G_1 \cup G_2$, of other subgraphs related to the spectral null equation.

5. Graph theory and spectral null codes

In this section we will present certain concepts and properties for spectral null codes and try to confirm and very them from a graph theoretical approach.

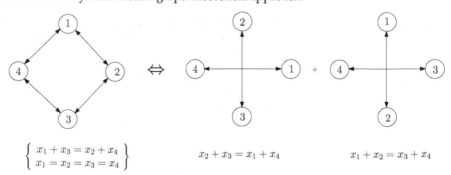

$$\left\{ \begin{array}{l} x_1 + x_3 = x_2 + x_4 \\ x_1 = x_2 = x_3 = x_4 \end{array} \right\} \qquad x_2 + x_3 = x_1 + x_4 \qquad x_1 + x_2 = x_3 + x_4$$

Fig. 5. All-zero equation representation for Graph $M = 4$

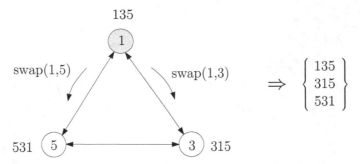

Fig. 6. Index-permutation sequences

5.1 Cardinalities approach

5.1.1 Hamming distance approach

The use of the Hamming distance [11] in this section is just to refer to the number of places that two permutation sequences representing the index-permutation symbols of each grouping A_i of the spectral null equation differ, and not in the study of the error correction properties of the spectral null codes.

To generate the permutation sequences, we start with any state representing an index-permutation symbol in each grouping as appearing in (1). A permutation sequence used as a starting point, contains the symbol from the start state followed by the rest of symbols from the other states taking into consideration the order of the symbols as appearing in (1). Fig. 6 shows the starting permutation sequence as 135. We swap the state-symbol with the following state-symbol in the permutation sequence based on the k-cube construction [12]. We end the swapping process at the last state in the graph. We do not swap symbols between the last state and the starting state for the reason to not disturb the obtained sequences at each state. As an example, for $M = 6$, Fig. 6 depicts the swaps and shows the resultant index-permutation codebooks for one grouping.

Definition 5.1. *The Hamming distance* $d_H(\mathbf{Y}^i, \mathbf{Y}^j)$ *is defined as the number of positions in which the two sequences* $\mathbf{Y}^i$ *and* $\mathbf{Y}^j$ *differ. We denote by* $\mathcal{H}_d(M, N)$ *the distance matrix, whose entries are the distances between index-permutation sequences from a spectral null code of length* $M = Nz$ *defined as follows:*

$$\mathcal{H}_d(M, N) = [h_{i,j}] \quad \text{with} \quad h_{i,j} = d_H(\mathbf{Y}^i, \mathbf{Y}^j). \tag{11}$$

Definition 5.2. *The Hamming distance between the same sequences or between sequences with non connected symbols is always equal to zero.*

Definition 5.3. *The sum on the Hamming distances in the* $\mathcal{H}_d(M, N)$ *distance matrix is*

$$|\mathcal{H}_d(M, N)| = \sum_{i=1}^{M} \sum_{j=1}^{M} h_{i,j}. \tag{12}$$

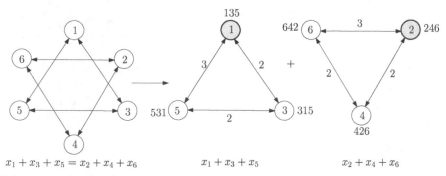

Fig. 7. Distances for Graph $M = 6$ with $N = 2$

In the following examples we consider different cases of number of groupings and number of elements in each grouping and we discuss their impact on the resultant Hamming distance and its relationship with the cardinalities of the spectral null codebooks.

Example 5.4. *We consider the case of $M = 6$ where the number of groupings is $N = 2$ and the number of variables in each grouping is $z = 3$. The corresponding spectral null equation is*

$$\overbrace{x_1 + x_3 + x_5}^{A_1} = \overbrace{x_2 + x_4 + x_6}^{A_2} \tag{13}$$

The equation (13) is presented by the graph in Fig. 7, where the index-permutation symbols are presented with their corresponding Hamming distances.

$$
\mathcal{H}_d(6,2) =
\begin{array}{c@{}c}
 & \begin{array}{cccccc} 135 & 315 & 513 & 246 & 426 & 624 \end{array} \\
\begin{array}{c} 135 \\ 315 \\ 513 \\ 246 \\ 426 \\ 624 \end{array} &
\left[\begin{array}{cccccc}
0 & 2 & 3 & 0 & 0 & 0 \\
2 & 0 & 2 & 0 & 0 & 0 \\
3 & 2 & 0 & 0 & 0 & 0 \\
0 & 0 & 0 & 0 & 2 & 3 \\
0 & 0 & 0 & 2 & 0 & 2 \\
0 & 0 & 0 & 3 & 2 & 0
\end{array} \right]
\end{array}
\tag{14}
$$

Each grouping in (13) is represented by a subgraph as depicted in Fig. 7. The Hamming distance matrix for all possible index-permutation sequences is presented in (14), where "0" represents the Hamming distance between same sequences or sequences with non connected symbols as defined in Definition 5.2. From Definition 5.3, we have,

$$|\mathcal{H}_d(6,2)| = 28.$$

Example 5.5. *For the case of $M = 6$ where $N = 3$ and $z = 2$, the corresponding spectral null equation is*

$$\overbrace{x_1 + x_4}^{A_1} = \overbrace{x_2 + x_5}^{A_2} = \overbrace{x_3 + x_6}^{A_3}. \tag{15}$$

The equation (15) is presented by the graph in Fig. 8. Using the concept of graph distance and the permutation sequences, we can have the distance values as depicted in Fig. 8.

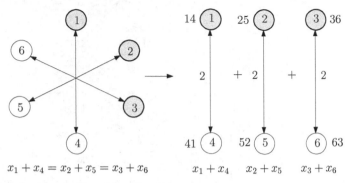

$$x_1 + x_4 = x_2 + x_5 = x_3 + x_6 \qquad x_1 + x_4 \quad x_2 + x_5 \quad x_3 + x_6$$

Fig. 8. Distances for Graph $M = 6$ for $N = 3$

The corresponding subgraphs for each grouping A_1, A_2 and A_3 are presented in Fig. 8.

$$\mathcal{H}_d(6,3) = \begin{array}{c} \\ 14 \\ 41 \\ 25 \\ 52 \\ 36 \\ 63 \end{array} \begin{array}{c} \begin{array}{cccccc} 14 & 41 & 25 & 52 & 36 & 63 \end{array} \\ \left[\begin{array}{cccccc} 0 & 2 & 0 & 0 & 0 & 0 \\ 2 & 0 & 0 & 0 & 0 & 0 \\ 0 & 0 & 0 & 2 & 0 & 0 \\ 0 & 0 & 2 & 0 & 0 & 0 \\ 0 & 0 & 0 & 0 & 0 & 2 \\ 0 & 0 & 0 & 0 & 2 & 0 \end{array} \right] \end{array} \qquad (16)$$

The Hamming distance matrix for all possible index-permutation sequences is presented in (16). From Definition 5.3, we have,

$$|\mathcal{H}_d(6,3)| = 12.$$

Comparing the two results we have,

$$|\mathcal{H}_d(6,2)| > |\mathcal{H}_d(6,3)|.$$

Example 5.6. *In this example we take the case of N not a prime number, where we have to suppose that $N = cd$, where c and d are integer factors of N. The equation, which leads to nulls, is*

$$\begin{aligned} A_u &= A_{u+vc}, \\ u &= 0,1,2,\ldots,c-1, \\ v &= 1,2,\ldots,d-1, \\ N &= cd, \end{aligned} \qquad (17)$$

We consider the case of $M = 8$, where N can be whether $N = 2$ or $N = 4$. The corresponding graph of each case is respectively depicted depicted in Fig. 9 as G_1 and G_2. From Definition 5.3, we have,

$$|\mathcal{H}_d(8,2)| = 40.$$

and

$$|\mathcal{H}_d(8,4)| = 16.$$

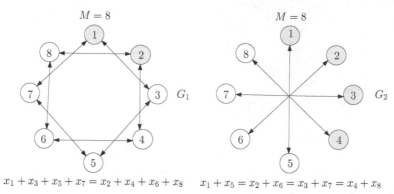

Fig. 9. Equation representation for Graph $M = 8$

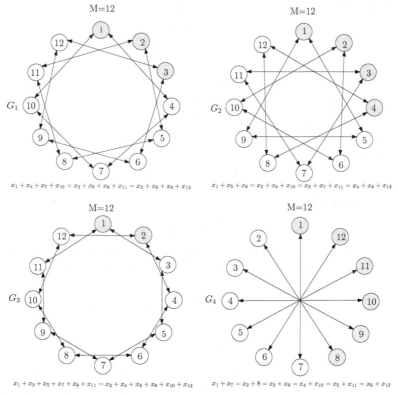

Fig. 10. Equation representation for Graph $M = 12$

Comparing the two results we have,

$$|\mathcal{H}_d(8,2)| > |\mathcal{H}_d(8,4)|.$$

Example 5.7. *In the case of $M = 12$, we have four combinations where the value of N could be $N = 4$, $N = 3$, $N = 2$ or $N = 6$ as depicted in (17). In each case we have a graph representing the spectral null equation as depicted in Fig. 10.*

From Definition 5.3, we have,

$$|\mathcal{H}_d(12,2)| = 64,$$

$$|\mathcal{H}_d(12,6)| = 24,$$

$$|\mathcal{H}_d(12,3)| = 60,$$

and

$$|\mathcal{H}_d(12,4)| = 56.$$

Comparing all the results we have,

$$|\mathcal{H}_d(12,2)| > |\mathcal{H}_d(12,6)|,$$

$$\text{and } |\mathcal{H}_d(12,3)| > |\mathcal{H}_d(12,4)|.$$

Theorem 5.8. *The sum on the Hamming distances for all index-permutation sequences is*

$$|\mathcal{H}_d(M,N)| = \begin{cases} 4N, & \text{for } z = 2, \\ 2N(3z - 2), & \text{for } z \geq 3. \end{cases}$$

Proof. Since the matrix $\mathcal{H}_d(M,N)$ is clearly symmetric, we can just prove half of the results of the theorem and then the final will be the double. For the case of $z = 2$ the proof is trivial since we swap only two symbols in each index-permutation sequence. Thus the sum on the distances is $4 \times N$. For the case of $z \geq 3$ we have a cycle graph [1]-[2], where the number of edges is equal to the number of vertices. Since we swap two symbols each time we move from one state to another, the distance at each edge is equal to two, except for the last edge connecting the first state to the last state where all symbols are swapped and the distance is equal to the length of the index-permutation sequences, which is z. The sum on the Hamming distances for a cycle graph for each grouping is $2 \times (z - 1) + z = 3 \times z - 2$. Thus the result on the sum of the Hamming distances in the matrix is $2 \times N \times (3 \times z - 2)$.   □

5.1.2 Graph-swap distance approach

The length of each grouping A_i, which is equal to the value of z plays an important role in cardinalities of the corresponding codebooks. We make use of the graph distance theory to see how z also plays an important role in the value of the graph distance.

Definition 5.9. *The graph-swap distance denoted by $\mathcal{G}_d$ between two index-permutation symbols represented by the vertices u and v of a finite graph is the minimum number of times of swaps that symbol u can take the position of symbol v in the graph.*

Definition 5.10. *The graph-swap distance between the same index-permutation symbol or between non connected symbols is always equal to zero.*

Definition 5.11. *We denote by $\mathcal{M}_{\mathcal{G}_d}(M,N)$ the graph-swap distance matrix, whose entries $m_{i,j}$ are the graph distances between two index-permutation symbols from a spectral null code of length $M = Nz$.*

Definition 5.12. *The sum on the graph-swap distances in the* $\mathcal{M}_{\mathcal{G}_d}(M, N)$ *distance matrix is*

$$|\mathcal{M}_{\mathcal{G}_d}(M, N)| = \sum_{i=1}^{M} \sum_{j=1}^{M} m_{i,j}. \tag{18}$$

Example 5.13. *We consider the case of $M = 8$ with $N = 2$ or $N = 4$, the corresponding graph-swap distance matrices are respectively as*

$$\mathcal{M}_{\mathcal{G}_d}(8,2) = \begin{array}{c} \\ 1 \\ 2 \\ 3 \\ 4 \\ 5 \\ 6 \\ 7 \\ 8 \end{array} \begin{array}{c} 1\ 2\ 3\ 4\ 5\ 6\ 7\ 8 \\ \left[\begin{array}{cccccccc} 0 & 0 & 1 & 0 & 2 & 0 & 1 & 0 \\ 0 & 0 & 0 & 1 & 0 & 2 & 0 & 1 \\ 1 & 0 & 0 & 0 & 1 & 0 & 2 & 0 \\ 0 & 1 & 0 & 0 & 0 & 1 & 0 & 2 \\ 2 & 0 & 1 & 0 & 0 & 0 & 1 & 0 \\ 0 & 2 & 0 & 1 & 0 & 0 & 0 & 1 \\ 1 & 0 & 2 & 0 & 1 & 0 & 0 & 0 \\ 0 & 1 & 0 & 2 & 0 & 1 & 0 & 0 \end{array} \right] \end{array}, \quad \text{and} \quad \mathcal{M}_{\mathcal{G}_d}(8,4) = \begin{array}{c} \\ 1 \\ 2 \\ 3 \\ 4 \\ 5 \\ 6 \\ 7 \\ 8 \end{array} \begin{array}{c} 1\ 2\ 3\ 4\ 5\ 6\ 7\ 8 \\ \left[\begin{array}{cccccccc} 0 & 0 & 0 & 0 & 1 & 0 & 0 & 0 \\ 0 & 0 & 0 & 0 & 0 & 1 & 0 & 0 \\ 0 & 0 & 0 & 0 & 0 & 0 & 1 & 0 \\ 0 & 0 & 0 & 0 & 0 & 0 & 0 & 1 \\ 1 & 0 & 0 & 0 & 0 & 0 & 0 & 0 \\ 0 & 1 & 0 & 0 & 0 & 0 & 0 & 0 \\ 0 & 0 & 1 & 0 & 0 & 0 & 0 & 0 \\ 0 & 0 & 0 & 1 & 0 & 0 & 0 & 0 \end{array} \right] \end{array}.$$

From Definition 5.12, we have $|\mathcal{M}_{\mathcal{G}_d}(8,2)| = 32$ and $|\mathcal{M}_{\mathcal{G}_d}(8,4)| = 8$. where we can see clearly that

$$|\mathcal{M}_{\mathcal{G}_d}(8,2)| > |\mathcal{M}_{\mathcal{G}_d}(8,4)|.$$

Theorem 5.14. *The sum on the graph distances for all index-permutation symbols is*

$$|\mathcal{M}_{\mathcal{G}_d}(M, N)| = \begin{cases} \left(\frac{z}{2}\right)^2 M, & \text{for } z \text{ even}, \\ \frac{z^2-1}{4} M, & \text{for } z \text{ odd}. \end{cases}$$

Proof. The graphs that we are using are cycle graphs. As long as we go through the edges of a graph the graph distance is incremented by one. When z is even, the first state has the farthest state to it located at $\frac{z}{2}$. So the graph distances from the first state to the $\frac{z}{2}$ state are in a numerical series of ratio one from one to $\frac{z}{2}$. From the state at the position $\frac{z}{2} - 1$ till the first state, the graph distances are in a numerical series of ratio one from one to $\frac{z}{2} - 1$. Adding the two series we get the final sum equal to $\left(\frac{z}{2}\right)^2 M$. Same analogy for the case of z as odd with a numerical series from one till $\frac{z-1}{2}$. $\square$

5.1.3 Adjacency-swap matrix approach

We introduce the adjacency-swap matrix inspired by graph theory as follows.

Definition 5.15. *The adjacency-swap matrix of index-permutation symbols is an $M \times M$ matrix $\mathcal{N}_{A_d}(M, N) = (n_{i,j})$ in which the entry $n_{i,j} = 1$ if there is a swap between an index symbol i and an index symbol j and is 0 if there is no swap between index symbol i and index symbol j as presented in each grouping of a spectral null equation.*

Example 5.16. *For the case of* $M = 6$ *with* $N = 2$ *or* $N = 3$, *the corresponding adjacency-swap matrices are*

$$
\mathcal{N}_{A_d}(6,2) = \begin{array}{c} \\ 1 \\ 2 \\ 3 \\ 4 \\ 5 \\ 6 \end{array} \begin{array}{c} 1\ 2\ 3\ 4\ 5\ 6 \\ \begin{bmatrix} 0\ 0\ 1\ 0\ 1\ 0 \\ 0\ 0\ 0\ 1\ 0\ 1 \\ 1\ 0\ 0\ 0\ 1\ 0 \\ 0\ 1\ 0\ 0\ 0\ 1 \\ 1\ 0\ 1\ 0\ 0\ 0 \\ 0\ 1\ 0\ 1\ 0\ 0 \end{bmatrix} \end{array}, \quad \text{and} \quad \mathcal{N}_{A_d}(6,3) = \begin{array}{c} \\ 1 \\ 2 \\ 3 \\ 4 \\ 5 \\ 6 \end{array} \begin{array}{c} 1\ 2\ 3\ 4\ 5\ 6 \\ \begin{bmatrix} 0\ 0\ 0\ 1\ 0\ 0 \\ 0\ 0\ 0\ 0\ 1\ 0 \\ 0\ 0\ 0\ 0\ 0\ 1 \\ 1\ 0\ 0\ 0\ 0\ 0 \\ 0\ 1\ 0\ 0\ 0\ 0 \\ 0\ 0\ 1\ 0\ 0\ 0 \end{bmatrix} \end{array}.
$$

We can see that $|\mathcal{N}_{A_d}(6,2)| = 12 > |\mathcal{N}_{A_d}(6,3)| = 6.$

M	N	z	$|\mathcal{C}_b(M,N)|$	$|\mathcal{H}_d(M,N)|$	$|\mathcal{M}_{\mathcal{G}_d}(M,N)|$	$|\mathcal{N}_{A_d}(M,N)|$
6	3	2	10	12	4	6
6	2	3	20	28	12	12
8	4	2	36	16	8	8
8	2	4	70	40	32	24
10	5	2	34	20	10	10
10	2	5	252	52	60	40
12	6	2	250	24	12	12
12	4	3	300	56	24	24
12	3	4	346	60	48	36
12	2	6	924	64	108	60
15	5	3	488	70	30	30
15	3	5	2252	78	90	60

Table 1. Graph Distances and Cardinalities of Different Codebooks

Theorem 5.17. *The total number of swaps in an adjacency-swap matrix is*

$$
|\mathcal{N}_{A_d}(M,N)| = (z-1)M
$$

Proof. The proof is trivial as per grouping we have z index-permutation symbols. Thus we have $z - 1$ ones in each row of the matrix $\mathcal{N}_{A_d}(M,N)$ which refer to the possible swaps of each symbol with others in the same grouping. The total number of swaps is $(z-1) \times M$. □

Table 1 presents few examples of the relationship between the cardinalities of spectral null codes denoted by $\mathcal{C}_b(M,N)$ and their correspondences of graph distances. It is clear from Table 1 that the cardinalities of different codebooks with the same length of codewords, increase when the number of swaps increases. This results is also verified in Table 1 based on the concept of distances from graph theory perspective.

5.2 Subsets approach

5.2.1 Subgraph theory

In this section we make use of one of the properties in graph theory related to the design of subgraphs as presented in Definition 3.1.

The elimination of states from any graph corresponding to the index-permutation symbols is in fact the same as eliminating the corresponding variables from the spectral null equation (1). The elimination of the variables is performed in such a way that the spectral null equation is always satisfied. This leads to the basic idea of eliminating an equivalent number of variable equal to N as a total number from different groupings in the spectral null equation. This is true when we eliminate only one variable from each grouping. In the case when we eliminate t variables with $1 < t < z$ from each grouping, we have a total number of eliminated variables of tN.

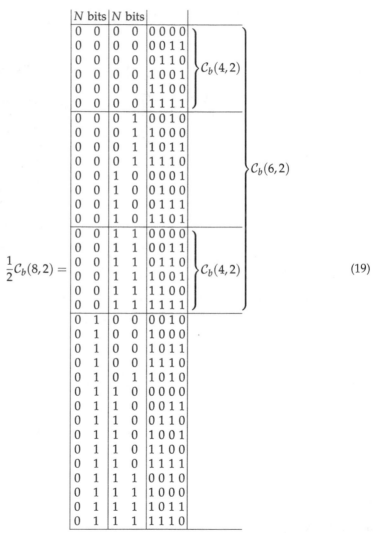

$$\frac{1}{2}C_b(8,2) =$$ (19)

N bits		N bits			
0	0	0	0	0 0 0 0	$C_b(4,2)$
0	0	0	0	0 0 1 1	
0	0	0	0	0 1 1 0	
0	0	0	0	1 0 0 1	
0	0	0	0	1 1 0 0	
0	0	0	0	1 1 1 1	
0	0	0	1	0 0 1 0	$C_b(6,2)$
0	0	0	1	1 0 0 0	
0	0	0	1	1 0 1 1	
0	0	0	1	1 1 1 0	
0	0	1	0	0 0 0 1	
0	0	1	0	0 1 0 0	
0	0	1	0	0 1 1 1	
0	0	1	0	1 1 0 1	
0	0	1	1	0 0 0 0	$C_b(4,2)$
0	0	1	1	0 0 1 1	
0	0	1	1	0 1 1 0	
0	0	1	1	1 0 0 1	
0	0	1	1	1 1 0 0	
0	0	1	1	1 1 1 1	
0	1	0	0	0 0 1 0	
0	1	0	0	1 0 0 0	
0	1	0	0	1 0 1 1	
0	1	0	0	1 1 1 0	
0	1	0	1	1 0 1 0	
0	1	1	0	0 0 0 0	
0	1	1	0	0 0 1 1	
0	1	1	0	0 1 1 0	
0	1	1	0	1 0 0 1	
0	1	1	0	1 1 0 0	
0	1	1	0	1 1 1 1	
0	1	1	1	0 0 1 0	
0	1	1	1	1 0 0 0	
0	1	1	1	1 0 1 1	
0	1	1	1	1 1 1 0	

Example 5.18. *We construct the code for the case of $M = 8$, with $N = 2$ and $z = 4$, which is represented by the codebook $C_b(8,2)$ in (19) (we present only the half of the codebook because of space*

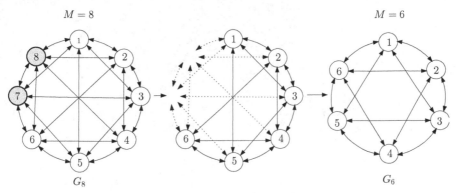

Fig. 11. Subgraph design from $M = 8$ to $M = 6$ with $N = 2$

limitation in the page) and which is designed from the spectral null equation presented as follows:

$$\overbrace{\underbrace{y_1 + y_3 + y_5 + y_7}_{z=4} = \underbrace{y_2 + y_4 + y_6 + y_8}_{z=4}}^{N=2}. \tag{20}$$

The corresponding graph for $\mathcal{C}_b(8,2)$ is G_8 as presented in Fig. 11.

From the spectral null equation (20) we eliminate the variables y_7 and y_8 using the addition property. Thus we get,

$$\overbrace{\underbrace{y_1 + y_3 + y_5}_{z=3} = \underbrace{y_2 + y_4 + y_6}_{z=3}}^{N=2}. \tag{21}$$

This resultant equation is the spectral null equation for the case of $M = 6$ with $N = 2$ and the corresponding codebook is denoted by $\mathcal{C}_b(6,2)$. Fig. 11 depicts the elimination of the states from a graph theory perspective.

Based on the same approach, we eliminate the variables y_5 and y_6 from the equation (21). The resultant spectral equation for the case of $M = 4$, with $N = 2$ and $z = 2$ is presented as follows:

$$\overbrace{\underbrace{y_1 + y_3}_{z=2} = \underbrace{y_2 + y_4}_{z=2}}^{N=2}. \tag{22}$$

The code generated from the spectral null equation (22) is denoted by the codebook $\mathcal{C}_b(4,2)$ as depicted in (19). The corresponding graph for $\mathcal{C}_b(4,2)$ is G_4 as presented in Fig. 12.

It is clear that from the codebook presented in (19), we have $\mathcal{C}_b(4,2) \subset \mathcal{C}_b(6,2) \subset \mathcal{C}_b(8,2)$ in terms of the existence of elements from the codebooks $\mathcal{C}_b(4,2)$ and $\mathcal{C}_b(6,2)$ in the codebook $\mathcal{C}_b(8,2)$, which is the same as for the subgraps where we have $G_4 \subset G_6 \subset G_8$.

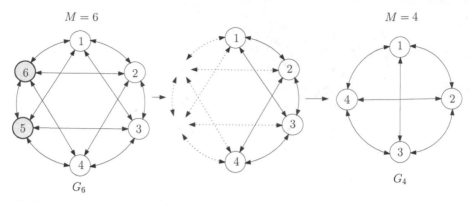

Fig. 12. Subgraph design from $M = 6$ to $M = 4$ with $N = 2$

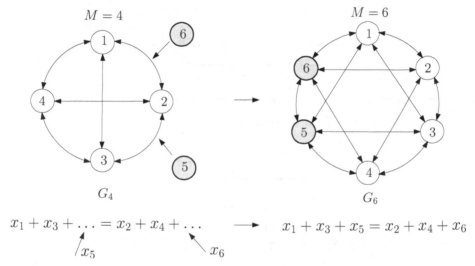

$$x_1 + x_3 + \ldots = x_2 + x_4 + \ldots \qquad \longrightarrow \qquad x_1 + x_3 + x_5 = x_2 + x_4 + x_6$$

Fig. 13. Supergraph design: From $M = 4$ to $M = 6$ with $N = 2$

5.2.2 Supergraph theory

The concept of supergraphs is totally opposite to what was introduced with the subgraphs. Although this concept is not treated in graph theory because of its complexity and the conditions that we should have to add vertices to any graph. This problem is already solved in the design of spectral null codes since we are dealing with spectral null equations where it is easy to add variables in all groupings in such a way the spectral null equations are satisfied. Thus it results in the addition of the corresponding states of the symbols in the corresponding permutation equation.

Definition 5.19. *A spectral null preserving supergraph is an extension of a graph with a multiple of N states, which always keeps the spectral null equation satisfied.*

Fig. 13 presents the mechanism of the addition of states to an existing graph. The example of a graph of six states, which is related to the case of $M = 6$, is actually an extension of the graph of four states which corresponds to the case of $M = 4$. An addition of a state corresponds to the addition of its corresponding variable in a way to keep the equation (1) satisfied.

6. Conclusion

Spectral shaping technique that design codes with certain power spectral density properties is used to construct codes called spectral null codes that can generate nulls at rational submultiples of the symbol frequency. These codes have great importance in certain applications like in the case of transmission systems employing pilot tones for synchronization and that of track-following servos in digital recording. these codes are not confined to magnetic recorders but they ware taken further to their utilization in write-once recording systems.

In this investigation we have shown how the use of graphs can give a new insight into the analysis and understanding the structure of the spectral null codes, where with incisive observations to spectral null codebooks, we could derive important properties that can be useful in the field of digital communications.

The relationship between the spectral null equations for our designed codes and the permutation sequences corresponding to the indices of the variables in those equations have lead to a very important derivation of certain properties based on graph theory approach.

The properties that we have presented could potentially lead to the discovery of other interesting properties for specific applications like those that we have investigated in [13].

The use of certain graph theory properties helped in understanding certain properties of spectral null codes. The introduction of the index-permutation sequences and the use of the concept of distances gave us an idea about the structure and the design conditions of spectral null codes.

7. References

[1] R. J. Wilson, *Graph theory and Combinatorics*. England: Pitman Advanced Publishing Program., 1979.

[2] J. L. Gross and J. Yellen, *Graph theory and its Applications*. USA: Chapman and Hall/CRC., 2006.

[3] K. A. S. Immink, "Spectral null codes," *IEEE Transactions on Magnetics*, vol. 26, no. 2, pp. 1130–1135, Mar. 1990.

[4] N. Hansen, "A head-positioning system using buried servos,Ť *IEEE Transactions on Magnetics*, vol. 17, no. 6, pp. 2735–2738, Nov. 1981.

[5] M. Haynes, "Magnetic recording techniques for buried servos,Ť *IEEE Transactions on Magnetics*, vol. 17, no. 6, pp. 2730–2734, Nov. 1981.

[6] C. Yeh and B. Parhami, "Parallel algorithms for index-permutation graphs. An extension of Cayley graphs for multiple chip-multiprocessors (MCMP)"*International Conference on Parallel Processing*,pp. 3–12, Sept. 2001.

[7] T. G. Swart, "Distance-Preserving Mappings and Trellis Codes with Permutation Sequences", Ph.D. dissertation, University of Johannesburg, Johannesburg, South Africa, Apr. 2006.

[8] E. Gorog, "Alphabets with desirable frequency spectrum properties," *IBM J. Res. Develop.*, vol. 12, pp. 234–241, May 1968.

[9] B. H. Marcus and P. H. Siegel, "On codes with spectral nulls at rational submultiples of the symbol frequency," *IEEE Trans. Inf. Theory*, vol. 33, no. 4, pp. 557–568, Jul. 1987.

[10] K. A. S. Immink, *Codes for mass data storage systems*, Shannon Foundation Publishers, The Netherlands, 1999.

[11] A. Viterbi and J. Omura, *Principles of Digital Communication and Coding*. McGraw-Hill Kogakusha LTD, Tokyo Japan, 1979.

[12] K. Ouahada and H. C. Ferreira, "A k-Cube Construction mapping mapping binary vector to permutation," in *Proceedings of the International Symposium on Information Theory*, South Korea, pp. 630–634, June 28–July 3, 2009.

[13] K. Ouahada, T. G. Swart, H. C. Ferreira and L. Cheng, "Binary permutation sequences as subsets of Levenshtein codes, run-length limited codes and spectral shaping codes," *Designs, Codes and Cryptography Journal*, vol. 48, no. 2, pp. 141–154, Aug. 2008.

2

Analysis of Modified Fifth Degree Chordal Rings

Bozydar Dubalski, Slawomir Bujnowski, Damian Ledzinski,
Antoni Zabludowski and Piotr Kiedrowski
University of Technology and Life Sciences, Bydgoszcz,
Poland

1. Introduction

Implementation of new telecommunications services has always been associated with the need to ensure network efficiency required to implement these services. Network efficiency can be described by a number of parameters such as: network bandwidth, propagation time, quality, reliability and fault tolerance. More and better performance, and thus network efficiency is achieved mainly by using more and more advanced technical and technological solutions. There were milestones solutions such as the use of coaxial transmission cables, optical fibers, and various techniques of multiplication like TDM or WDM (Newton, 1996). Significant impact on the way to deliver services had wireless transmission, which has found widespread use in communication networks since the end of last century.

In addition to technical and technological solutions to improve network efficiency by using system solutions such as: protocols or topology (topology control) appropriate for the type of connection or service. Type of system solutions in the network is closely related to the technology involved in the network, so it can be said that technology determines the solutions. Examples may be the different topological approaches as a result or a consequence of the limiters for a specific technology; e. g.:

- Networks based on SONET/SDH have limitations as to the path length (number of nodes in the path) as result of synchronization signals distribution,
- In turn, networks based on WDM technology, where network nodes are OADM (ALU: Alcatel-Lucent, 2011) multiplexers there are restrictions as to the length of the path (in the literal sense) associated with the phenomenon of dispersion,
- Recent example is the WSN networks, which are increasingly common application in various areas of life, such as the implementation of communication solutions for Smart Grid (Al-Karaki, 2004), which dealt with the Authors of this chapter, in the case of WSN network must resolve a number of problems associated with reliable transmission over a large area using short-range devices.

From the above examples it follows that in order to provide high efficiency network technology solutions are not always sufficient and require additional system solutions, which should always go hand in hand with these technological ones. Therefore, proposed in this publication the solutions are always up to date.

2. Background

A critical issue in designing telecommunications systems is choosing the interconnection network topology as it has the biggest impact on efficiency, speed, and reliability of the entire system (Bhuyan, 1987). Nowadays, analysis of regular network structures is one of the most important issues in telecommunications and computer science.

These networks can be model by symmetric digraphs, i.e., a directed graph G with vertex set $V(G)$ and edge set $E(G)$, such that, if $[v_i, v_j]$ is in $E(G)$, then $[v_j, v_i]$ is also in $E(G)$. So any edge of digraph connecting vertices v_i and v_j can be replaced by two directed edges $[v_i, v_j]$ and $[v_j, v_i]$ (Narayanan et al., 2001).

It is obvious that the best service and reliability parameters one can obtain by forming complete networks (described by a complete graph), but only small networks can be built in this way. In (Kocis, 1992) a survey of known topologies has been presented. Among the analyzed topologies that would be used in designing the distributed structures, the authors of this publication have chosen rings as they are very simple and extensible. They are characterized by connectivity equal to 2 (damage of one edge or node ensures possibility of transmission), are not expensive (number of edges is equal to the number of nodes), are regular and symmetric, but possess poor transmission parameters.

Halfway between the complete graph and the ring is the chordal ring structure (Arden & Lee, 1981). The chordal ring is a ring with additional chords. It is defined by pair (p, Q), where p is the number of nodes of the ring and Q is the set of chords. Each chord connects every pair of nodes of the ring that are at distance q_i in the ring.

The application of this type of structure is useful due to its simplicity, clear topology, resistance to damages, simplicity of routings, and good extension (Kocis, 1992).

The application of chordal rings in computer systems (Mans, 1999), TDM networks (communication between distributed switching modules) (Bujnowski, 2003), core optical networks (Freire & da Silva, 1999, 2001a, 2001b; Liestman et al., 1998; Narayanan & Opatrny, 1999; Narayanan et al., 2001), and optical access networks (Pedersen, 2005; Pedersen et al., 2004a, 2004b, 2005; Bujnowski et al., 2003) has been analyzed. The authors of this publication, in their earlier works on modeling of telecommunication and computer networks, present an analysis of chordal rings (Bujnowski et al. 2004a, 2004b, 2005).

In the beginning the general definition of chordal ring will be giving.

Definition 1. A chordal ring is a ring with additional edges called chords. A chordal ring is defined by the pair (p, Q), where p denotes the number of nodes of the ring and Q denotes the set of chord lengths $Q \subseteq \{1, 2, ..., \lfloor p/2 \rfloor\}$. Since it is a ring, every node is connected to exactly two other nodes (i.e. assume a numbering of the nodes $1, 2, ..., p$ – then node i is connected to node $i-1$ and $i+1$ (mod p). Node 0 is connected to p and 1). Each chord of length $q \in Q$ connects every two nodes of the ring that are at distance q. The chordal ring will be further denoted as $G(p; 1, q_1, ..., q_i)$, $q_1 < ... < q_i$. In general, the degree of chordal rings is $2i$, unless there is a chord of length $p/2$. In this case p should be even and rings' degree is $2i - 1$ (Gavoille, n.d).

In the papers (Bujnowski et al., 2008a, 2009b, 2010; Dubalski et al., 2007, 2008; Pedersen et al., 2009) the authors have previously analysed the transmission properties of third, fourth and sixth degree chordal rings and modified graphs of these types. These topologies are the subject of many publications of the researchers from Putra University (Farah et al, 2008, 2010a, 2010b; Azura et al., 2008, 2010; Farah et al. 2010, 2011).

In this publication the survey of the chordal rings consisting of fifth degree nodes (Fig. 1) will be presented. Until now this type of the regular structures is not widely examined, so authors decided to focus on it (Dubalski, 2010).

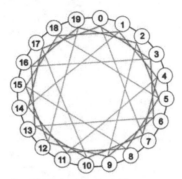

Fig. 1. An example of chordal ring fifth nodal degree

Average distance and diameter was chosen in order to provide a general and simple indication of transmission properties of the analyzed topologies. These follow standard definitions as summarized below. For more basic definitions of e.g. paths and path lengths, please refer to any basic graph theory book, such as (Distel, 2010).

Definition 2. The diameter $D(G)$ is the largest path length among all of the shortest length of the paths between any pair of nodes. It is defined as follows:

$$D(G) = \max_{v_i v_j} \{d_{\min}(v_i, v_j)\} \tag{1}$$

where v_i means the number of the node, d_{min} minimal distance (number of edges) between i-th and j-th node.

Definition 3. The average path length d_{av} between all pairs of nodes is defined by the formula:

$$d_{av} = \frac{1}{p(p-1)} \sum_{i=0}^{p-1} \sum_{j=0}^{p-1} d_{\min}(v_i, v_j) \tag{2}$$

where $d_{min}(v_i, v_j)$ is the minimal number of edges between a source node v_i and every other chosen node $v_{j,}$, and p denotes the number of nodes.

A Reference Graph (a virtual example is shown in Fig. 2) can be determined, which presents a reference for all regular graphs of degree 5. It represents lower bounds for average distance and diameter for all these graphs, but since it is a "virtual graph" these bounds may not always be achievable.

The Reference Graph possesses parameters as follows:

1. The number of nodes p_{dr} in d-th layer is determined by formula:

$$p_{1r} = 5$$
$$p_{dr} = 20 \cdot 2^{2(d-2)} \; when \quad d > 1 \tag{3}$$

2. Total number of nodes $p_{D(G)r}$ versus graph diameter is described by expression:

$$p_{D(G)r} = \frac{5 \cdot 4^{D(G)r} - 2}{3} \tag{4}$$

3. Value of diameter versus total number of nodes can be calculated following formula:

$$D(G) = \left\lceil \log_4 \left(\frac{3p_r + 2}{5} \right) \right\rceil \tag{5}$$

4. Average path length d_{avr} in function of diameter is equal to:

$$d_{avr} = \frac{1 + \left(3 \cdot D(G)_r - 1\right) \cdot 4^{D(G)_r}}{3 \cdot \left(4^{D(G)_r} - 1\right)} \tag{6}$$

5. This graph is symmetrical, its all parameters are equal regardless from which node they are calculated.

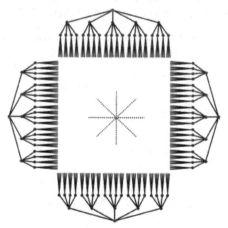

Fig. 2. General diagram of virtual infinite Reference Graph

Only one Reference Graph fifth nodal degree exists in reality, it is the complete graph consisting of 6 nodes.

Two other reference graphs, named as Ideal and Optimal graphs, are also useful for determining average distance and diameter of the chordal rings. They provide theoretical values, which in the following will be compared to values obtained in the real graphs. As for the reference graph mentioned above, the optimal and ideal graphs do not always exist.

In order to determine parameters of the theoretical calculated reference topologies of chordal rings two types of these structures were defined. The first one is called the ideal graph and the second one - optimal graph. In fact these graphs exist only in particular cases, but they are useful as reference models for evaluation expected parameters of tested graphs.

Definition 4. The ideal chordal ring with degree $D(G)$ is the regular graph with total number of nodes p_i given by the formula:

$$p_i = 1 + \sum_{d=1}^{D(G)-1} |p_d| + |p_{D(G)}| \tag{7}$$

where p_d means the number of nodes that belong to the d-th layer (the layer is the subset of nodes that are at a distance d from the source node), while $p_{D(G)}$ denotes the number of the remaining nodes which appear in the last layer. For ideal rings, for every n and $m < D(G)$ $p_n \cap p_m = \varnothing$. If for certain $D(G)$ the subset $p_{D(G)}$ of chordal ring reaches the maximal possible value, then such a ring is called the optimal ring (optimal graph).

For ideal chordal ring the average path length d_{avi} is expressed as:

$$d_{avi} = \frac{\sum_{d=1}^{d(G)-1} d|p_d| + D(G)|p_{d(G)}|}{p_i - 1} \tag{8}$$

whereas for the optimal graph the average path length d_{avo} is equal to:

$$d_{avo} = \frac{\sum_{d=1}^{d(G)} d\, p_d}{p_o - 1} \tag{9}$$

where d – layer number, p_d – number of nodes in d-th layer, p_o – number of nodes in optimal graph.

Optimal graphs were used to calculate the formulas describing parameters of each type of analyzed chordal ring, whereas ideal rings were served to compare calculated theoretically and obtained in reality parameters of analyzed structures.

The basic topology of fifth degree chordal rings in Fig. 3 is shown. The definition, short presentation and author's consideration concerned of this structure are given below.

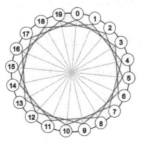

Fig. 3. Basic chordal ring fifth nodal degree CHR5(20; 3,10)

Definition 5. The basic chordal ring fifth nodal degree called CHR5 is an undirected graph, based on a cycle with additional connections (chords). It is denoted by CHR5(p; q_1,q_2) where p must be even and means number of nodes creating the ring, chord length $q_1 > p/2$ is odd, even too, chord length q_2 is equal to $p/2$. The values of p and q_1 must be prime each other (Bujnowski, 2011).

In order to calculate the diameters and average path lengths appearing in optimal graphs it is necessary to evaluate the maximal number of nodes appearing in each layer. In the table 1 the numbers of nodes in the first successive layers of virtual optimal ring are shown (d denotes the layer number, p_d - the number of nodes appearing in d-th layer).

d	1	2	3	4	5	6	7	8
p_{do}	5	12	20	28	36	44	52	60

Table 1. Maximal number of nodes in the layers

If $d > 1$ the power of these sets is described by formula:

$$p_{do} = 4(2d-1) \tag{10}$$

Using the formula given above, the total number of nodes p_o in the optimal graph with diameter $D(G)$ can be calculated ($D(G) > 1$):

$$p_o = 4D(G)^2 + 2 . \tag{11}$$

The total number of nodes in optimal graphs versus its diameter is shown in table 2.

$d(G)$	1	2	3	4	5	6	7	8
p_{do}	6	18	38	66	102	146	198	258

Table 2. Total numbers of nodes forming optimal graphs versus diameter

The average path length in optimal graphs is given by formula:

$$d_{avo} = \frac{8D(G)^3 + 6D(G)^2 - 2D(G) + 3}{3\left(4D(G)^2 + 1\right)} \tag{12}$$

Only one optimal graph exists in reality. It is the complete graph which possesses 6 nodes, but the ideal chordal rings can be found. Whereas it founded two groups of ideal graphs consisting of p nodes, which can be described by formulas given below.

The graphs belonging to the first group are defined as follows:

$$p_i = 4D(G)^2 \quad (D(G) > 1) \tag{13}$$

so

$$D(G) = \frac{\sqrt{p_i}}{2} \tag{14}$$

In this case a chord length q_1 of ideal graphs is equal to:

$$q_1 = 2D(G) - 1 \quad or \quad q_1 = 2D(G) + 1$$
$$q_1 = \sqrt{p_i} - 1 \quad or \quad q_1 = \sqrt{p_i} + 1 \tag{15}$$

these both graphs are isomorphic each other.

The average path length can be express by formula:

$$d_{avi} = \frac{8D(G)^3 + 3D(G)^2 - 8D(G) + 3}{3\left(4D(G)^2 - 1\right)} \tag{16}$$

The graphs belonging to the second group are described as follows:

If $D(G) > 2$ then:

$$p_i = 4D(G)^2 - 4D(G) = 4D(G)\left[D(G) - 1\right] \tag{17}$$

So when the number of nodes is equal to p_i then

$$D(G) = \frac{1 + \sqrt{1 + p_i}}{2} \tag{18}$$

The lengths of chords used to construct ideal graphs can be calculated using formulas:

$$q_1 = 2D(G) - 1 \quad or \quad q_1 = \sqrt{1 + p_i} \tag{19}$$

When the number of nodes creating chordal ring is given by equation:

$$p_i = 4(9i^2 + 9i + 2) \quad or \quad p_i = 6(6i^2 + 10i + 4) \quad where \quad i \in (1, 2, \cdots, n)$$
$$then \quad q_1 = 2D(G) + 1 \quad or \quad q_1 = \sqrt{p_i + 1} + 2$$
$$if \quad p_i = 4(9i^2 + 9i + 2) \quad then \quad q_1 = \frac{4D(G)^2 - 2D(G) - 3}{3} \tag{20}$$
$$if \quad p_i = 6(6i^2 + 10i + 4) \quad then \quad q_1 = \frac{4D(G)^2 - 6D(G) + 3}{3}$$

The average path length of all these graphs is described by formula:

$$d_{avi} = \frac{8D(G)^3 - 6D(G)^2 - 8D(G) + 3}{3\left(4D(G)^2 + 4D(G) - 1\right)} \tag{21}$$

Unfortunately the parameters of CHR5 graphs are considerably different of Reference Graph parameters, what is shown in fig. 4 and 5 given above.

It follows from the difference of number of nodes appearing in successive layers and thus the difference of total number of nodes appearing in dependence of its diameter as well.

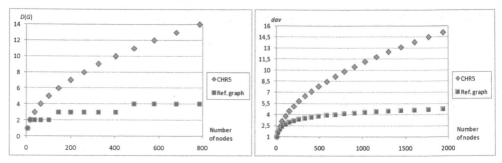

Fig. 4. Comparison of diameter and average path length of Reference Graphs and CHR5

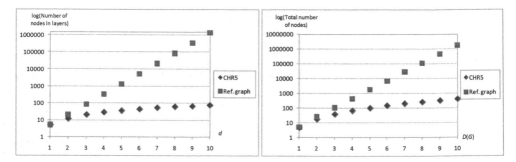

Fig. 5. Differences of number of nodes in successive layers and total number of nodes in Reference Graphs and CHR5

The aim of authors of this publication was to find structures possessing basic parameters which values would be closer to reference graph parameters.

3. Analysis of modified graphs fifth degree

The authors prepared two programs which were used to make it possible to examine the analysed graphs - "Program Graph Finder" and "Find the best distribution of nodes in the layers". The first one - "Program Graph Finder" was used in the first stage of analysis for quite simple topologies, the second one "Find the best distribution of nodes in the layers" – for more complicated structures, when the number of variables describing the way of connections is greater than 4.

The real values of parameters of modified chordal rings were calculated using these programs and compared to those obtained in a theoretical way.

In the following sections, an analysis of 15 different regular structures based on chordal rings is presented. Each of the type of graphs is defined, examples are given, the distribution of nodes in different layers is analyzed, and the ideal and optimal graphs are compared to real graphs. Also, basing on the analysis of nodes in different layers, the average distance and diameter can be calculated as a function of the number of nodes.

The graphs are divided into 3 groups, each consisting of 5 types of graphs. The first group of graphs needs to have a number of nodes divisible by two, and the second group of graphs a number of nodes divisible by 4. The third group of graphs also has a number of nodes divisible by 4, but for these no mathematical expressions of node distribution (and thus the average distance and diameter) were found.

3.1 First group of chordal rings

As previously mentioned, for each type of graph we present:

- Definitions
- Descriptions
- Distribution of nodes
- Expressions for key parameters
- Comparisons of parameters for real and theoretical graphs.

Graph CHR5_a.

Definition 6. The modified fifth degree chordal rings called CHR5_a (Fig. 6) is denoted by CHR5_a(p; q_1,q_2), where p is even and means number of nodes; q_1, q_2 are chords. Chords q_1 and q_2 are odd and $< p/2$. Chord q_1 generates a Hamiltonian cycle. whereas q_2 is odd too and $< p/2$. Each even node i_{2k} is connected to five other nodes: i_{2k-1}, i_{2k+1}, $i_{2k-q1(\mathrm{mod}\ p)}$, $i_{2k+q1(\mathrm{mod}\ p)}$, $i_{2k+q2(\mathrm{mod}\ p)}$, while odd node i_{2k+1} is connected to i_{2k}, i_{2k+2}, $i_{2k+1-q1(\mathrm{mod}\ p)}$, $i_{2k+1+q1(\mathrm{mod}\ p)}$ and $i_{2k+1-q2(\mathrm{mod}\ p)}$ ($0 \le k < p/2$). The values of p and q_1 must be prime each other (this ensures that the Hamiltonian cycle is created). ☑

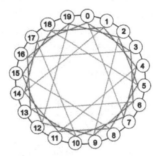

Fig. 6. Modified chordal ring CHR5_a(20; 3,7)

In table 3 the numbers of nodes in the layers of optimal rings as the function of node degree are shown.

d	1	2	3	4	5	6	7	8
p_{do}	5	16	33	58	89	128	173	226

Table 3. Maximal number of nodes in the successive layers

When d (layer number) is odd, then the power of these sets is described by the expression:

$$p_{do\ odd} = \frac{1}{2}(7d^2 + 3) \tag{22}$$

and when d is even:

$$p_{do\,even} = \frac{1}{2}(7d^2 + 4) \tag{23}$$

The general expression has the following form:

$$p_{do} = \frac{1}{2}(7d^2 + 4 - d(\mathrm{mod}\,2)) \tag{24}$$

The total number of nodes p_o forming an optimal graph which possesses diameter $D(G)$ is expressed as:

$$p_{o\,odd} = \frac{7}{12}\left(2D(G)^3 + 3D(G)^2 + 4D(G)\right) + \frac{3}{4}$$

$$p_{o\,even} = \frac{7}{12}\left(2D(G)^3 + 3D(G)^2 + 4D(G)\right) + 1$$

$$p_o = \frac{7}{12}\left(2D(G)^3 + 3D(G)^2 + 4D(G)\right) + 1 - \frac{D(G)(\mathrm{mod}\,2)}{4} \tag{25}$$

which confirms the results obtained by constructing the possible graphs, as shown in Table 4.

$D(G)$	1	2	3	4	5	6	7	8
p_o	6	22	55	113	202	330	503	729

Table 4. Diameters and total numbers of nodes in virtual, optimal graphs

The average path length in optimal graphs can be calculated as:

$$d_{avo} = \frac{3}{2}\frac{7d(G)^4 + 14d(G)^3 + 14d(G)^2 + 8d(G) - 2d(G)d(G)(\mathrm{mod}\,2) - d(G)\,\mathrm{mod}\,2}{14d(G)^3 + 21d(G)^2 + 28d(G) - 3d(G)\,\mathrm{mod}\,2} \tag{26}$$

Fig. 7 shows a comparison of diameters and average path lengths between theoretical and real graphs.

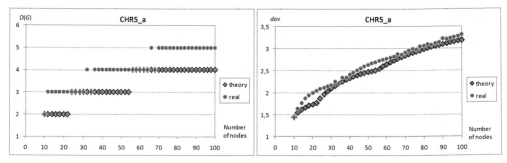

Fig. 7. Comparison of diameter and average path length of theoretical and real graphs CHR5_a

All graphs of this type are symmetrical, the values of their basic parameters do not depend on the number of source node.

Graph CHR5_b.

Definition 6. The modified fifth degree chordal ring called CHR5_b is denoted by $CHR5\_b(p; q_1, q_2, q_3)$ where p is even and means number of nodes; q_1, q_2, q_3 are chords, where chord q_1 and q_2 possess even lengths, whereas the length of q_3 is odd. The values of p and q_1, q_2, q_3 must be lower than $p/2$. Each even node i_{2k} is connected to five other nodes: i_{2k-1}, i_{2k+1}, $i_{2k-q1(\mathrm{mod}\ p)}$, $i_{2k+q1(\mathrm{mod}\ p)}$, $i_{2k+q3(\mathrm{mod}\ p)}$, while odd node i_{2k+1} is connected to i_{2k}, i_{2k+2}, $i_{2k+1-q2(\mathrm{mod}\ p)}$, $i_{2k+1+q2(\mathrm{mod}\ p)}$ and $i_{2k+1-q3(\mathrm{mod}\ p)}$ $(0 \le k < p/2)$.

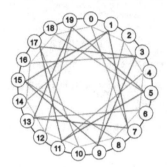

Fig. 8. Modified chordal ring CHR5_b(20; 2,6,7)

Fig. 8 shows an example of CHR5_b. In table 5, the numbers of nodes in the layers of an optimal graph is shown.

d	1	2	3	4	5	6	7	8
p_{do}	5	20	61	140	267	454	713	1056

Table 5. Maximal number of nodes in the layers

When d is bigger than 2, the maximal number of nodes which can appear in the successive layers is described by:

$$p_{do} = 2d^3 + 5d - 8 \tag{27}$$

The total number of nodes p_o in the optimal graph with diameter $D(G) > 1$ is given by:

$$p_o = \frac{1}{2}\left(D(G)^4 + 2D(G)^3 + 6D(G)^2 - 11D(G) + 18\right) \tag{28}$$

This was also confirmed by constructing the possible graphs. These results can be seen in Table 6.

$d(G)$	1	2	3	4	5	6	7	8
p_{do}	6	26	87	227	494	948	1661	2717

Table 6. Total numbers of nodes in optimal graphs versus diameter

The average path length in optimal graphs can be expressed as:

$$d_{avo} = \frac{12D(G)^5 + 30D(G)^4 + 70D(G)^3 - 45D(G)^2 - 97D(G) + 300}{15\left(D(G)^4 + 2D(G)^3 + 6D(G)^2 - 11D(G) + 16\right)} \qquad (29)$$

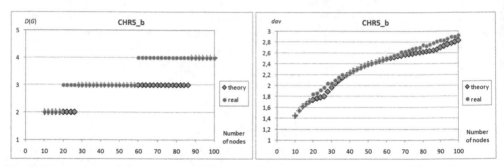

Fig. 9. Comparison of diameter and average path length of theoretical and real graphs CHR5_b

Fig. 9 shows a comparison of the diameter and average path length between theoretical and real graphs.

Not all these graphs are symmetric, but most of the graphs possessing parameters equal or close to ideal graphs are symmetric. In table 7 examples of the real and ideal chordal rings are presented.

Number of nodes	q_1	q_2	q_3	d_{av}
8	2	2	3	1.2857143
10	2	4	3	1.4444444
12	2	2	5	1.5454545
14	2	2	7	1.6153846
16	2	6	7	1.6666666
18	2	4	7	1.7058823
38	6	12	9	2.1891892
40	6	18	9	2.2307692
42	4	8	19	2.2682927
44	6	18	15	2.3023255
46	4	8	19	2.3333333
48	10	22	7	2.3617022
48	14	22	17	2.3617022
50	4	8	21	2.3877552
52	10	14	17	2.4117646
54	4	14	23	2.4339623
56	10	22	5	2.4545455
58	4	14	23	2.4736843

Table 7. Examples of ideal graphs CHR5_b

Graph CHR5_c

Definition 7. The modified fifth degree chordal ring called CHR5_c is denoted by CHR5_c(p; q_1, q_2, q_3), where p is even and means number of nodes; q_1, q_2, q_3 are chords, all chords possess odd lengths less then $p/2$. The values of p and q_1, q_2, q_3 must be prime each other. Each even node i_{2k} is connected to five other nodes: i_{2k-1}, i_{2k+1}, $i_{2k+q1(\text{mod } p)}$, $i_{2k+q2(\text{mod } p)}$, $i_{2k+q3(\text{mod } p)}$, while odd node i_{2k+1} is connected to i_{2k}, i_{2k+2}, $i_{2k+1-q1(\text{mod } p)}$, $i_{2k+1-q2(\text{mod } p)}$ and $i_{2k+1-q3(\text{mod } p)}$ where ($0 \le k < p/2$). ☑

Fig. 10 shows an example of CHR5_c. In Table 8 the numbers of nodes in the layers of an optimal graph is shown.

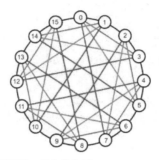

Fig. 10. Modified chordal ring CHR5_c(16; 3,5,,7)

d	1	2	3	4	5	6	7	8
p_{do}	5	20	50	110	200	340	550	850

Table 8. Number of nodes appearing in successive layers

In the case when d – number of layer is bigger than 2 the number of nodes in the layers can be described by the following expression:

$$p_{do} = 5\left(\frac{2}{3}d^3 - 5d^2 + \frac{67}{3}d - 30\right) \tag{30}$$

The total number of nodes p_o in the optimal graph with diameter $D(G) > 1$ is given by:

$$p_o = \frac{1}{6}\left(5D(G)^4 - 40D(G)^3 + 265D(G)^2 - 590D(G) + 516\right) \tag{31}$$

In table 9 the total number of nodes in virtual, optimal graphs, as described by the above expression, is shown.

$D(G)$	1	2	3	4	5	6	7	8
p_o	6	26	76	186	386	726	1276	2126

Table 9. Total numbers of nodes in optimal graphs

The average path length in optimal graphs can be calculated using the expression:

$$d_{avo} = \frac{8D(G)^5 - 55D(G)^4 + 310D(G)^3 - 305D(G)^2 - 678D(G) + 1260}{2\left(5D(G)^4 - 40D(G)^3 + 265D(G)^2 - 590D(G) + 510\right)} \qquad (32)$$

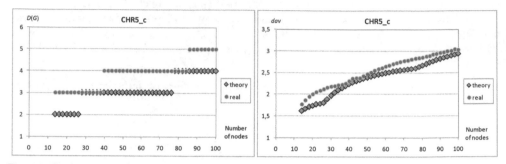

Fig. 11. Comparison of diameter and the average path length of theoretical and real CHR5_c graphs

Fig. 11 shows a comparison of diameter and the average path length between theoretical and real graphs. All graphs of this type are symmetrical, but they couldn't find any ideal graph.

Graph CHR5_d.

Definition 8. The modified fifth degree chordal ring called CHR5_d is denoted by CHR5_d(p; $q_1,q_2,p/2$), where p means the number of nodes and is positive and even; q_1, q_2 are chords which possess odd lengths less then $p/2$. The values of p and q_1, q_2 must be prime each other. Each even node i_{2k} is connected to five other nodes: i_{2k-1}, i_{2k+1}, $i_{2k+q1(mod\ p)}$, $i_{2k+q2(mod\ p)}$, $i_{2k+p/2(mod\ p)}$, while odd node i_{2k+1} is connected to i_{2k}, i_{2k+2}, $i_{2k+1-q1(mod\ p)}$, $i_{2k+1-q2(mod\ p)}$ and $i_{2k+1+p/2(mod\ p)}$ where $(0 \le k < p/2)$.

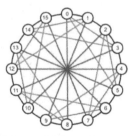

Fig. 12. Example of modified chordal ring CHR5_d(16; 3,5,,8)

Fig. 12 shows an example of CHR5_d. In table 10, the numbers of nodes in the layers of an optimal graph are shown. It should be noted that there are two different number of nodes in layers, depending on whether the total number of nodes is divisible by 4 or not.

d	1	2	3	4	5	6	7	8
$p_o = 0$ (mod 4)	5	16	36	66	106	156	216	286
$p_o \ne 0$ (mod 4)	5	18	39	72	113	166	227	300

Table 10. Maximal number of nodes in the successive layers

When the number of layer d is bigger than 1, the number of nodes in the layers can be described by the following expression:

$$\text{If } p = 0 \ (\text{mod } 4) \text{ then } \quad p_{do} = 5d^2 - 5d + 6$$
$$\text{if } p \neq 0 \ (\text{mod } 4) \text{ and layer number is even then } \quad p_{do} = 5d^2 - 3d + 4 \quad (33)$$
$$\text{if } p \neq 0 \ (\text{mod } 4) \text{ and layer number is odd then } \quad p_{do} = 5d^2 - 3d + 3$$

The total number of nodes p_o in the optimal graph depending on the diameter is given by:

$$p_o = \frac{D(G)\left(4D(G)^2 + 13\right)}{3} \quad \text{if } p_o = 0 \bmod 4$$

$$p_o = \frac{D(G)\left(10D(G)^2 + 6D(G) + 17\right)}{6} + 1 \quad \text{if } p \neq 0 \bmod 4 \text{ and } D(G) \text{ is even} \quad (34)$$

$$p_o = \frac{D(G)\left(10D(G)^2 + 6D(G) + 17\right)}{6} + \frac{1}{2} \quad \text{if } p \neq 0 \bmod 4 \text{ and } D(G) \text{ is odd}$$

In Table 11 the total number of nodes in virtual optimal graphs described by formula given above is shown.

$D(G)$	1	2	3	4	5	6	7	8
$p_o = 0 \ (\text{mod } 4)$	6	22	58	124	230	386	602	888
$p_o \neq 0 \ (\text{mod } 4)$	6	24	63	135	248	414	641	941

Table 11. Total numbers of nodes forming optimal graphs versus diameter

The average path length in optimal graphs can be calculated using the expressions:

$$\text{If } p_o = 0 \bmod 4$$
$$d_{avo} = \frac{D(G)\left(2D(G)^4 + 5D(G)^3 + 12D(G)^2 + 10D(G) + 1\right)}{2\left(5D(G)^3 + 13D(G) - 3\right)} \quad (35)$$

$$\text{if } p_o \neq 0 \bmod 4$$
$$d_{avo} = \frac{6D(G)\left(5D(G)^2 - 3D(G) + 3 + (D(G) + 1)\bmod 2\right)}{D(G)\left(10D(G)^2 + 6D(G) + 17\right) + 3 + 3(D(G) + 1)\bmod 2}$$

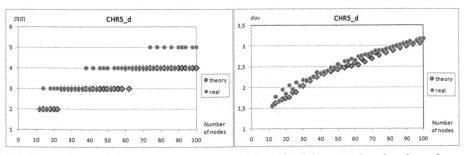

Fig. 13. Comparison of diameter and average path length of theoretical and real graphs CHR5_d

Fig. 13 shows a comparison of diameter and average path length between theoretical and real graphs.

All this type of chordal rings are symmetrical. Ideal graphs are only for the cases where the number of nodes is divisible by 4. Examples are given in Table 12.

Number of nodes	q_1	q_2	$p/2$	d_{av}
12	3	5	6	1,545455
16	3	5	8	1,666667
28	7	11	14	2,037037
32	5	13	16	2,16129
36	5	13	18	2,257143
40	7	17	20	2,333333
44	5	13	22	2,395349
52	5	17	26	2,490196
76	13	21	38	2,893333
80	7	25	40	2,949367
84	9	23	42	3,00000
88	7	27	44	3,045977
96	7	29	48	3,126316

Table 12. Examples of ideal graphs CHR5_d

Graph CHR5_e.

Definition 9. The modified fifth degree chordal ring called CHR5_e is denoted by CHR5_e(p; $q_1, q_2, p/2$), where p means the number of nodes and is positive and even; q_1, q_2 are chords which possess even lengths less then $p/2$. The values of $p/2$ and q_1, q_2 must be prime each other. Each even node i_{2k} is connected to five other nodes: i_{2k-1}, i_{2k+1}, $i_{2k+q1(\text{mod } p)}$, $i_{2k-q1(\text{mod } p)}$, $i_{2k+p/2(\text{mod } p)}$, while odd node i_{2k+1} is connected to i_{2k}, i_{2k+2}, $i_{2k+1+q2(\text{mod } p)}$, $i_{2k+1-q2(\text{mod } p)}$ and $i_{2k+1+p/2(\text{mod } p)}$ where $(0 \le k < p/2)$.

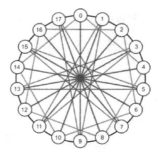

Fig. 14. Example modified chordal ring CHR5_e(18; 4,8,9)

Fig. 14 shows an example of CHR5_e. In Table 13 the number of nodes appearing in the successive layers of optimal graphs is shown. It should be observed that there are two different number of nodes in layers, depending on the total number of nodes are they divisible by 4 or not.

d	1	2	3	4	5	6	7	8
$p_o = 0 \pmod 4$	5	16	42	88	152	232	328	440
$p_o \neq 0 \pmod 4$	5	18	48	96	160	240	336	448

Table 13. Maximal number of nodes in the successive layers

When the number of layers d - is bigger than 1 the number of nodes in the layers can be described by the following expressions:

$$if \; p = 0 \,(\mathrm{mod}\,4) \; and \; d \rangle 2 \; then \quad p_{do} = 8d(d-1)$$
$$if \; p \neq 0 \,(\mathrm{mod}\,4) \; and \; d \rangle 3 \; then \quad p_{do} = 8d(d-1)-8$$
(36)

The total number of nodes p_o in the optimal graph depending on the diameter is given by:

$$p_o = \frac{8D(G)^3 - 32d(G) + 72}{3} = \frac{8D(G)\left(D(G)^2 - 4\right)}{3} + 24 \quad if \; p_o = 0 \bmod 4 \; and \; D(G) \rangle 2$$
(37)

$$p_o = \frac{8D(G)^3 - 8d(G) + 24}{3} = \frac{8D(G)\left(D(G)^2 - 1\right)}{3} + 8 \quad if \; p_o \neq 0 \bmod 4 \; and \; D(G) \rangle 1$$

In Table 14 the total number of nodes in virtual optimal graphs described by the above expressions is shown.

$D(G)$	1	2	3	4	5	6	7	8
$p_o = 0 \pmod 4$	6	22	64	152	304	536	864	1304
$p_o \neq 0 \pmod 4$	6	24	72	168	328	568	904	1352

Table 14. Total numbers of nodes forming optimal graphs versus diameter

The average path length in optimal graphs can be calculated using this expression:

$$When \; p_o = 0 \bmod 4 \; and \; D(G) \rangle 2$$
$$d_{avo} = \frac{6D(G)^4 + 4D(G)^3 - 18D(G)^2 - 16D(G) + 105}{8D(G)^3 - 32D(G) + 69}$$
(38)
$$when \; p_o \neq 0 \bmod 4 \; and \; D(G) \rangle 1$$
$$d_{avo} = \frac{6D(G)^4 + 4D(G)^3 - 6D(G)^2 - 4D(G) + 27}{8D(G)^3 - 8D(G) + 21}$$

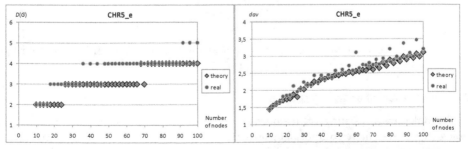

Fig. 15. Comparison of diameter and average path length of theoretical and real graphs CHR5_e

Fig. 15 shows diameter and average path lengths in theoretical and real graphs with up to 100 nodes.

Some but not all of these graphs are symmetric, and as illustrated in Fig. 15 the non symmetric graphs generally have parameters closer to those of ideal graphs. Only a few ideal graphs are found, of which some examples are shown in Table 15.

Number of nodes	q_1	q_2	$p/2$	d_{av}
10	2	4	5	1,444444
12	2	2	6	1,545455
14	2	4	7	1,615385
16	2	6	8	1,666667
28	6	10	14	2,037037
32	6	10	16	2,161290
34	4	10	17	2,151515
38	4	8	19	2,243243
42	4	8	21	2,317073

Table 15. Examples of ideal graphs CHR5_e

To sum up, in the first group of analyzed graphs the best parameters have CHR5_b graphs. They possess minimal diameter and average path length in comparison to the other analyzed chordal rings, and the parameters of the real graphs are close or equal to parameters of ideal graphs. Additionally, most of the best graphs are symmetric, what is also an advantage for the application in real networks.

Fig. 16 shows the comparisons of the real graphs in the first group.

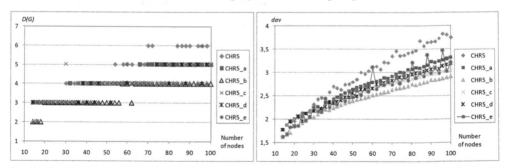

Fig. 16. Comparison of diameter and average path length of all real modified chordal rings belonging to the first group of analyzed graphs.

3.2 Second group of analyzed graphs

The chordal rings consisting of $4i$ nodes ($i = 2, 3, 4, \ldots,$) belong to this group. These topologies are often more complicated, since they are less symmetric. Basing on patterns for ideal and optimal graphs it is possible to derive expressions for average distance and diameter for all of the different topologies in this group of graphs.

Graph CHR5_f.

Definition 10. The modified fifth degree chordal ring called CHR5_f is denoted by CHR5_f(p; q_1,q_2,q_3, $p/2$), where p means the number of nodes and is positive and divisible by 4; q_1, q_2, q_3, are chords which possess even lengths less then $p/2$. The values of $p/4$ and q_1, q_2, q_3 must be prime each other. Each even node i_{2k} is connected to five other nodes: i_{2k-1}, i_{2k+1}, $i_{2k+q1(\text{mod } p)}$, $i_{2k-q1(\text{mod } p)}$, $i_{2k+p/2(\text{mod } p)}$, while odd node i_{2k+1} is connected to i_{2k}, i_{2k+2}, $i_{2k+1+q2(\text{mod } p)}$, $i_{2k+1-q2(\text{mod } p)}$ and $i_{2k+1+p/2(\text{mod } p)}$ and node i_{2k-1} is connected to i_{2k}, i_{2k-2}, $i_{2k-1+q3(\text{mod } p)}$, $i_{2k-1-q3(\text{mod } p)}$ and $i_{2k-1+p/2(\text{mod } p)}$ where ($0 \le k < p/2$).

An example is shown in Fig. 17.

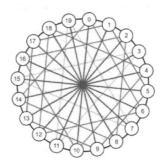

Fig. 17. Example of modified chordal ring CHR5_f(20; 4,6,,8,10)

This structure is more complicated since the number of nodes appearing in successive layers depends on whether the total number of nodes is divisible by 8 or not, and also on whether it seen from odd or even node number in the graph. This creates multiple cases, which also complicates deriving the basic parameters. Table 16 shows the experimentally obtained results, which are the basis for further analysis.

d	Number of nodes	1	2	3	4	5	6	7	8
$p_{do\ p\ =\ 0\ (mod\ 8)}$	Even	5	16	44	112	248	488	888	1496
	Odd	5	16	46	116	262	536	984	1640
$p_{do\ p\ \neq 0\ (mod\ 8)}$	Even	5	16	48	136	312	616	1096	1784
	Odd	5	18	58	152	340	668	1172	1884

Table 16. Maximal number of nodes in the successive layers

The number of nodes in the layers can be described by the following formula:

$$When\ p = 0\ (\text{mod }8)\ and\ d \rangle 5\ then$$

$$p_{do\ even} = 5\frac{1}{3}d^3 - 8d^2 - 173\frac{1}{3} + 664 \quad p_{do\ odd} = 5\frac{1}{3}d^3 - 8d^2 - 125\frac{1}{3} + 424$$

$$when\ p \neq 0\ (\text{mod }8)\ and\ d \rangle 4\ then$$

$$p_{do\ even} = 5\frac{1}{3}d^3 - 8d^2 - 93\frac{1}{3} + 312 \quad p_{do\ odd} = 5\frac{1}{3}d^3 - 8d^2 - 69\frac{1}{3} + 220$$

(39)

In Table 17 the total number of nodes in optimal graphs given by above expression is shown.

$d(G)$	Number of nodes	1	2	3	4	5	6	7	8
$p_{do\ p\ =\ 0\ (mod\ 8)}$	even	6	22	66	178	426	914	1802	3298
	odd	6	22	68	184	446	982	1966	3606
$p_{do\ p\ \neq 0\ (mod\ 8)}$	even	6	22	70	206	518	1134	2230	4014
	odd	6	24	82	234	574	1242	2414	4298

Table 17. Total numbers of nodes forming optimal graphs versus diameter

The total number of nodes p_o in the optimal graph as a function of the diameter is given by:

$$When\ p = 0\ (mod\,8)\ and\ d \rangle 4$$

$$p_{o\,even} = 1\frac{1}{3}D(G)^4 - 89\frac{1}{3}D(G)^2 + 576D(G) - 1054$$

$$p_{o\,odd} = 1\frac{1}{3}D(G)^4 - 65\frac{1}{3}D(G)^2 + 360D(G) - 554 \tag{40}$$

$$when\ p \neq 0\ (mod\,8)\ and\ d \rangle 3$$

$$p_{o\,even} = 1\frac{1}{3}D(G)^4 - 49\frac{1}{3}D(G)^2 + 264D(G) - 402$$

$$p_{o\,odd} = 1\frac{1}{3}D(G)^4 - 49\frac{1}{3}D(G)^2 + 264D(G) - 400$$

The average path length in optimal graphs can be calculated using expressions:

$$When\ p_o = 0\ mod\ 8\ and\ D(G) \rangle 2$$

$$d_{avo} = \frac{6D(G)^4 + 4D(G)^3 - 18D(G)^2 - 16D(G) + 105}{8D(G)^3 - 32D(G) + 69} \tag{41}$$

$$when\ p_o \neq 0\ mod\ 8\ and\ D(G) \rangle 1$$

$$d_{avo} = \frac{6D(G)^4 + 4D(G)^3 - 6D(G)^2 - 4D(G) + 27}{8D(G)^3 - 8D(G) + 21}$$

Fig. 19 shows diameter and average path lengths in theoretical and real graphs with up to 100 nodes.

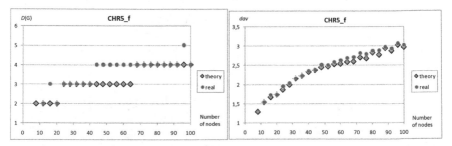

Fig. 18. Comparison of diameter and average path length of theoretical and real graphs CHR5_f

Only three ideal graphs found, which is presented in Table 18.

Number of nodes	q_1	q_2	q_3	$p/2$	d_{av}
12	2	4	4	6	1,545
20	6	4	4	10	1,737
32	6	12	12	16	2,161

Table 18. All founded ideal graphs CHR5_f with up to 100 nodes.

Graph CHR5_g.

Definition 11. The modified fifth degree chordal ring called CHR5_g is denoted by CHR5_g(p; q_1,q_2,q_3), where p means the number of nodes, is positive and divisible by 4; q_1 is chord has odd length; q_2, q_3 are chords which possess even lengths and less then $p/2$. The values of p and q_1 must be prime each other. Even node $i_{2k=0(mod4)}$ is connected to five other nodes: i_{2k-1}, i_{2k+1}, $i_{2k+q1(mod\ p)}$, $i_{2k-q1(mod\ p)}$ and $i_{2k+q2(mod\ p)}$, when $i_{2k=2(mod4)}$ than this node is connected to i_{2k-1}, i_{2k+1}, $i_{2k+q1(modp)}$, $i_{2k-q1(mod\ p)}$ and $i_{2k-q2(modp)}$; while odd node $i_{(2k+1)=1(mod4)}$ is connected to i_{2k}, i_{2k+2}, $i_{2k+1+q1(modp)}$, $i_{2k+1-q1(modp)}$, $i_{2k+1+q3(modp)}$; and any node $i_{(2k+1)=3(mod4)}$ is connected to i_{2k}, i_{2k+2}, $i_{2k+1+q1(modp)}$, $i_{2k+1-q1(modp)}$, $i_{2k+1-q3(modp)}$. ☑

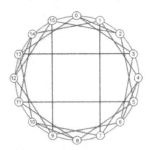

Fig. 19. Example of modified chordal ring CHR5_g(16; 3,2,6)

An example is shown in Fig. 20. The number of nodes in the layers of an optimal graph is given in table 19.

d	1	2	3	4	5	6	7	8
p_{do}	5	16	42	102	183	302	491	704

Table 19. Maximal number of nodes in the layers

In the case when the number of layer is bigger than 1, the number of nodes in the layers can be described by the following formula:

$$\text{When} \quad d = 0 \,(mod\,3) \text{ then } p_{do} = \frac{4}{3}d^3 + \frac{8}{3}d - 2$$

$$\text{when} \quad d = 1 \,(mod\,3) \text{ and } d \rangle 2 \text{ then } p_{do} = \frac{4}{3}d^3 + \frac{2}{9}d^2 + \frac{29}{9}d + \frac{2}{9} \qquad (42)$$

$$\text{when} \quad d = 2 \,(mod\,3) \text{ and } d \rangle 4 \text{ then } p_{do} = \frac{4}{3}d^3 - \frac{2}{9}d^2 + \frac{41}{9}d - \frac{8}{9}$$

The total number of nodes p_o in the optimal graph can be calculated using the expressions:

$$\text{When} \quad d = 0 \,(\text{mod}\,3) \; \text{then} \quad p_o = \frac{1}{3}D(G)^4 + \frac{2}{3}D(G)^3 + 2D(G)^2 + \frac{2}{3}D(G) - 1$$

$$\text{when} \quad d = 1\,(\text{mod}\,3) \; \text{and} \; d \rangle 2 \; \text{then} \quad p_o = \frac{1}{3}D(G)^4 + \frac{2}{3}D(G)^3 + \frac{20}{9}D(G)^2 + \frac{5}{9}D(G) + \frac{2}{9} \quad (43)$$

$$\text{when} \quad d = 2\,(\text{mod}\,3) \; \text{and} \; d \rangle 2 \; \text{then} \quad p_o = \frac{1}{3}D(G)^4 + \frac{2}{3}D(G)^3 + 2D(G)^2 + \frac{4}{3}D(G) + \frac{2}{3}$$

In Table 20 the total number of nodes in optimal graphs described by formula given above is shown.

$D(G)$	1	2	3	4	5	6	7	8
p_o	6	22	64	166	349	651	1142	1846

Table 20. Diameters and total numbers of nodes in optimal graphs

The average path length in the optimal graphs as a function of its diameter can be calculated using equations (44).

When $D(G) = 0\,(\text{mod}\,3)$

$$d_{avo} = \frac{64{,}8\left(\dfrac{D(G)}{3}\right)^5 + 54\left(\dfrac{D(G)}{3}\right)^4 + \dfrac{124}{3}\left(\dfrac{D(G)}{3}\right)^3 - 10\left(\dfrac{D(G)}{3}\right)^2 - \dfrac{47}{15}\dfrac{D(G)}{3} - 4}{\dfrac{1}{3}\left(\dfrac{D(G)}{3}\right)^4 + \dfrac{2}{3}\left(\dfrac{D(G)}{3}\right)^3 + 2\left(\dfrac{D(G)}{3}\right)^2 + \dfrac{2}{9}D(G) - 2}$$

when $D(G) = 1\,(\text{mod}\,3)$

$$d_{avo} = \frac{64{,}8\left(\dfrac{D(G)-1}{3}\right)^5 - 54\left(\dfrac{D(G)-1}{3}\right)^4 + \dfrac{124}{3}\left(\dfrac{D(G)-1}{3}\right)^3 - 14\left(\dfrac{D(G)-1}{3}\right)^2}{\dfrac{1}{3}\left(\dfrac{D(G)-1}{3}\right)^4 + \dfrac{2}{3}\left(\dfrac{D(G)-1}{3}\right)^3 + \dfrac{20}{9}\left(\dfrac{D(G)-1}{3}\right)^2 + \dfrac{5}{27}(D(G)-1) - \dfrac{7}{9}} +$$

$$+ \frac{-\dfrac{43}{15}\dfrac{D(G)-1}{3} - 4}{\dfrac{1}{3}\left(\dfrac{D(G)-1}{3}\right)^4 + \dfrac{2}{3}\left(\dfrac{D(G)-1}{3}\right)^3 + \dfrac{20}{9}\left(\dfrac{D(G)-1}{3}\right)^2 + \dfrac{5}{27}(D(G)-1) - \dfrac{7}{9}}$$

when $D(G) = 2\,(\text{mod}\,3)$

$$d_{avo} = \frac{64{,}8\left(\dfrac{D(G)-2}{3}\right)^5 + 162\left(\dfrac{D(G)-2}{3}\right)^4 + \dfrac{574}{3}\left(\dfrac{D(G)-2}{3}\right)^3 + 117\left(\dfrac{D(G)-2}{3}\right)^2}{\dfrac{1}{3}\left(\dfrac{D(G)-2}{3}\right)^4 + \dfrac{2}{3}\left(\dfrac{D(G)-2}{3}\right)^3 + 2\left(\dfrac{D(G)-2}{3}\right)^2 + \dfrac{4}{9}(D(G)-2) - \dfrac{1}{3}} +$$

$$+ \frac{-\dfrac{523}{15}\dfrac{D(G)-2}{3} + 1}{\dfrac{1}{3}\left(\dfrac{D(G)-2}{3}\right)^4 + \dfrac{2}{3}\left(\dfrac{D(G)-2}{3}\right)^3 + 2\left(\dfrac{D(G)-2}{3}\right)^2 + \dfrac{4}{9}(D(G)-2) - \dfrac{1}{3}}$$

(44)

Fig. 20 shows diameter and average path lengths in theoretical and real graphs with up to 100 nodes.

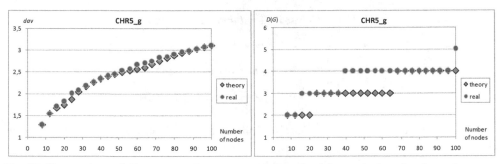

Fig. 20. Comparison of diameter and average path length of theoretical and real graphs CHR5_g

Graph CHR5_h.

Definition 12. The modified fifth degree chordal ring called CHR5_h is denoted by CHR5_h(p; q_1,q_2,q_3,q_4,q_5), where p means the number of nodes and is positive and divisible by 4; q_1, q_2, q_3, q_4 are chords which possess even lengths less then $p/2$, $q_5=p/2$. Even node i_{2k} is connected to five other nodes: i_{2k-1}, i_{2k+1}, $i_{2k+q5(\mathrm{mod}\ p)}$ and to $i_{2k+q1(\mathrm{mod}\ p)}$, $i_{2k-q1(\mathrm{mod}\ p)}$ when number of node is equal to 0 mod 4 or to $i_{2k+q2(\mathrm{mod}\ p)}$, $i_{2k-q2(\mathrm{mod}\ p)}$ when number of node is equal to 2 mod 4, while odd node i_{2k+1} is connected to i_{2k}, i_{2k+2}, $i_{2k+1+q5(\mathrm{mod}\ p)}$ and $i_{2k+1+q3(\mathrm{mod}\ p)}$, $i_{2k+1-q3(\mathrm{mod}\ p)}$ when number of node is equal to 1 mod 4 or to $i_{2k+1+q4\ (\mathrm{mod}\ p)}$, $i_{2k+1-q4\ (\mathrm{mod}\ p)}$ when number of node is equal to 3 mod 4.

An example is shown in Fig. 21, and the number of nodes in the layers of an optimal graph is shown in tables 21 and 22.

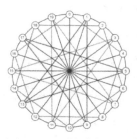

Fig. 21. Example of modified chordal ring CHR5_h(20; 4,4,8,8,10)

d	1	2	3	4	5	6	7	8
p_{do}	5	18	64	196	524	1244	2636	5068

Table 21. The number of nodes appearing in layers when p is not divided by 8

d	1	2	3	4	5	6	7	8
p_{do}	5	16	48	136	358	868	1908	3804

Table 22. The number of nodes appearing in layers when p is divided by 8

The number of nodes in the layers, as shown in Tables 21 and 22, can be described by the following expression:

$$
When \quad d > 5 \text{ and } p \text{ is not divisible by 8 then}
$$
$$
p_{do} = \frac{8}{3}d^4 - \frac{16}{3}d^3 + 154\frac{2}{3}d^2 + 1133\frac{8}{3}d - 2292
$$
$$
when \quad d > 7 \text{ and } p \text{ is divisible by 8 then} \tag{45}
$$
$$
p_{do} = \frac{8}{3}d^4 - \frac{16}{3}d^3 - 266\frac{2}{3}d^2 + 2285\frac{1}{3}d - 5604
$$

In Table 23 the total number of nodes in optimal graphs described by expression given above is shown.

$D(G)$	1	2	3	4	5	6	7	8
p_o is not divisible by 8	6	24	88	284	808	2052	4688	9756
p_o is divisible by 8	6	22	70	206	564	1432	3340	7144

Table 23. Total numbers of nodes forming optimal graphs versus diameter

$$
When \quad D(G) > 4 \text{ and } p \text{ is not divisible by 8 then}
$$
$$
p_o = \frac{8}{15}D(G)^5 - \frac{160}{3}D(G)^3 + 488d^2 + 1751\frac{1}{5}d + 2364
$$
$$
when \quad D(G) > 6 \text{ and } p \text{ is divisible by 8 then} \tag{46}
$$
$$
p_o = \frac{8}{15}D(G)^5 - \frac{272}{3}D(G)^3 - 1008\frac{2}{3}D(G)^2 + 4505\frac{13}{15}D(G) + 7624
$$

The average path length in optimal graphs:

$$
When \quad D(G) > 4 \text{ and } p \text{ is not divisible by 8}
$$
$$
d_{avo} = \frac{\frac{4}{9}D(G)^6 + \frac{4}{15}D(G)^5 - 40\frac{2}{9}D(G)^4 + 298\frac{2}{3}D(G)^3 - 618\frac{2}{9}D(G)^2 - 956\frac{14}{15}D(G)}{\frac{8}{15}D(G)^5 - \frac{160}{3}D(G)^3 + 488d^2 + 1751\frac{1}{5}d + 2363} +
$$
$$
+ \frac{3905}{\frac{8}{15}D(G)^5 - \frac{160}{3}D(G)^3 + 488d^2 + 1751\frac{1}{5}d + 2363}
$$
$$
When \quad D(G) > 4 \text{ and } p \text{ is divisible by 8} \tag{47}
$$
$$
d_{avo} = \frac{\frac{4}{9}D(G)^6 + \frac{4}{15}D(G)^5 - 68\frac{2}{3}D(G)^4 + 626\frac{2}{3}D(G)^3 + 1726\frac{2}{9}D(G)^2 - 2420\frac{14}{15}D(G)}{\frac{8}{15}D(G)^5 - \frac{272}{3}D(G)^3 - 1008\frac{2}{3}D(G)^2 + 4505\frac{13}{15}d + 7623} +
$$
$$
+ \frac{14695}{\frac{8}{15}D(G)^5 - \frac{272}{3}D(G)^3 - 1008\frac{2}{3}D(G)^2 + 4505\frac{13}{15}d + 7623}
$$

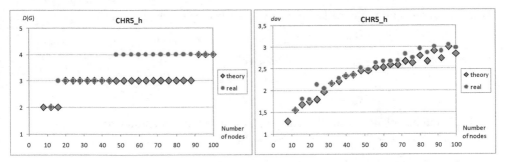

Fig. 22. Comparison of diameter and average path length of theoretical and real graphs CHR5_h

Fig. 22 shows diameter and average path lengths in theoretical and real graphs with up to 100 nodes.

There are only three chordal rings which possess basic parameters equal to parameters of theoretical graphs: These are: CHR5_h(12; 4,4,4,6), CHR5_h(40; 4,12,4,12,20) and CHR5_h(44; 4,8,12,16,22).

Graph CHR5_i.

Definition 13. The modified fifth degree chordal ring called CHR5_i is denoted by CHR5_i(p; q_1,q_2,q_3,q_4), where p means the number of nodes and is positive and divisible by 4; q_1, q_2, q_3, q_4 are chords which possess: q_1, odd length and q_2, q_3, q_4 - even lengths less then $p/2$. Each node is connected to five other nodes. Even node i_{2k} is connected to i_{2k-1}, i_{2k+1}, $i_{2k+q1(mod\ p)}$, $i_{2k-q1(mod\ p)}$ and $i_{2k+q?(mod\ p)}$, while odd node $i_{(?k+1)=1(mod4)}$ is connected to i_{2k}, i_{2k+2}, $i_{2k+1-q2(mod\ p)}$, $i_{2k+1+q3(mod\ p)}$, $i_{2k+1-q3(mod\ p)}$ and odd node $i_{(2k+1)=3(mod4)}$ is connected to i_{2k}, i_{2k+2}, $i_{2k+1-q2(mod\ p)}$, $i_{2k+1+q4(mod\ p)}$, $i_{2k+1-q4(mod\ p)}$.

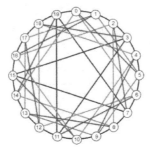

Fig. 23. Example of modified chordal ring CHR5_i(20; 5,6,4,8)

An example of a CHR5_i is shown in Fig. 23. The distribution of nodes in the layers depends on whether the graph is seen from odd or even node in Table 24 is given.

d	1	2	3	4	5	6	7	8	Node number
p_{do}	5	20	63	190	465	1010	2001	3594	even
	5	20	69	196	493	1094	2141	3790	odd

Table 24. Maximal number of nodes in the layers

The distribution of nodes in the layers can be described by the expression:

$$When \quad d > 4 \quad and \quad node \quad number \quad is \quad even$$

$$p_{do} = d^4 + 6d^2 - 192d + 650$$

$$when \quad d > 4 \quad and \quad node \quad number \quad is \quad odd$$

$$p_{do} = d^4 + 6d^2 - 136d + 398$$

(48)

The total number of nodes in optimal graphs calculated depending on the source node number, given in Table 25, can be expressed as follows:

$$When \quad D(G) > 3 \quad and \quad node \quad number \quad is \quad even$$

$$p_{o\ even} = \frac{D(G)^5}{5} + \frac{D(G)^4}{2} + \frac{7}{3}D(G)^3 - 93D(G)^2 + 554\frac{29}{30}D(G) - 935$$

$$when \quad D(G) > 3 \quad and \quad node \quad number \quad is \quad odd$$

$$p_{oodd} = \frac{D(G)^5}{5} + \frac{D(G)^4}{2} + \frac{7}{3}D(G)^3 - 65D(G)^2 + 330\frac{29}{30}D(G) - 475$$

(49)

$D(G)$	1	2	3	4	5	6	7	8	Node number
p_o	6	26	89	279	744	1754	3755	7349	even
	6	26	95	291	784	1878	4019	7809	odd

Table 25. Total numbers of nodes in the optimal graphs

The average path length in optimal graphs is equal to:

$$d_{avo} = \frac{d_{avoeven} + d_{avoodd}}{2} =$$

$$= \frac{\frac{D(G)^6}{6} + \frac{D(G)^5}{2} + 1\frac{11}{12}D(G)^4 - 42\frac{1}{3}D(G)^3 + 132\frac{5}{12}D(G)^2 + 176\frac{1}{3}D(G) - 764}{2\left(\frac{D(G)^5}{5} + \frac{D(G)^4}{2} + \frac{7}{3}D(G)^3 - 93D(G)^2 + 554\frac{29}{30}D(G) - 936\right)} +$$

$$+ \frac{\frac{D(G)^6}{6} + \frac{D(G)^5}{2} + 1\frac{11}{12}D(G)^4 - 61\frac{1}{3}D(G)^3 + 230\frac{5}{12}D(G)^2 + 293D(G) - 1646}{2\left(\frac{D(G)^5}{5} + \frac{D(G)^4}{2} + \frac{7}{3}D(G)^3 - 65D(G)^2 + 330\frac{29}{30}D(G) - 476\right)}$$

(50)

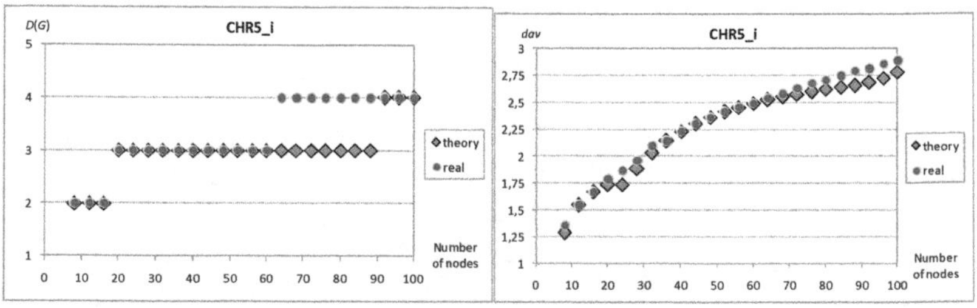

Fig. 24. Comparison of diameter and average path length of theoretical and real graphs CHR5_i

Fig. 24 shows diameter and average path lengths in theoretical and real graphs with up to 100 nodes.

For graphs with less than 72 nodes it is possible to find real graphs with parameters close to those of ideal graphs. However the difference becomes bigger for larger graphs. The differences seem to come from the different path lengths calculated from odd and even nodes. The ideal graphs found are shown in table 26.

Number of nodes	q_1	q_2	q_3	q_4	d_{av}
12	5	2	4	4	1,545
16	7	2	4	4	1,667
36	13	10	16	16	2,143
40	9	6	12	12	2,231
44	9	6	12	20	2,302
48	5	10	20	20	2,362
52	5	10	20	20	2,412

Table 26. Ideal graphs CHR5_i

Graph CHR5_j.

Definition 14. The modified fifth degree chordal ring called CHR5_j is denoted by CHR5_j(p; q_1,q_2,q_3,q_4), where p is the number of nodes. It must be positive and divisible by 4. Chords q_1 and q_2 have odd lengths; q_3 and q_4 possess even lengths, all chords lengths are less then $p/2$. Each node is connected to five other nodes. Even nodes i_{2k} are connected to i_{2k-1}, i_{2k+1} and to $i_{2k+q1(\text{mod } p)}$, $i_{2k+q2(\text{mod } p)}$ and $i_{2k+q3(\text{mod } p)}$ when $2k = 0$ (mod4) or to $i_{2k-q3(\text{mod } p)}$ when $2k = 2$ (mod4); while odd nodes $i_{(2k+1)}$ are connected to i_{2k}, i_{2k+2}, $i_{2k+1-q1(\text{mod } p)}$, $i_{2k+1-q2(\text{mod } p)}$ and to $i_{2k+1+q4(\text{mod } p)}$ when $2k+1=1$ (mod4) or to $i_{2k+1-q4(\text{mod } p)}$ when $2k+1 = 3$ (mod4).

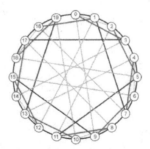

Fig. 25. Example of modified chordal ring CHR5_j(20; 3,9,2,6)

Based on going through all the real graphs, the distribution of nodes in the layers of CHR5_j is as follows (Table 27):

d	1	2	3	4	5	6	7	8
P_{do}	5	20	62	174	375	718	1303	2136

Table 27. Maximal number of nodes in the successive layers

The distribution of nodes in the layers of an optimal graph is described by expression:

When $d \rangle 2$ and $d = 0 \pmod 3$ then

$$p_{do} = \frac{77}{162}\left(\frac{d}{3}\right)^4 + 27\frac{1}{2}\left(\frac{d}{3}\right)^2 - 4$$

when $d \rangle 2$ and $d = 1 \pmod 3$ then

$$p_{do} = \frac{77}{162}\left(\frac{d-1}{3}\right)^4 + 52\frac{1}{3}\left(\frac{d-1}{3}\right)^3 + 53\frac{1}{2}\left(\frac{d-1}{3}\right)^2 + 24\frac{2}{3}\left(\frac{d-1}{3}\right) + 5 \qquad (51)$$

when $d \rangle 2$ and $d = 2 \pmod 3$ then

$$p_{do} = \frac{77}{162}\left(\frac{d-2}{3}\right)^4 + 101\frac{2}{3}\left(\frac{d-2}{3}\right)^3 + 129\frac{1}{2}\left(\frac{d-2}{3}\right)^2 + 83\frac{1}{3}\left(\frac{d-2}{3}\right)^2 + 22$$

In Table 28 the total number of nodes in virtual, optimal graphs is shown as function of diameter.

$d(G)$	1	2	3	4	5	6	7	8
p_{do}	6	26	88	262	637	1355	2658	4794

Table 28. Total numbers of nodes in optimal graphs versus diameter

The total number of nodes in optimal graphs calculated as a function of its diameter can be expressed as follows:

When $D(G) \rangle 2$ and $D(G) = 0 \pmod 3$ then

$$p_o = 23\frac{1}{10}\left(\frac{D(G)}{3}\right)^5 + 19\frac{1}{4}\left(\frac{D(G)}{3}\right)^4 + 31\frac{2}{3}\left(\frac{D(G)}{3}\right)^3 +$$

$$+ 12\frac{3}{4}\left(\frac{D(G)}{3}\right)^2 + 2\frac{7}{30}\left(\frac{D(G)}{3}\right) - 1$$

when $D(G) \rangle 2$ and $D(G) = 1 \pmod 3$ then

$$p_o = 23\frac{1}{10}\left(\frac{D(G)-1}{3}\right)^5 + 57\frac{3}{4}\left(\frac{D(G)-1}{3}\right)^4 + 84\left(\frac{D(G)-1}{3}\right)^3 + \qquad (52)$$

$$+ 66\frac{1}{4}\left(\frac{D(G)-1}{3}\right)^2 + 26\frac{8}{9}\left(\frac{D(G)-1}{3}\right) + 4$$

when $D(G) \rangle 2$ and $D(G) = 2 \pmod 3$ then

$$p_o = 23\frac{1}{10}\left(\frac{D(G)-2}{3}\right)^5 + 96\frac{1}{4}\left(\frac{D(G)-2}{3}\right)^4 + 185\frac{2}{3}\left(\frac{D(G)-2}{3}\right)^3 +$$

$$+ 195\frac{3}{4}\left(\frac{D(G)-2}{3}\right)^2 + 110\frac{7}{30}\left(\frac{D(G)-2}{3}\right) + 26$$

The average path length in optimal graphs is described by (53):

When $D(G) \rangle 2$ *and* $D(G) = 0 \pmod 3$ *then*

$$d_{avo} = 57\frac{3}{4}\left(\frac{D(G)}{3}\right)^6 + 57\frac{3}{4}\left(\frac{D(G)}{3}\right)^5 + 77\frac{5}{12}\left(\frac{D(G)}{3}\right)^4 +$$

$$+39\frac{7}{12}\left(\frac{D(G)}{3}\right)^3 + 4\frac{251}{300}\left(\frac{D(G)}{3}\right)^2 - 2\frac{101}{300}\left(\frac{D(G)}{3}\right) - 4$$

when $D(G) \rangle 2$ *and* $D(G) = 1 \pmod 3$ *then*

$$d_{avo} = 57\frac{3}{4}\left(\frac{D(G)-1}{3}\right)^6 - 173\frac{1}{4}\left(\frac{D(G)-1}{3}\right)^5 + 272\frac{11}{12}\left(\frac{D(G)-1}{3}\right)^4 +$$

$$-261\frac{3}{4}\left(\frac{D(G)-1}{3}\right)^3 + 146\frac{1}{3}\left(\frac{D(G)-1}{3}\right)^2 - 42\left(\frac{D(G)-1}{3}\right) + 1$$

(53)

when $D(G) \rangle 2$ *and* $D(G) = 2 \pmod 3$ *then*

$$d_{avo} = 57\frac{3}{4}\left(\frac{D(G)}{3}\right)^6 - 57\frac{3}{4}\left(\frac{D(G)}{3}\right)^5 + 77\frac{5}{12}\left(\frac{D(G)}{3}\right)^4 +$$

$$-42\frac{11}{12}\left(\frac{D(G)}{3}\right)^3 + 16\frac{5}{6}\left(\frac{D(G)}{3}\right)^2 - 2\frac{1}{3}\left(\frac{D(G)}{3}\right) - 4$$

Fig. 27 shows diameter and average path lengths in theoretical and real graphs with up to 100 nodes.

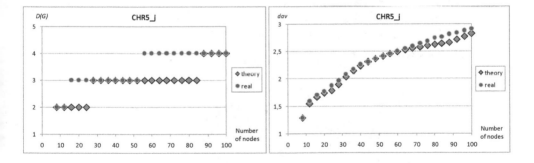

Fig. 26. Comparison of diameter and average path length of theoretical and real graphs CHR5_j

There are a few graphs having values of basic parameters equal to those of ideal graphs. For example: CHR5_j(44; 5,17,14,22), CHR5_j(48; 5,17,14,22), CHR5_j(52; 5,15,18,26), CHR5_j(56; 7,19,10,22).

Among all graphs belonging to the second group the best parameters (minimal diameter and minimal average path length given the number of nodes) were found in CHR5_i but other in minimal degree are slightly different it (especially from CHR5_j). In Fig. 27 the comparison of the second group of graphs is shown.

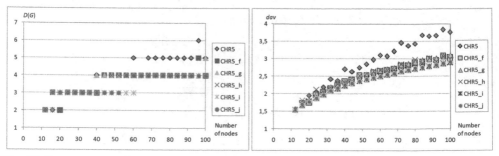

Fig. 27. Comparison of basic parameters of the chordal rings belonging to the second group of graphs

3.3 Analysed graphs – Third group

There are a number of other topologies, for which we have not found any nice expressions for the distribution of nodes in layers, and thus no expressions of the average distance and diameter could be derived. Due to the good basic parameters the topologies have been described, but further research is needed in order to provide more precise descriptions.

Graph CHR5_k.

Definition 15. The modified fifth degree chordal ring called CHR5_k is denoted by CHR5_k(p; q_1,q_2,q_3,q_4,q_5,q_6), where p is the number of nodes. It must be positive and divisible by 4. All chords have even lengths less than p/2. Each node is connected to five other nodes. Even nodes i_{2k} are connected to i_{2k-1}, i_{2k+1} and to $i_{2k+q1(mod\ p)}$, $i_{2k-q1(mod\ p)}$, $i_{2k+q5(mod\ p)}$ when 2k = 0 (mod4) or to $i_{2k+q2(mod\ p)}$, $i_{2k-q2(mod\ p)}$, $i_{2k-q5(mod\ p)}$ when 2k = 2 (mod4); while odd nodes $i_{(2k+1)}$ are connected to i_{2k}, i_{2k+2} and to $i_{2k+1+q3(mod\ p)}$, $i_{2k+1-q3(mod\ p)}$, $i_{2k+1+q6(mod\ p)}$ when 2k+1 = 1 (mod4) or to $i_{2k+1+q4(mod\ p)}$, $i_{2k+1-q4(mod\ p)}$, $i_{2k+1-q6(mod\ p)}$ when 2k+1 = 3 (mod4).

An example is shown in Fig. 28.

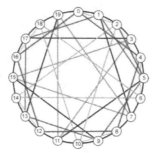

Fig. 28. Example of modified chordal ring CHR5_k(20; 4,8,4,8,2,6)

In Table 29 the distribution of nodes in layers is shown, based on observations of all graphs.

d	1	2	3	4	5	6	7	8
p_{do}	5	20	80	284	895	2520	6333	14334

Table 29. Maximal number of nodes in the successive layers

Using the results shown in Table 29, the counted total number of nodes in virtual optimal graphs is presented in table 30.

$D(G)$	1	2	3	4	5	6	7	8
p_o	6	26	106	390	1285	3805	10138	24472

Table 30. Total numbers of nodes forming optimal graphs versus diameter

In Fig. 29 comparison of diameter and average path length of theoretical and real graphs CHR5_k is shown.

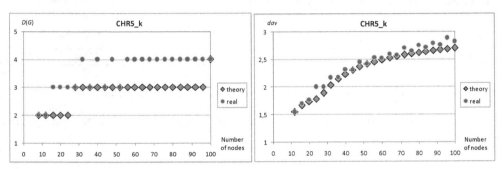

Fig. 29. Comparison of diameter and average path length of theoretical and real graphs CHR5_k

Only three graphs with average distance and diameter equal to those of ideal graphs can be found. These are CHR5_k(12; 4,4,4,2,6), CHR5_k(44; 4,8,12,16,18,22), CHR5_k(52; 4,8,12,16,22,26).

Graph CHR5_l.

Definition 16. The modified fifth degree chordal ring called CHR5_l is denoted by CHR5_l(p; q_1,q_2,q_3,q_4,q_5), where p means the number of nodes and is positive and divisible by 4. Chord q_1 has odd length, other chords have even length. The lengths of all chords are less than $p/2$. Each node is connected to five other nodes. Even nodes i_{2k} are connected to i_{2k-1}, i_{2k+1} and to $i_{2k+q1(\text{mod } p)}$ and to $i_{2k+q2(\text{mod } p)}$, $i_{2k+q3(\text{mod } p)}$ when $2k = 0$ (mod4) or to $i_{2k-q2(\text{mod } p)}$, $i_{2k-q3(\text{mod } p)}$ when $2k = 2$ (mod4); while odd nodes $i_{(2k+1)}$ are connected to i_{2k}, i_{2k+2}, $i_{2k+1-q1(\text{mod } p)}$, and to $i_{2k+1+q4(\text{mod } p)}$, $i_{2k+1+q5(\text{mod } p)}$ when $2k+1 = 1$ (mod4) or to $i_{2k+1-q4(\text{mod } p)}$, $i_{2k+1-q5(\text{mod } p)}$ when $2k+1 = 3$ (mod4).

An example is shown in Fig. 30.

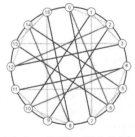

Fig. 30.Example of modified chordal ring CHR5_l(16; 7,6,2,2,6)

In Table 31 the distribution of nodes in the layers is shown.

d	1	2	3	4	5	6	7	8
p_{do}	5	20	71	228	555	1216	2442	4458

Table 31. Maximal number of nodes in the successive layers

Using the results shown in Table 31, the total number of nodes in virtual optimal graphs calculated in this way was presented in Table 32.

$D(G)$	1	2	3	4	5	6	7	8
p_o	6	26	97	325	880	2096	4538	8996

Table 32. Total numbers of nodes forming optimal graphs versus diameter

In Fig. 31 comparison of diameter and average path length of theoretical and real graphs CHR5_l is shown.

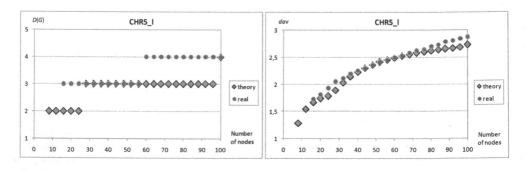

Fig. 31. Comparison of diameter and average path length of theoretical and real graphs CHR5_l

Examples of ideal graphs include: CHR5_l(16; 3,2,2,6,6), CHR5_l(44; 11,6,14,22,18), CHR5_l(48; 7,10,18,14,22), CHR5_l(52; 11,6,18,26,22), CHR5_l(56; 11,14,6,22,26).

Graph CHR5_m.

Definition 17. The modified fifth degree chordal ring called CHR5_m is denoted by CHR5_m(p; q_1,q_2,q_3,q_4), where p means the number of nodes and is positive and divisible by 4. All chords have even length less than $p/2$. Each node is connected to five other nodes. Even nodes i_{2k} are connected to i_{2k-1}, i_{2k+1}, $i_{2k+q1(\text{mod } p)}$, $i_{2k-q1(\text{mod } p)}$ and to $i_{2k+q2(\text{mod } p)}$ when $2k = 0$ (mod4) or to $i_{2k-q2(\text{mod } p)}$ when $2k = 2$ (mod4); while odd nodes $i_{(2k+1)}$ are connected to i_{2k}, i_{2k+2}, $i_{2k+1+q3(\text{mod } p)}$, $i_{2k+1-q3(\text{mod } p)}$, and to $i_{2k+1+q4(\text{mod } p)}$ when $2k+1 = 1$ (mod4) or to $i_{2k+1-q4(\text{mod } p)}$ when $2k+1 = 3$ (mod4).

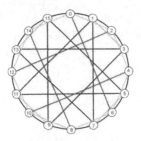

Fig. 32.Example of modified chordal ring CHR5_m(16; 2,6,6,2)

An example of CHR5_m is given in Fig. 32.

In Table 33 the distribution of nodes in layers is shown. The total number of nodes in optimal graphs, calculated based on these results, are shown in Table 34.

d	1	2	3	4	5	6	7	8
p_{do}	5	20	71	210	511	1064	1997	3440

Table 33. Maximal number of nodes in the layers

$D(G)$	1	2	3	4	5	6	7	8
p_o	6	26	97	307	818	1882	3879	7319

Table 34. Total numbers of nodes forming optimal graphs versus diameter

In Fig. 34 comparison of diameter and average path length of theoretical and real graphs CHR5_m is shown.

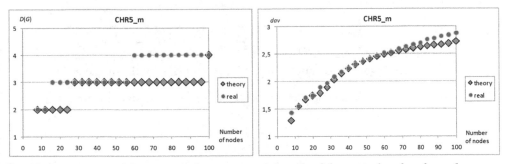

Fig. 33.Comparison of diameter and average path length of theoretical and real graphs CHR5_m

Examples of ideal chordal rings with up to 100 nodes include: CHR5_m(12; 2,2,6,6), CHR5_m(40; 6,14,10,18), CHR5_m(44; 6,14,10,18), CHR5_m(48; 10,22,6,14), CHR5_m(56; 18,26,14,6), CHR5_m(40; 6,14,10,18).

Graph CHR5_n.

Definition 18. The modified fifth degree chordal ring called CHR5_n is denoted by CHR5_n(p; q_1,q_2,q_3,q_4,q_5,q_6), where p means the number of nodes and is positive and divisible by 4. All chords have even length less than $p/2$. Each node is connected to five other nodes. Even nodes i_{2k} are connected to i_{2k-1}, i_{2k+1} and to $i_{2k+q1(\mathrm{mod}\ p)}$, $i_{2k+q2(\mathrm{mod}\ p)}$, $i_{2k+q3(\mathrm{mod}\ p)}$ when $2k = 0$ (mod4) or to $i_{2k-q1(\mathrm{mod}\ p)}$, $i_{2k-q2(\mathrm{mod}\ p)}$, $i_{2k-q3(\mathrm{mod}\ p)}$ when $2k = 2$ (mod4); while odd nodes $i_{(2k+1)}$ are connected to i_{2k}, i_{2k+2} and to $i_{2k+1+q4(\mathrm{mod}\ p)}$, $i_{2k+1+q5(\mathrm{mod}\ p)}$,$i_{2k+1+q6(\mathrm{mod}\ p)}$ when $2k+1 = 1$ (mod4) or to $i_{2k+1-q4(\mathrm{mod}\ p)}$, $i_{2k+1-q5(\mathrm{mod}\ p)}$,$i_{2k+1-q6(\mathrm{mod}\ p)}$ when $2k+1 = 3$ (mod4).

An example of a CHR5_n graph is shown in Fig. 34.

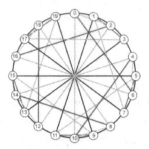

Fig. 34. Example of modified chordal ring CHR5_n(20; 2,6,10,2,6,10)

In Table 35 the distribution of nodes in layers is presented.

d	1	2	3	4	5	6	7	8
p_{do}	5	20	77	272	764	1916	4268	8696

Table 35. Maximal number of nodes in the layers

The total number of nodes in optimal graphs calculated based on the results given in Table 35 is shown in Table 36.

$D(G)$	1	2	3	4	5	6	7	8
p_o	6	26	103	375	1139	3055	7323	16019

Table 36. Total numbers of nodes in optimal graphs as a function of the diameter

In Fig. 36 shows the comparison of diameter and average path length of theoretical and real graphs CHR5_n.

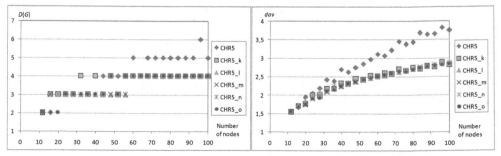

Fig. 35. Comparison of diameter and average path length of theoretical and real graphs CHR5_n

Going through all nodes with up to 100 nodes only one ideal graph was found, namely CHR5_n(52; 10,6,14,18,26,22) with 52 nodes.

Graph CHR5_o.

Definition 19. The modified fifth degree chordal ring called CHR5_o is denoted by CHR5_o(p; q_1,q_2,q_3,q_4,q_5) where p means the number of nodes and is positive and divisible by 4. All chords have even length less than $(p/2+1)$. Each node is connected to five other nodes. Even nodes i_{2k} are connected to i_{2k-1}, i_{2k+1} and to $i_{2k+q1(\bmod\ p)}$, $i_{2k-q1(\bmod\ p)}$, $i_{2k+q2(\bmod\ p)}$ when $2k = 0$ (mod4) or to $i_{2k-q2(\bmod\ p)}$ when $2k = 2$ (mod4); while odd nodes $i_{(2k+1)}$ are connected to i_{2k}, i_{2k+2}, $i_{2k+1+q5(\bmod\ p)}$ and to $i_{2k+1+q3(\bmod\ p)}$, $i_{2k+1-q3(\bmod\ p)}$ when $2k+1 = 1$ (mod4) or to $i_{2k+1+q4(\bmod\ p)}$, $i_{2k+1-q4(\bmod\ p)}$ when $2k+1 = 3$ (mod4).

An example of a CHR5_o graph is shown in Fig. 36.

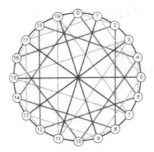

Fig. 36. Example of modified chordal ring CHR5_o(20; 6,2,4,8,10)

In Table 37 the distribution of nodes in layers is shown. Based on these numbers Table 38 is derived, showing the numbers of nodes in optimal graphs as a function of the diameter.

d	1	2	3	4	5	6	7	8	Node number
p_{do}	5	20	73	244	699	1726	3779	7498	even
	5	20	78	254	719	1778	3893	7696	odd

Table 37. Maximal number of nodes in the layers

$d(G)$	1	2	3	4	5	6	7	8	Node number
p_{do}	6	26	99	343	1042	2768	6547	14045	even
	6	26	104	358	1077	2855	6748	14444	odd

Table 38. Total numbers of nodes forming optimal graphs versus diameter

A number of ideal graphs can be found, such as CHR5_o(12; 6,2,2,4,4), CHR5_o(16; 2,6,6,4,4), CHR5_o(20; 10,10,6,4,4), CHR5_o(36; 10,18,14,4,8), CHR5_o(44; 6,14,10,20,4), CHR5_o(52; 6,10,14,20,24).

Fig. 37 shows the comparison of diameter and average path length of theoretical and real CHR5_o graphs.

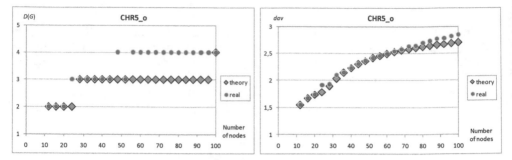

Fig. 37.Comparison of diameter and average path length of theoretical and real graphs CHR5_o

As for the previous group of graphs, the chordal rings of this group are compare with respect to the basic parameters. The comparison is given in Fig. 38.

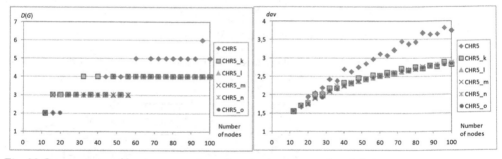

Fig. 38.Comparison of basic parameters of analyzing group chordal rings

From Fig. 38 it follows that all graphs belonging to third group of chordal rings have very similar properties when the number of nodes is smaller than 100.

4. Analysis of obtained results

Based on the obtained results for all 15 groups of graphs presented, the values of minimum diameter and average paths lengths can be compared. Despite the differences found between theoretical and real parameters the comparisons will be based on the theoretical values.

First, Fig. 39 presents the average path lengths as a function of the diameter in the graphs. This does not take into account the number of nodes in the graphs, which vary significantly between the different graphs as can be seen in Table 39 and Fig. 40. It can be seen that for a given number of nodes, CHR5_k has the smallest diameter.

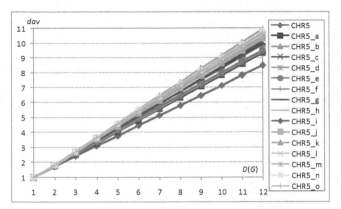

Fig. 39. Comparison of the calculated average paths length in the function of the diameter of graphs

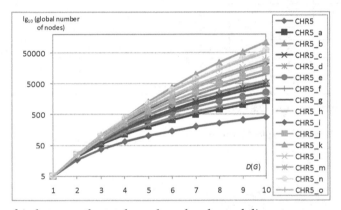

Fig. 40. Relationship between the total number of nodes and diameters

In the following section, a comparison of the average path length as a function of the number of nodes is presented. In order to compare graphs with different numbers of nodes, a "Normalized estimator of average path length" (E_{nav}) is introduced as follows:

$$E_{nav} = \frac{d_{avD(G)} \sum\limits_{d=1}^{d(G)} p_{dr}}{d_{avrD(G)} \sum\limits_{d=1}^{d(G)} p_d} \qquad (54)$$

Where $d_{avD(G)}$ means the average path length of a particular graph when its diameter is equal to $D(G)$, and $d_{avrD(G)}$ means the average path length of the reference graph, $\sum\limits_{d=1}^{d(G)} p_{dr}$ is the total number of nodes in relation graph, $\sum\limits_{d=1}^{d(G)} p_d$ - total number of nodes in the particular graph.

Graph	$D(G)$									
	1	2	3	4	5	6	7	8	9	10
CHR5	6	18	38	66	102	146	198	258	326	402
CHR5_a	6	22	55	113	202	330	503	729	1014	1366
CHR5_b	6	26	87	227	494	948	1661	2717	4212	6254
CHR5_c	6	26	76	186	386	726	1276	2126	3386	5186
CHR5_d	6	23	61	130	239	400	622	915	1288	1753
CHR5_e	6	23	68	160	316	552	884	1328	1900	2616
CHR5_f	6	23	72	201	491	1068	2103	3804	6411	10196
CHR5_g	6	22	64	166	349	651	1142	1846	2840	4228
CHR5_h	6	23	79	245	686	1742	4014	8450	16430	29842
CHR5_i	6	26	92	285	764	1816	3887	7579	13674	23158
CHR5_j	6	26	88	262	637	1355	2658	4794	8148	13240
CHR5_k	6	26	106	390	1285	3805	10138	24472	54108	110878
CHR5_l	6	26	97	325	880	2096	4538	8996	16706	29420
CHR5_m	6	26	97	307	818	1882	3879	7319	12876	21406
CHR5_n	6	26	103	375	1139	3055	7323	16019	32520	62092
CHR5_o	6	26	102	351	1060	2812	6648	14245	28120	51864

Table 39.Total number of nodes versus diameters.

The results are shown in Table 40 and Fig. 41.

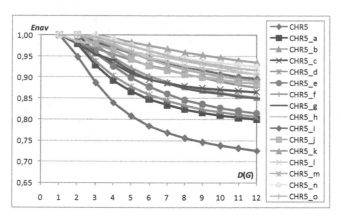

Fig. 41. Distribution of the normalized estimator of average path length versus graph diameter

From Fig. 41 it can be seen that he CHR5_k has the relatively shortest average path length.

Not surprisingly, the distributions of the number of nodes in the layers have a great impact on the two basic parameters. This can be seen also from expressions (7) and (9). The difference in the distribution for all graphs is shown in Fig. 42.

Graph	$D(G)$											
	1	2	3	4	5	6	7	8	9	10	11	12
CHR5	1,000	0,948	0,886	0,840	0,807	0,785	0,768	0,756	0,746	0,738	0,732	0,726
CHR5_a	1,000	0,979	0,928	0,892	0,866	0,848	0,834	0,824	0,816	0,810	0,804	0,800
CHR5_b	1,000	1,000	0,977	0,947	0,922	0,903	0,889	0,878	0,870	0,863	0,857	0,853
CHR5_c	1,000	1,000	0,958	0,932	0,909	0,894	0,885	0,879	0,875	0,871	0,868	0,865
CHR5_d	1,000	0,985	0,938	0,903	0,877	0,858	0,844	0,833	0,824	0,818	0,811	0,806
CHR5_e	1,000	0,985	0,957	0,926	0,898	0,877	0,860	0,847	0,836	0,828	0,820	0,814
CHR5_f	1,000	0,982	0,967	0,954	0,941	0,928	0,918	0,908	0,899	0,891	0,883	0,876
CHR5_g	1,000	0,979	0,953	0,940	0,914	0,895	0,885	0,874	0,866	0,861	0,855	0,851
CHR5_h	1,000	0,985	0,978	0,969	0,961	0,954	0,946	0,939	0,931	0,925	0,914	0,908
CHR5_i	1,000	1,000	0,984	0,971	0,957	0,945	0,935	0,925	0,916	0,909	0,901	0,896
CHR5_j	1,000	1,000	0,978	0,964	0,943	0,926	0,916	0,907	0,899	0,894	0,889	0,885
CHR5_k	1,000	1,000	1,000	0,992	0,983	0,975	0,967	0,959	0,953	0,946	0,941	0,936
CHR5_l	1,000	1,000	0,990	0,981	0,961	0,947	0,937	0,928	0,921	0,916	0,911	0,907
CHR5_m	1,000	1,000	0,990	0,974	0,957	0,941	0,929	0,919	0,910	0,903	0,897	0,892
CHR5_n	1,000	1,000	0,997	0,990	0,975	0,964	0,954	0,946	0,939	0,934	0,929	0,925
CHR5_o	1,000	1,000	0,995	0,985	0,973	0,962	0,952	0,943	0,935	0,928	0,922	0,917
E_{nav}												

Table 40. Distribution of the normalized estimator of the average path length versus the graph diameter

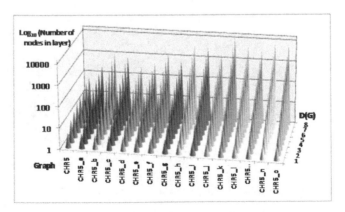

Fig. 42. Distribution of nodes numbers in the layers for all graphs

In the charts given below the comparison of the maximum number of nodes appearing in the layers of all analyzed graphs is presented.

On the basis of Fig. 42 and Fig. 43 it seems to be sufficient to analyze the distributions of numbers occurring in the first few layers to select a graph having the best basic parameters. This obviously reduces the time and effort for comparisons. The results again confirm that CHR5_k seems to be superior in terms of these basic parameters.

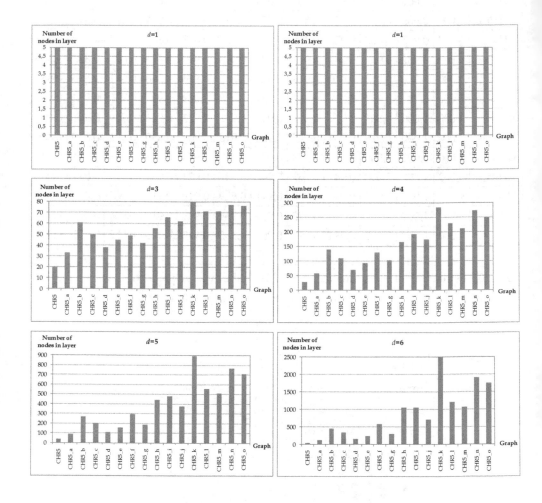

Fig. 43. Distribution of the number nodes in the first six layers

In order to make an objective assessment of the CHR5_k parameters, they were compared to the parameters of the Reference Graph as previously described. Table 41 and Fig. 44 show the distribution of nodes in different layers of these two graphs, as a function of their diameters.

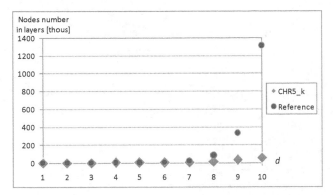

Fig. 44. Comparison of nodes number in successive layers

In Table 41 the total number of nodes in CHR5_k and reference graphs is compared for different diameters.

$D(G)$	1	2	3	4	5	6	7	8	9	10
CHR5_k	5	20	80	284	895	2520	6333	14334	29636	56770
Reference Graph	5	20	80	320	1280	5120	20480	81920	327680	1310720
Total number of nodes										

Table 41. Total number of nodes in Reference Graph and CHR5_k

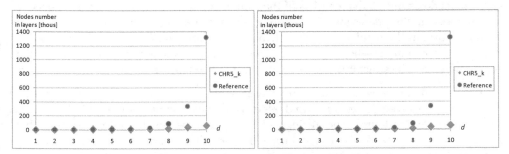

Fig. 45. Comparison the number of nodes in layers and total number of nodes versus layer number

In Fig. 45 a comparison of node numbers in successive layers and total number of nodes as a function of number layer in ideal CHR5_k and Reference Graph is shown.

Table 42 and Fig. 46 show a comparison of the average length as a function of the both graphs diameter, taking into account the total number of nodes corresponding to this diameter. As in the previous comparisons E_{nav} (54) is used.

$D(G)$	1	2	3	4	5	6	7	8	9	10
CHR5_k	1,0000	1,0000	1,0000	0,9920	0,9829	0,9747	0,9669	0,9595	0,9527	0,9465
E_{nav}										

Table 42. The average path length as a function of the diameter

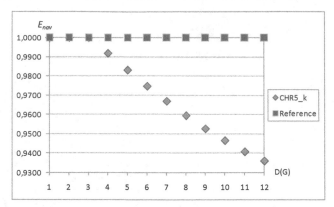

Fig. 46. Comparison of average path length as a function of the graphs diameter

5. Conclusions

In this publication the analysis of a different construction of 5th degree chordal rings was presented. The authors' main aim was to find structures which have the minimal diameter and minimal average of path in respect to number of nodes which create these graphs.

Presented considerations in the paper have rather theoretical nature, without a strict reference to practical applications. It is difficult to imagine a real regular WAN communication network that consists of thousands of nodes. However the interconnection structures connecting thousands of microprocessors, or sensors require the construction of such networks as the regular ones. The main objective function for regular interconnection structures is to minimize the network diameter or average path length. So, this is the main reason why such the structures were analyzed and studied in our paper.

In this regard the program was worked out which allows to calculate the analyzed parameters. It allowed describing virtual reference graphs namely optimal and ideal graphs. In this way we also found chordal rings which possess the smallest difference regards to average path length and diameter, which Reference Graphs have. They examined many types of structures and concluded that parameters of the real graphs are slightly different of the theoretically calculated graph parameter values. The obtained results became the basis for preparing the general formulas for determining these parameters without the need of simulation. As a side-result of the paper, we have shown that these reference graphs can be used for obtaining fairly good estimation of distance parameters in a simple manner. Additionally, they concluded that it is enough to inspect the maximal number of nodes which can appear in first few layers in aim to choose the best topology.

This publication presents the results of analysis of the modified chordal rings fifth degree. This analysis was carried out for 15 graphs divided into 3 groups. Each of the group included 5 types of graphs. Since the graphs were analyzed are regular graphs odd degree, hence all the graphs have to have an even number of nodes. The nodes number of all graphs belonging to the first group is divisible by two, the second and the third by four (it follows from used method of their construction).

For each group of graphs their analysis based on results obtained thanks to the application of testing programs constructed by authors. It made possible to carry out a distribution of maximal number of nodes in layers, to count total number of nodes in virtual, optimal graphs. Based on obtained results, for the first two groups of graphs they found strict mathematical expressions describing the distribution of nodes in layers, the total number of nodes, the average path length for optimal graphs, whilst for the third group such formulas were not found. Additionally the prepared programs allowed us to compare the basic parameters of real and theoretically constructed graphs.

Among the all analyzed graphs, the structures CHR5_f – CHR5_o have the most acceptable basic parameters. Those graphs however have a fundamental limitation: the network should consist of nodes with nodes number has to be divisible by 4. The parameters of these networks in more or less deviate from the parameters of the reference graph (graph ideal), and what's involved, they are usually asymmetrical (depending on the choice of the source node, obtained values differ). Also, computational process is rather complex, and takes long time.

From the point of view of application, according to the authors, the most appropriate structure of the regular network topology are graphs CHR5_b. These chordal rings are symmetrical, their parameters, if are not equal to the parameters of optimal graphs, they are very close to them; they are simple and easy to design and implement. The main limitation is the fact that the number of nodes, creating these networks, has to be even, but this follows from the assumption that the structure has to be regular, and the degree of nodes is five. Fig. 47 shows the comparison of the best two structures.

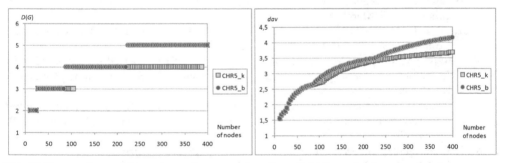

Fig. 47.Comparison of diameter and average length path both analyzing chordal rings versus the number of nodes

As the justification of this last conclusion the presented above diagrams can be used. In Fig. 47 we show the comparison of basic parameters of chordal rings CHR5_b and the best graph - CHR5_k. It can be observed that in up to 84 nodes, theoretically calculated parameters of both types of rings are identical, and up to 224 nodes - not much different from each another. Thus, taking into account the advantages of CHR5_b structure described before, it should be used to construct regular networks possessing not a huge number of nodes or in a large network consisting of a few identical regular structures.

Future work can focus on both theoretical and practical aspects. For the theoretical aspects, it would be a big help to find more precise and reachable bounds. This would make it possible to assess graph types, and to know how close to optimal they are. Moreover, a more thorough study of how precise distance estimates can be given using ideal and optimal graph would be interesting. Such a study could also cover other types of graphs. Another direction for further research would be to study new groups of graphs.

More practical aspects could deal with analyzing how well the good theoretical properties translate into good network properties. This could be done through simulation of different network configuration and traffic scenarios, and/or by studying how feasible the graphs are for assignment of physical, optical or logical links. In order to demonstrate that the topologies are useful in real-world settings, case studies would be a good place to start.

6. Acknowledgment

The authors would like to thank Associate Professor Jens Myrup Pedersen and Assistant Professor Tahir Riaz from Aalborg University, Department of Electronic Systems, and Professor Dr. Mohamed Othman from Universiti Putra Malaysia, Department of Communication Technology and Networks, for a fruitful collaboration with the work on analyzing chordal rings and other network topologies. Their inputs for this chapter have been particularly valuable.

7. References

A.N. Al-Karaki & A.E.Kamal. (2004). *Routing Techniques in Wireless Sensor Networks: a Survey*, IEEE Wireless Comm., pp. 6-28, Dec. 2004

ALU: Alcatel-Lucent 1830. (2011). *Photonic Service Switch 4 (Pss-4) Release 1.5.0/1.5.1 User Guide*, Issue 2, March 2011

W. Arden & H. Lee. (1981). *Analysis of Chordal Ring Network*. IEEE Transactions on Computers Vol. 30 No. 4 pp. 291-295, 1981

R. N. F. Azura; M. Othman; M.. H. Selamat & P. Y. Hock. 2008. *Modified Degree Six Chordal Rings Network Topology*. Proceedings of Simposium Kebangsaan Sains Matematik Ke-16, 3-5, pp. 515-522, Jun 2008

R. N. F. Azura; M. Othman; M.. H. Selamat & P. Y. Hock. (2010). *On properties of modified degree six chordal rings networks*. Malaysian Journal of Mathematical Sciences. 4(2), pp. 147-157.

L. N. Bhuyan. (1987). *Interconnection Networks for Parallel and Distributed Processing*. IEEE Computer Vol. 20 No. 6, pp. 9-12, 1987

S. Bujnowski. (2003). *Analysis & Synthesis Homogeneous Structure Networks Connecting Communications Modules*. PhD Thesis, UTP, Bydgoszcz, 2003.

S. Bujnowski; B. Dubalski & A. Zabłudowski. (2003). *Analysis of Chordal Rings*. Mathematical Techniques and Problems in Telecommunications. Centro International de Matematica, Tomar 2003, pp. 257-279

S. Bujnowski; B. Dubalski & A. Zabłudowski. (2004). *The Evaluation of Transmission Ability of 3rd Degree Chordal Rings with the Use of Adjacent Matrix*. The Seventh INFORMS Telecommunications Conference, pp. 219-221. Boca Raton 2004

S. Bujnowski; B. Dubalski & A. Zabłudowski. (2004). *Analysis of 4th Degree Chordal Rings*. Proceedings of International Conference on the Communications in Computing. Las Vegas, 2004, pp. 318 – 324

S. Bujnowski; B. Dubalski & A. Zabłudowski. (2005). *Analysis of Transmission Properties of 3rd Degree Chordal Rings*. Kwartalnik Elektroniki i Telekomunikacji. 2005, 51, z.4, pp. 521 – 539

S. Bujnowski; B. Dubalski; J. M. Pedersen & A. Zabłudowski. (2008). *Struktury topologiczne CR3m oraz NdRm*. Przegląd Telekomunikacyjny LXXXI, nr 8/9, 2008, pp. 1133 – 1141 (in Polish)

S. Bujnowski; B. Dubalski; J. M. Pedersen & A. Zabłudowski. (2009). *Analysis of Regular Structures Third Degree Based on Chordal Rings*. Image Processing and Communications. Vol. 14, nr 1, pp. 13-24, 2009 - 4

S. Bujnowski; B. Dubalski; A. Zabłudowski; D. Ledziński; T. Marciniak & J. M. Pedersen. (2010). *Comparison of Modified 6 Degree Chordal Rings*. Image Processing and Communications. Challenges 2. 2010, ISSN 1867-5662, pp. 435 – 446

S. Bujnowski; B. Dubalski; A. Zabłudowski; J. M. Pedersen & T. Riaz. (2011). *Analysis of Degree 5 Chordal Rings For Network Topologies*. Image Processing & Communications. Challenge 3, 2011, Springer. ISSN 1867-5662, pp. 445-459

R. Diestel, Graph Theory, 4th Edition. Springer-Verlag, Heidelberg. Graduate Texts in Mathematics, Volume 173. July 2001.

B. Dubalski; S. Bujnowski; A. Zabłudowski & J. M. Pedersen. (2007). *Introducing Modified Degree 4 Chordal Rings with Two Chord Lengths*. Proceedings of the Fourth IASTED Asian Conference "Communication Systems and Networks". Phuket 2007, ISBN CD: 978-0-88986-658-4, pp. 561-574 - 2

B. Dubalski; A. Zabludowski; S. Bujnowski & J. M. Pedersen. (2008). *Comparison of Modified Chordal Rings Fourth Degree to Chordal Rings Sixth Degree*. Proceedings of Electronics in Marine, ELMAR 2008, Zadar, Croatia. Volume 2, pp. 597-600

B. Dubalski; A. Zabłudowski; D. Ledziński; J. M. Pedersen & T. M. Riaz. (2010). *Evaluation Of Modified Degree 5 Chordal Rings for Network Topologies*. Proceedings of 2010 Australasian Telecommunication Networks and Applications Conference. Auckland – New Zealand, ISBN 978-1-4244-8171-2, pp. 66-71

R. N. Farah; M. Othman; H. Selemat & P.Y. Hock. (2008). *Analysis of Modified Degree Six Chordal Rings and Traditional Chordal Rings Degree Six Interconnection Network*. International Conference of Electronics Design, Penang, Malaysia, 2008

R. N. Farah & M. Othman. (2010). *In Modified Chordal Rings Degree Six Geometrical Representation Properties*. Proceedings of Fundamental Science Congress, Kuala Lumpur, Malaysia, May 18-19, 2010.

R. N. Farah; M. Othman & M. H. Selamat. (2010). *Combinatorial properties of modified chordal rings degree four networks*. Journal of Computer Science. 6(3), pp. 279-284.

R. N. Farah; M. Othman & M. H. Selamat (2010). *An optimum free-table routing algorithms of modified and traditional chordal rings networks of degree four*. Journal of Material Science and Engineering. 4(10), pp. 78-89.

R. N. Farah; M. Othman; M. H. Selamat & M. Rushdan. (2010). *In Layers Shortest Path of Modified Chordal Rings Degree Six Networks*. Proceedings of International Conference on Intelligent Network and Computing, Kuala Lumpur, Malaysia, November 26-28, 2010.

R. N. Farah; N. Irwan; M. Othman; M. H. Selamat & M. Rushdan. (2011). *In An Efficient Broadcasting Schemes for Modified Chordal Rings Degree Six Networks*. Proceedings of

International Conference on Information and Industrial Electronics, Chengdu, China, January 13-14, 2011

M. M. Freire & H.J.A. da Silva. (1999). *Assessment of Blocking Performance in Bidirectional WDM Ring Networks with Node-to-Node and Full-Mesh Connectivity.* European Conference on Networks and Optical Communications (NOC'99), Delft, 1999

M. M. Freire & H.J.A. da Silva. (2001). *Influence of Wavelength on Blocking Performance of Wavelength Routed Chordal Ring Networks.* Proceedings of 3rd Conference on Telecommunications, Figueira da Foz, 2001

M. M. Freire & H.J.A. da Silva. (2001). *Performance Comparison of Wavelength Routing Optical Networks with Chordal Ring and Mesh-Torus Topologies.* ICN (1), pp. 358-367, 2001

C. Gavoille. (n.d). *A Survey on Internal Routing.*
http://deptinfo.labri.ubordeaux.fr/~gavoille/article/ survey /node28.html

G. Kotsis. (1992). *Interconnection Topologies and Routing for Parallel Processing Systems.* ACPC, Technical Report Series, ACPC/TR92-19, 1992

A.L. Liestman; J. Opatrny & M. Zaragoza. (1998). *Network Properties of Double and Triple Fixed Step Graphs.* International Journal of Foundations of Computer Science 9, pp. 57-76, 1998

B. Mans. (1999). *On the Interval Routing of Chordal Rings.* ISPAN '99 IEEE - International Symposium on Parallel Architectures, Algorithms and Networks, Fremantle, Australia, 1999, pp. 16-21

L. Narayanan & J. Opatrny. (1999). *Compact Routing on Chordal Rings of Degree Four.* Algorithmica, Vol. 23, pp. 72-96, 1999

L. Narayanan; J. Opatrny & D. Sotteau. (2001). *All-to-All Optical Routing in Chordal Rings of Degree 4.* Algorithmica Vol. 31, pp. 155-178, 2001

H. Newton. (1996). *Newton's Telecom Dictionary.* 11th Edition. A Flatiron/Publishing, Inc. Book, 1996,

J. M. Pedersen; A. Patel; T. P. Knudsen & O. B. Madsen. (2004). *Generalized Double Ring Network Structures.* Proc. of SCI 2004, The 8th World Multi-Conference On Systemics, Cybernetics and Informatics. Vol. 8, pp. 47-51. Orlando, USA, July 2004.

J. M. Pedersen; T. P. Knudsen & O. B. Madsen. (2004). *Comparing and Selecting Generalized Double Ring Network Structures.* Proc. of IASTED CCN 2004, The Second IASTED International Conference Communication and Computer Networks, pp. 375-380. Cambridge, USA, November 2004.

J. M. Pedersen. (2005). *Structural Quality of Service in Large-Scale Networks.* PhD. Thesis Aalborg University, April 2005

J. M. Pedersen; M. T. Riaz & O. Brun Madsen. (2005). *Distances in Generalized Double Rings and Degree Three Chordal Rings.* Proc. of IASTED PDCN 2005. IASTED International Conference on Parallel and Distributed Computing and Networks, pp. 153-158. Innsbruck, Austria, February 2005.

J. M. Pedersen; J. M. Gutierrez; T. Marciniak; B. Dubalski & A. Zabłudowski. (2009). *Describing N2R Properties Using Ideal Graphs.* Advances in Mesh Networks, MESH 2009. The Second International Conference on Advances in Mesh Networks. ISBN: 978-0-7695-3667-5, pp. 150-154

3

Poly-Dimension of Antimatroids

Yulia Kempner[1] and Vadim E. Levit[2]

[1]*Holon Institute of Technology*
[2]*Ariel University Center of Samaria*
Israel

1. Introduction

A partial cube is a graph that can be isometrically embedded into a hypercube. In other words, a partial cube is a subgraph of a hypercube that preserves distances - the distance between any two vertices in the subgraph is the same as the distance between those vertices in the hypercube. Partial cubes were first introduced by Graham and Pollak (Graham & Pollak, 1971) as a model for communication networks and were extensively studied afterwards.

Many important families of combinatorial structures can be represented as subgraphs of partial cubes, for instance, antimatroids. An antimatroid is an accessible set system closed under union. There are two equivalent definitions of antimatroids, one as set systems and the other as languages (Björner & Ziegler, 1992; Korte et al., 1991). An algorithmic characterization of antimatroids based on the language definition was introduced in (Boyd & Faigle, 1990). Later, another algorithmic characterization of antimatroids, which depicted them as set systems was developed in (Kempner & Levit, 2003). Dilworth (Dilworth, 1940) was the first to study antimatroids using another axiomatization based on lattice theory, and they have been frequently rediscovered in other contexts. An antimatroid can be viewed as a special case of either greedoids or semimodular lattices, and as a generalization of partial orders and distributive lattices. While classical examples of antimatroids connect them with posets, chordal graphs, convex geometries etc., a game theory gives a framework, in which antimatroids are considered as permission structures for coalitions (Algaba et al., 2004; 2010). There are also rich connections between antimatroids and cluster analysis (Kempner & Muchnik, 2003). Glasserman and Yao (Glasserman & Yao, 1994) used antimatroids to model the ordering of events in discrete event simulation systems. In mathematical psychology, antimatroids are used to describe feasible states of knowledge of a human learner (Cosyn & Uzun, 2009; Eppstein et al., 2008; Falmagne & Doignon, 2011).

The notion of "antimatroid with repetition" was conceived by Bjorner, Lovasz and Shor (Björner et al., 1991) as an extension of the notion of antimatroid in the framework of non-simple languages. Further they were investigated by the name of "poly-antimatroids" (Kempner & Levit, 2007; Nakamura, 2005), where the set system approach was used.

An antimatroid with the ground set of size n may be isometrically embedded into the **hypercube** $\{0,1\}^n$. A poly-antimatroid with the same ground set is isometrically embedded into n-dimensional integer **lattice** $\mathbb{Z}^n$. In this research we concentrate on interrelations between antimatroids and poly-antimatroids and prove that these two structures are isometrically isomorphic.

The *poly-dimension* of an antimatroid is the minimum dimension d such that the antimatroid is isometrically isomorphic to some d-dimensional poly-antimatroid. This definition is a direct analog of the lattice dimension of graphs (Eppstein, 2005). In this paper the exact characterization of antimatroids of poly-dimension 1 and 2 is given and a conjecture concerning antimatroids of any poly-dimension d is suggested.

This chapter is organized as follows.

Section 1 contains an extended introduction to the theory of antimatroids.

Section 2 gives basic information about distances on graphs and isometric isomorphisms. We concentrate on interrelations between antimatroids and poly-antimatroids. In particular, we construct an isometric isomorphism between these structures.

In Section 3 we introduce the *poly-dimension* of an antimatroid and prove our main theorem characterizing antimatroids of poly-dimension 2. In addition we present a linear labeling algorithm that isometrically embeds such an antimatroid into the integer lattice $\mathbb{Z}^2$.

Section 4 discusses an open problem.

2. Preliminaries

Let E be a finite set. A *set system* over E is a pair $(E, \mathcal{S})$, where $\mathcal{S}$ is a family of sets over E, called *feasible* sets. We will use $X \cup x$ for $X \cup \{x\}$, and $X - x$ for $X - \{x\}$.

Definition 2.1. *(Korte et al., 1991) A finite non-empty set system $(E, \mathcal{S})$ is an antimatroid if*

$(A1)$ *for each non-empty $X \in \mathcal{S}$, there exists $x \in X$ such that $X - x \in \mathcal{S}$*

$(A2)$ *for all $X, Y \in \mathcal{S}$, and $X \nsubseteq Y$, there exists $x \in X - Y$ such that $Y \cup x \in \mathcal{S}$.*

Any set system satisfying $(A1)$ is called *accessible*.

In addition, we use the following characterization of antimatroids.

Proposition 2.2. *(Korte et al., 1991) For an accessible set system $(E, \mathcal{S})$ the following statements are equivalent:*

(i) $(E, \mathcal{S})$ *is an antimatroid*

(ii) $\mathcal{S}$ *is closed under union $(X, Y \in \mathcal{S} \Rightarrow X \cup Y \in \mathcal{S})$*

An "antimatroid with repetition" was invented by Bjorner, Lovasz and Shor (Björner et al., 1991) by studying the set of configurations of the Chip Firing Game (CFG). Further it was investigated by the name of "poly-antimatroid" as a generalization of the notion of the antimatroid for multisets. A *multiset* A over E is a function $f_A : E \to N$, where $f_A(e)$ is a number of repetitions of an element e in A. A *poly-antimatroid* is a finite non-empty multiset system $(E, \mathcal{F})$ that satisfies the antimatroid properties $(A1)$ and $(A2)$. So antimatroids may be considered as a particular case of poly-antimatroids. An example of an antimatroid $(\{x, y, z\}, \mathcal{S})$ and a poly-antimatroid $(\{x, y\}, \mathcal{F})$ is illustrated in Figure 1.

Definition 2.3. *A multiset system $(E, \mathcal{F})$ satisfies the chain property if for all $X, Y \in \mathcal{F}$, and $X \subset Y$, there exists a chain $X = X_0 \subset X_1 \subset ... \subset X_k = Y$ such that $X_i = X_{i-1} \cup x_i$ and $X_i \in \mathcal{F}$ for $0 \leq i \leq k$. We call the system a chain system.*

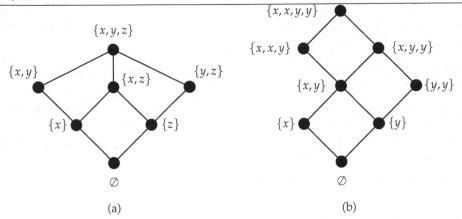

Fig. 1. (a) Antimatroid. (b) Poly-antimatroid.

It is easy to see that this chain property follows from $(A2)$, but these properties are not equivalent. If $\varnothing \in \mathcal{F}$, then accessibility follows from the chain property. In general case, there are accessible set systems that do not satisfy the chain property. Indeed, consider the following example illustrated in Figure 2 (a). Let $E = \{1, 2, 3\}$ and $\mathcal{F} = \{\varnothing, \{1\}, \{2\}, \{2, 3\}, \{1, 2, 3\}\}$). It is easy to check that the set system is accessible, but there are no chain from $\{1\}$ to $\{1, 2, 3\}$. Vice versa, it is possible to construct a system, that satisfies the chain property and it is not accessible. For example, $\mathcal{F} = \{\{1\}, \{3\}, \{1, 2\}, \{2, 3\}, \{1, 2, 3\}\}$ in Figure 2(b).

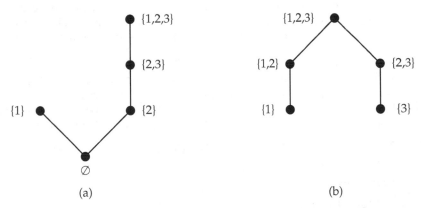

Fig. 2. (a) Accessible system but not chain system. (b) Chain system but not accessible.

In fact, if we have an accessible set system satisfying the chain property, then the same system but without the empty set (or without all subsets of cardinality less then some k) is not accessible, but still satisfies the chain property.

Examples of chain systems include poly-antimatroids, convex geometries, matroids and other hereditary systems (matchings, cliques, independent sets of a graph).

Regular set systems have been defined by Honda and Grabisch (Honda & Grabisch, 2008).

Definition 2.4. *A set system* $(E, \mathcal{S})$ *is a regular set system if* $\varnothing, E \in \mathcal{S}$ *and all maximal chains have length n, where* $n = |E|$.

Equivalently, under the condition $\varnothing, E \in \mathcal{S}$, $(E, \mathcal{S})$ is a regular set system if and only if $|A - B| = 1$ for any $A, B \in \mathcal{S}$ such that A is a cover of B.

Actually, a regular set system is a chain system with an additional constraint demanding from both the empty set $\varnothing$ and the ground set E to belong to the set system $\mathcal{S}$.

An algorithmic characterization of chain systems with empty set $\varnothing$, called *strongly accessible set system*, was presented in (Boley et al., 2010).

Augmenting systems were introduced in game theory as a restricted cooperation model. The model is a weakening of the antimatroid structure and strengthening of the chain system.

Definition 2.5. *(Bilbao, 2003) A set system* $(E, \mathcal{S})$ *is a augmenting system if*

1. $\varnothing \in \mathcal{S}$

2. *it satisfies the chain property*

3. *for all* $X, Y \in \mathcal{S}$ *with* $X \cap Y \neq \varnothing$, *we have* $X \cup Y \in \mathcal{S}$.

An augmenting system is an antimatroid if and only if it is closed under union (Bilbao, 2003).

A comparative study of various families of set systems are given in (Grabisch, 2009).

Antimatroids have already been investigated within the framework of lattice theory by Dilworth (Dilworth, 1940). The feasible sets of an antimatroid ordered by inclusion form a lattice, with lattice operations: $X \vee Y = X \cup Y$, and $X \wedge Y$ is the maximal feasible subset of the set $X \cap Y$ called a *basis*. Since an antimatroid is closed under union, it has only one basis.

A finite lattice L is *semimodular* if whenever $x, y \in L$ both cover $x \wedge y$, $x \vee y$ covers x and y.

A finite lattice L is called *join-distributive* (Björner & Ziegler, 1992) if for any $x \in L$ the interval $[x, y]$ is Boolean, where y is the join of all elements covering x. Such lattices are appeared under several different names, e.g., locally free lattices (Korte et al., 1991) and upper locally distributive lattices (ULD) (Felsner & Knauer, 2009; Magnien et al., 2001; Monjardet, 2003).

Proposition 2.6. *(Björner & Ziegler, 1992) For an accessible set system* $(E, \mathcal{S})$ *the following statements are equivalent:*

(i) $(E, \mathcal{S})$ *is an antimatroid.*

(ii) $(\mathcal{S}, \subseteq)$ *is a join-distributive lattice.*

(iii) $(\mathcal{S}, \subseteq)$ *is a semimodular lattice.*

In fact, the two concepts - antimatroids and join-distributive lattices - are essentially equivalent.

Theorem 2.7. *((Björner & Ziegler, 1992; Korte et al., 1991) A finite lattice is join-distributive if and only if it is isomorphic to the lattice of feasible sets of some antimatroid.*

It is easy to see that feasible sets of a poly-antimatroid ordered by inclusion form a join-distributive lattice as well. This kind of findings was discussed in (Magnien et al., 2001), where it was proved that any CFG is equivalent to a simple CFG (antimatroid), i.e., the lattices of their configuration spaces are isomorphic.

3. Isometry

For each graph $G = (V, E)$ the distance $d_G(u, v)$ between two vertices $u, v \in V$ is defined as the length of a shortest path joining them.

If G and H are arbitrary graphs, then a mapping $f : V(G) \to V(H)$ is an *isometric embedding* if $d_H(f(u), f(v)) = d_G(u, v)$ for any $u, v \in V(G)$.

Let $E = \{x_1, x_2, ... x_d\}$. Define a graph $H(E)$ as follows: the vertices are the finite subsets of E, two vertices A and B are adjacent if and only if the symmetric difference $A \triangle B$ is a singleton set. Then $H(E)$ is the *hypercube* Q_n on E (Djokovic, 1973). The hypercube can be equivalently defined as the graph on $\{0, 1\}^d$ in which two vertices form an edge if and only if they differ in exactly one position.

The shortest path distance $d_H(A, B)$ on the hypercube $H(E)$ is the Hamming distance between A and B that coincides with the symmetric difference distance: $d_H(A, B) = |A \triangle B|$.

A graph G is called a *partial cube* if it can be isometrically embedded into a hypercube $H(E)$ for some set E.

A family of sets S is *well-graded* (Doignon & Falmagne, 1997) if any two sets $P, Q \in S$ can be connected by a sequence of sets $P = R_0, R_1, ..., R_n = Q$ formed by single-element insertions and deletions ($|R_i \triangle R_{i+1}| = 1$), such that all intermediate sets in the sequence belong to S and $|P \triangle Q| = n$. This sequence of sets is called a *tight path*.

Any set system (E, S) defines an undirected graph $G_S = (S, E_S)$, where $E_S = \{\{P, Q\} \in S : |P \triangle Q| = 1\}$.

Theorem 3.1. *(Ovchinnikov, 2008) The graph G_S defined by a set system (E, S) is a partial cube on E if and only if the family S is well-graded.*

Proposition 3.2. *A family S of every antimatroid (E, S) is well-graded.*

Proof. We prove that there is a tight path between each $P, Q \in S$. If $P \subset Q$, then the existence of such path follows immediately from the chain property. If $P \not\subset Q$, then there exist chains from both P and Q to $P \cup Q$ that forms the tight path. $\square$

Thus each antimatroid is a partial cube and may be represented as a graph G_S that is a subgraph of the hypercube $H(E)$ or as a set of points of the hypercube $\{0, 1\}^d$. For example, see an antimatroid in Figure 3. The distance in an antimatroid (E, S) considered as a graph coincides with the Hamming distance between sets, i.e., $d_S(A, B) = |A \triangle B|$ for any $A, B \in S$.

A poly-antimatroid $(E, \mathcal{F})$ may be represented as a set of points in the digital space $\mathbb{Z}^d$, since each $A \in \mathcal{F}$ is defined by the sequence $(f_A(x_1), f_A(x_2), ..., f_A(x_d))$ that may be denoted as a point $(a_1, a_2, ..., a_d)$ in $\mathbb{Z}^d$. For example, see a two-dimensional poly-antimatroid in Figure 4.

The symmetric difference distance can be generalized to multisets by summing the absolute difference of the multiplicities of corresponding elements

$$|A \triangle B| = \sum |f_A(x_i) - f_B(x_i)|$$

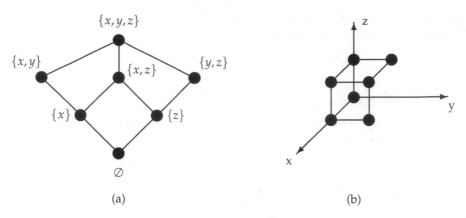

Fig. 3. Representation of an antimatroid: (a) as a graph of a family of subsets and (b) as a set of points in $\mathbb{Z}^3$.

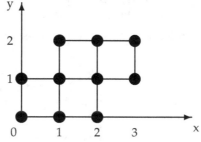

Fig. 4. A two-dimensional poly-antimatroid.

that coincides with the L_1 distance $(||A - B||_1)$ on the digital space $\mathbb{Z}^d$.

A multiset system $(E, \mathcal{F})$ defines an undirected graph $G_{\mathcal{F}}$ in the same way as a set system. So each poly-antimatroid $(E, \mathcal{F})$ may be represented as a graph $G_{\mathcal{F}}$.

These representations of poly-antimatroids are isometrically isomorphic.

Proposition 3.3. *The distance in a poly-antimatroid $(E, \mathcal{F})$ considered as a graph $G_{\mathcal{F}}$ coincides with the symmetric difference distance, i.e., $d_{\mathcal{F}}(A, B) = |A \triangle B|$ for any $A, B \in \mathcal{F}$.*

Proof. For any $A, B \in \mathcal{F}$ there is a multiset $A \cup B \in \mathcal{F}$. Then there is a path in the poly-antimatroid from A to $A \cup B$ (by adding successively all the elements from $B - A$) and from $A \cup B$ to B (by removing successively all the elements from $A - B$). Therefore, $d_{\mathcal{F}}(A, B) \leq |A \triangle B|$. On the other hand, in any path from A to B every two adjacent vertices differ by only one element, and, consequently, $d_{\mathcal{F}}(A, B) \geq |A \triangle B|$.   □

Thus a poly-antimatroid with the ground set of size d may be isometrically embedded into the d-dimensional lattice $\mathbb{Z}^d$.

Since each antimatroid is a poly-antimatroid, one may prove that every poly-antimatroid is isometrically isomorphic to some antimatroid.

Consider a poly-antimatroid $(E, \mathcal{F})$ given as a set of points in the digital space $\mathbb{Z}^d$. We use the Djokovic technique (Djokovic, 1973) elaborated by Eppstein (Eppstein, 2005) to embed a finite subset of the lattice $\mathbb{Z}^d$, and, therefore, a poly-antimatroid $(E, \mathcal{F})$, into a hypercube.

Let $\lambda_i = \max\{a_i | A \in \mathcal{F}\}$ and $\tau = \sum\limits_{i=1}^{d} \lambda_i$.

Consider the following mapping $\mu : \mathbb{Z}^d \to \{0,1\}$.

Define $\mu(A) = (\mu_1(A), \mu_2(A), ..., \mu_\tau(A))$, where for each k such that $\sum\limits_{i=1}^{j-1} \lambda_i < k \le \sum\limits_{i=1}^{j} \lambda_i$

$$\mu_k(A) = \begin{cases} 1 & a_j \ge k - \sum\limits_{i=1}^{j-1} \lambda_i \\ 0 & otherwise \end{cases}.$$

It is clear that $\mu(0, 0, ..., 0) = (0, 0, ..., 0)$.

For example, consider the poly-antimatroid depicted in Figure 4. Mapping μ embeds the poly-antimatroid into a hypercube $\{0,1\}^5$ such that $\mu(1,0) = (1,0,0,0,0)$, $\mu(0,1) = (0,0,0,1,0)$, $\mu(2,1) = (1,1,0,1,0)$ and so on. See an isometrically isomorphic antimatroid in Figure 5.

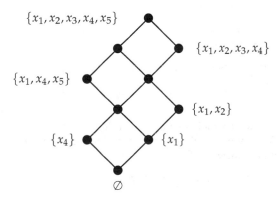

$\{x_1, x_2, x_3, x_4, x_5\}$

$\{x_1, x_2, x_3, x_4\}$

$\{x_1, x_4, x_5\}$

$\{x_1, x_2\}$

$\{x_4\}$ $\{x_1\}$

$\varnothing$

Fig. 5. An isometrically isomorphic antimatroid.

Lemma 3.4. *(Eppstein, 2005) The map μ is an isometry from a poly-antimatroid $(E, \mathcal{F})$ into a hypercube.*

Note that isometry μ is full-dimensional, i.e., each coordinate μ_i takes on both value 0 and 1 for at least one point.

It remains to show that $\mu(\mathcal{F})$ forms an antimatroid. We consider now a hypercube on some set $U = \{u_1, u_2, ..., u_\tau\}$. Then $\widehat{\mu}(A) = \{u_i | \mu_i(A) = 1\}$. Let $S = \{X \subseteq U : \exists A \in \mathcal{F}, X = \widehat{\mu}(A)\}$.

Proposition 3.5. *For every poly-antimatroid $(E, \mathcal{F})$, the set system $(U, S = \widehat{\mu}(\mathcal{F}))$ is an antimatroid.*

Proof. At first, $\emptyset \in S$. Indeed, $\emptyset \in \mathcal{F}$, i.e., $(0,0,...,0) \in \mathcal{F}$ and $\widehat{\mu}(0,0,...,0) = \emptyset$. Then S is accessible, since the map μ is distance-preserving. To see it, consider a non-empty $X \in S$. There is a set $A \in \mathcal{F}$ such that $X = \widehat{\mu}(A)$. Let $A = (a_1, a_2, ..., a_d)$. Since $(E, \mathcal{F})$ is a poly-antimatroid, there is $j \in \{1, 2, ..., d\}$ such that $B = (a_1, a_2, ..., a_j - 1, ..., a_d) \in \mathcal{F}$. Then

$\mu(A)$ and $\mu(B)$ differs only in the position $k = \sum\limits_{i=1}^{j-1} \lambda_i + a_j$, i.e., $\widehat{\mu}(A) - \widehat{\mu}(B) = \{u_k\}$.

Now let us prove, that the family S is closed under union, i.e., $X, Y \in S \Rightarrow X \cup Y \in S$. Since $X, Y \in S$ there are two points $A, B \in \mathcal{F}$ such that $X_1 = \widehat{\mu}(A)$ and $X_2 = \widehat{\mu}(B)$. Let $A = (a_1, a_2, ..., a_d)$ and $B = (b_1, b_2, ..., b_d)$. Then, $A \cup B = (c_1, c_2, ..., c_d)$, where $c_i = \max(a_i, b_i)$. Hence $\mu(A \cup B) = (v_1, v_2, ..., v_\tau)$, where $v_i = \max(\mu_i(A), \mu_i(B))$. So, $\widehat{\mu}(A \cup B) = \widehat{\mu}(A) \cup \widehat{\mu}(B)$, which implies $X \cup Y = \widehat{\mu}(A \cup B) \in S$. $\square$

So we can say that the two structures - a poly-antimatroid $(E, \mathcal{F})$ and an antimatroid $(U, \widehat{\mu}(\mathcal{F}))$ are essentially identical.

4. Poly-dimension

Clearly, an antimatroid is a poly-antimatroid as well. The question is what is a minimal dimension d such that an antimatroid $(U, \mathcal{S})$ may be represented by an isomorphic poly-antimatroid $(\widehat{U}, \mathcal{F})$ in the space $\mathbb{Z}^d$.

We will call this minimal dimension $\pi(\mathcal{S})$ the *poly-dimension* of the antimatroid $(U, \mathcal{S})$.

Each antimatroid $(U, \mathcal{S})$ may be considered also as a directed graph $G = (V, E)$ with $V = \mathcal{S}$ and $(A, B) \in E \Leftrightarrow \exists c \in B$ such that $A = B - c$.

Denote *in-degree* of the vertex A as $deg_{in}(A)$, and *out-degree* as $deg_{out}(A)$, where

$$deg_{in}(A) = |\{c : A - c \in \mathcal{S}\}|, deg_{out}(A) = |\{c : A \cup c \in \mathcal{S}\}|$$

Consider antimatroids for which their maximum in-degree and maximum out-degree is at most p, and there is at least one feasible set for which in-degree or out-degree equals p. We will call such antimatroids *p-antimatroids*.

Proposition 4.1. *The poly-dimension of an 1-antimatroid is one.*

Proof. It is easy to see that 1-antimatroid is a chain $\emptyset \subseteq \{a_1\} \subseteq ... \subseteq \{a_1, a_2, ..., a_n\}$, that may be represented as a poly-antimatroid $\emptyset \subseteq \{x\} \subseteq ... \subseteq \{x, x, ..., x\}$ belonging to the axe x. The "only if" part of the proof is clear. $\square$

To find the poly-dimension of a 2-antimatroid we show that 2-antimatroids have the special structure of "sub-grid".

A *poset antimatroid* is a particular case of antimatroid, which is formed by the lower sets of a poset (partially ordered set). The poset antimatroids can be characterized as the unique antimatroids which are closed under intersection (Korte et al., 1991).

Let us prove that any 2-antimatroid is a poset antimatroid.

An *endpoint* of a feasible set X is an element $e \in X$ such that $X - e$ is a feasible set too. A feasible set X is an *e-path* if it has a single endpoint e.

It is easy to see that an e-path is a minimal feasible set containing e.

The following lemma gives the characterization of poset antimatroids in terms of e-paths. We present a proof for the sake of completeness.

Lemma 4.2. *(Algaba et al., 2004) An antimatroid (U, S) is a poset antimatroid if and only if every $i \in U$ has a unique i-path in S.*

Proof. Let (U, S) be a poset antimatroid and let $A, B, A \neq B$ be two distinct i-paths for $i \in U$. Then $i \in A \cap B \in S$. Since each i-path is a minimal feasible set containing i, then $A \not\subseteq B$ and $B \not\subseteq A$. Hence the chain property implies the existence of an element $j \in A - (A \cap B)$ such that $A - j \in S$. This is a contradiction with A being an i-path.

Now suppose that every $i \in U$ has a unique i-path in S. Let $X, Y \in S$. If $X \cap Y = \emptyset$ then $X \cap Y \in S$ by definition of an antimatroid.

If $X \cap Y \neq \emptyset$ then for every $i \in U$ there exists an i-path $H_1^i \subseteq X$ and $H_2^i \subseteq Y$, because each i-path is a minimal feasible set containing i. By assumption $H_1^i = H_2^i = H^i$. Thus, for all $i \in X \cap Y$, an i-path $H^i \subseteq X \cap Y$, and so $X \cap Y = \bigcup_{i \in X \cap Y} H^i$. Therefore, $X \cap Y \in S$, since every antimatroid is closed under union. Hence, (U, S) is a poset antimatroid. $\square$

Lemma 4.3. *Any 2-antimatroid (U, S) has a unique i-path for each $i \in U$.*

Proof. Suppose the opposite. Let $A, B \in S$ be two different i-paths. Note that $A \not\subseteq B$ and $B \not\subseteq A$, since each i-path is a minimal set containing i.

Consider the set $Z = A \cup B$. From the chain property it follows that there is $b \in B - A$, such that $Z - b \in S$. On the other hand, there is $a \in A - B$ with $Z - a \in S$. Since $(A - i) \cup (B - i) = Z - i \in S$, we have that $deg_{in}(Z) \geq 3$. $\square$

Lemmas 4.2 and 4.3 imply the following.

Proposition 4.4. *Any 2-antimatroid is a poset antimatroid.*

Denote $C_k = \{X \in S : |X| = k\}$ a family of feasible sets of cardinality k.

Definition 4.5. *A lower zigzag is a sequence of feasible sets $P_0, P_1, ..., P_{2m}$ such that any two consecutive sets in the sequence differ by a single element and $P_{2i} \in C_k$, and $P_{2i-1} \in C_{k-1}$ for all $0 \leq i \leq m$.*

In the same way we define an *upper zigzag* in which $P_{2i-1} \in C_{k+1}$.

Each zigzag $P_0, P_1, ..., P_{2m}$ is a path connecting P_0 and P_{2m}, and so the distance on the zigzag $d(P_0, P_{2m}) = 2m$ is always no less than the distance $d_S(P_0, P_{2m})$ on an antimatroid (U, S).

In Figure 6 we can see two sets ($A = \{1, 2, 3, 5\}$ and $B = \{1, 3, 4, 5\}$) that are connected by an unique lower zigzag, such that the distance on the zigzag is 4, while $|A \triangle B| = 2$. Note, that the distance on the upper zigzag is indeed 2. For two sets $X = \{1, 2, 5\}$ and $Y = \{3, 4, 5\}$ both distances on the lower zigzag and on the upper zigzag are equal to 6, while $|X \triangle Y| = 4$.

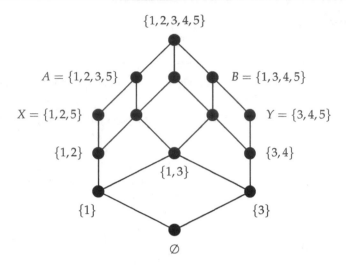

Fig. 6. An antimatroid without distance preserving zigzags.

In order to get the distance on a zigzag equal to the set distance, an antimatroid has to be a poset antimatroid.

Theorem 4.6. *In each poset antimatroid* $(U, \mathcal{S})$ *every two feasible sets* A, B *of the same cardinality* k *can be connected both by a lower and an upper zigzag in such a way that the distance between these sets* $d_{\mathcal{S}}(A, B)$ *coincides with the distance on the zigzags.*

Proof. We use induction on k. For every pair of one-element feasible sets $\{x\}$ and $\{y\}$ there is a lower zigzag (via $\varnothing$) and an upper zigzag (via $\{x, y\}$) such that the distance between these sets $d_{\mathcal{S}}(\{x\}, \{y\}) = 2$ on the antimatroid coincides with distances on the zigzags.

Assume that the hypothesis of the theorem is correct for all sets $X \in \mathcal{S}$ with $|X| < k$.

Consider two sets $A, B \in C_k$ and let $d_{\mathcal{S}}(A, B) = |A \triangle B| = 2m$. Since $(U, \mathcal{S})$ is a poset antimatroid, the set $A \cap B \in \mathcal{S}$. Hence, by the chain property, there is $a \in A - B$ such that $A - a \in \mathcal{S}$, and $b \in B - A$ with $B - b \in \mathcal{S}$. These two sets belong to C_{k-1}, so there are an upper zigzag $A - a = P_1, P_2, ..., P_{2l-1} = B - b$ connecting $A - a$ with $B - b$ (see Figure 7). Since $d_{\mathcal{S}}(A - a, B - b) = |A \triangle B| - 2 = 2m - 2$, the distance on the upper zigzag between $A - a$ and $B - b$ is $2m - 2$ by the induction hypothesis, i.e., $2l - 1 = 2m - 1$. It is easy to see that neither A nor B belongs to the upper zigzag $A - a = P_1, P_2, ..., P_{2m-1} = B - b$, because otherwise it implies that $d_{\mathcal{S}}(A, B) \le 2m - 2$. Hence exists a lower zigzag $A = P_0, A - a = P_1, P_2, ..., P_{2m-1} = B - b, P_{2m} = B$ connecting A and B such that the distance on the zigzag between A and B equals $2m$.

Let $P^i = P_{i-1} \cup P_{i+1}$ for each $i = 1, 3, ..., 2m - 1$. All obtained sets are different, because otherwise $d_{\mathcal{S}}(A, B) < 2m$. So there is the upper zigzag by length $2m$ connecting A and B. $\quad\square$

Note that if for a zigzag $P_0, P_1, ..., P_{2m}$ the distance on the zigzag $d(P_0, P_{2m}) = 2m$ is equal to the distance $d_{\mathcal{S}}(P_0, P_{2m}) = |P_0 \triangle P_{2m}|$ then the zigzag preserves the distance for each pair P_i, P_j, i.e., $d_{\mathcal{S}}(P_i, P_j) = d(P_i, P_j) = |j - i|$.

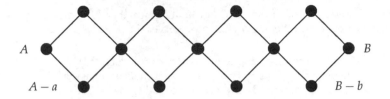

Fig. 7. Zigzags of an antimatroid.

For poset antimatroids there are distance preserving zigzags connecting two given sets, but these zigzags are not obliged to connect all feasible sets of the same cardinality. In Figure 8 we can see that there is a poset antimatroids, for which it is not possible to build two distance preserving zigzags connecting all feasible sets of the same cardinality.

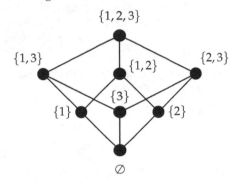

Fig. 8. A poset antimatroid without total zigzags.

It is worth mentioning that in 2-antimatroids all feasible sets of the same cardinality may be connected by one distance preserving zigzag.

Proposition 4.7. *In 2-antimatroid (U, S) all feasible sets C_k of the same cardinality k can be connected both by a lower and an upper zigzag in such a way that the distance between any two sets $d_S(P_i, P_j)$ belonging to the same zigzag coincides with distance on the zigzag $d(P_i, P_j) = |j - i|$.*

Proof. We proceed by induction on cardinality k. For one-elements sets it is obvious. Assume that it is correct for all sets $A \in S$ with $|A| < k$.

Each set from C_k is obtained from some set of cardinality $k - 1$. Since an antimatroid is closed under union and its maximum in-degree and out-degree is 2, the $n + 1$ sets of cardinality $k - 1$ connected by an upper zigzag form n sets from C_k. In addition to these n sets, each of two extreme sets (P_0 and P_{2n}) can form one more set of cardinality k that does not belong to the upper zigzag. Since each out-degree is less than 3 there are no other sets of cardinality k. So, the set C_k is connected by a lower zigzag. Since an antimatroid is closed under union and its maximum in-degree is 2, all feasible sets of the same cardinality k are also connected by an upper zigzag.

Now prove that the distance on the zigzags coincides with the distance on the antimatroid. Since maximum in-degree and out-degree are 2, the obtained lower zigzag is only lower zigzag connecting P_0 and P_{2n}. Then from Theorem 4.6 it follows that it is a distance preserving zigzag. Similarly, we obtain that the upper zigzag preserves the distance as well.   □

For each edge $(A, B) \in S$ there exists a single element $c \in B$ such that $A = B - c$, and so we can label each edge by such element. Then, each 2-antimatroid consists of quadrilaterals with equal labels on opposite pairs of edges, since if it contains the vertices X, $X \cup a$, $X \cup b$ for some $X \in S$, and $a, b \in U$, it also contains $X \cup \{a, b\}$. This property, called by $\cup$-coloring, was invented in (Felsner & Knauer, 2009) as the characterization of join-distributive lattices.

Proposition 4.8. *There exists a labeling of edges of a 2-antimatroid by two labels x or y such that opposite pairs of edges in each quadrilateral have equal labels.*

Proof. To obtain such labeling we consequently scan a 2-antimatroid by layers C_k beginning from the empty set ($k = 0$). The first zigzag of edges we mark by two labels x and y in turn, beginning from x or from y arbitrary. From the Proposition 4.7 follows that the next zigzag can be marked alternate by two labels (x, y) in such a way that the opposite pairs of edges in each quadrilateral have equal labels (see Figure 9).

If we reach the layer C_k with one element, the next zigzag is labeled alternate by two labels without dependence on the previous layers.   □

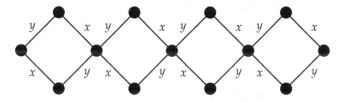

Fig. 9. A zigzag labeling.

It follows from accessibility that for each $A \in S$ there is a shortest path connecting A with $\emptyset$. The length of such path is equal to the cardinality of A. Let $M(A)$ be a multiset containing all labels on this shortest path. Obviously, $M(A) \in \mathbb{Z}^2$. It is possible that there are some shortest paths between A and $\emptyset$. Prove that all such paths result in the same multiset.

Lemma 4.9. *In every 2-antimatroid (U, S) for each $A \in S$ all shortest paths connecting A with $\emptyset$ forms the same multiset $M(A)$.*

Proof. We use induction on cardinality k. For one-elements sets it is obviously. Assume it is correct for all sets $A \in S$ with $|A| < k$. Consider $Y \in C_k$. If Y is obtained from only one set of cardinality $k - 1$, then $M(Y)$ is unique. If $Y = A \cup a = B \cup b$, then there is $X = A \cap B \in S$ such that $A = X \cup b$ and $B = X \cup a$, since every 2-antimatroid is a poset antimatroid. Let a label of $(X, X \cup b)$ be x, and a label of $(X, X \cup a)$ be y. Thus $M(A) = M(X) \cup x$ and $M(B) = M(X) \cup y$, and so $M(Y) = M(X) \cup \{x, y\}$. Therefore, by induction hypothesis, all shortest path connecting Y with $\emptyset$ forms the same $M(Y)$.   □

Proposition 4.10. *In every 2-antimatroid (U, S) the mapping $M : (U, S) \to \mathbb{Z}^2$, where each $A \in S$ corresponds to $M(A)$, is an isometry.*

Proof. We have to prove that for all $A, B \in \mathcal{S}$ the distance $d_{\mathcal{S}}(A, B)$ on the antimatroid $(U, \mathcal{S})$ is equal to the distance $\|M(A) - M(B)\|_1$ on $\mathbb{Z}^2$, i.e., it holds

$$|A \triangle B| = |M(A) \triangle M(B)|.$$

If $A \subseteq B$, then there is a shortest path connecting B with $\emptyset$ via A, and so $|A \triangle B| = |B - A| = |M(B) - M(A)| = |M(A) \triangle M(B)|$.

Consider two incomparable sets $A, B \in \mathcal{S}$. Since a 2-antimatroid is a poset antimatroid, $A \cap B \in \mathcal{S}$. Then there are a shortest path P_A connecting A with $\emptyset$ via $A \cap B$, and a shortest path P_B connecting B with $\emptyset$ via $A \cap B$. Since $|A \triangle B| = |A - A \cap B| + |B - A \cap B|$, remain to prove that the part of P_A from A till $A \cap B$ is labeled by only one label (for example x) and the part of P_B from B till $A \cap B$ is labeled by only one but another label (y).

Indeed, from $A \cap B$ go out two different edges marked by two different labels, where one belongs to P_A and the second belongs to P_B. Let the label of the first edge was x and the label of the second edge was y. Denote obtained sets as A_1 and B_1. Note that $A_1 = (A \cap B) \cup a$ and $B_1 = (A \cap B) \cup b$, where $a \in A - B$ and $b \in B - A$. If $A_1 \neq A$ then from A_1 go out two edges. The first edge $(A_1, A_1 \cup b)$ is labeled by y and does not belong to P_A since $b \in B - A$. The second edge (A_1, A_2) is labeled by x and belongs to P_A. The same is correct for each set on path P_A from A till $A \cap B$. So the part of P_A from A till $A \cap B$ is labeled by label x only. By the same way we obtain that the part of P_B from B till $A \cap B$ is labeled by y only.   □

Thus we have the following.

Theorem 4.11. *The poly-dimension of a 2-antimatroid is two.*

5. Open problems

It is clear that the poly-dimension of a d-antimatroid is at least d.

Conjecture 5.1. *The poly-dimension of an antimatroid equals d if and only if it is a d-antimatroid.*

6. References

Algaba, E.; Bilbao, J.M.; van den Brink, R. & Jimenez-Losada A. (2004). Cooperative Games on Antimatroids. *Discrete Mathematics*, Vol.282, 1-15.

Algaba, E.; Bilbao, J.M. & Slikker, M. (2010). A Value for Games Restricted by Augmenting Systems. *SIAM J. Discrete Mathematics*, Vol.24, No.3, 992-1010

Bilbao, J.M. (2003). Cooperative Games under Augmenting Systems. *SIAM J. Discrete Mathematics*, Vol.17, No.1, 122-133.

Björner, A.; Lovász, L. & Shor, P.R. (1991). Chip-firing Games on Graphs. *European Journal of Combinatorics*, Vol.12, 283-291.

Björner, A. & Ziegler, G.M. (1992). Introduction to Greedoids, In: *Matroid applications*, N. White, (Ed.), Cambridge Univ.Press, Cambridge, UK.

Boley, M; Horváth,T; Poigné,A & Wrobel,S. (2010). Listing Closed Sets of Strongly Accessible Set Systems with Applications to Data Mining. *Theoretical Computer Science - TCS* , Vol. 411, No. 3, 691-700.

Boyd, E.A. & Faigle, U. (1990). An Algorithmic Characterization of Antimatroids. *Discrete Applied Mathematics*, Vol.28, 197-205.

Cosyn, E. & Uzun, H.B. (2009). Note on Two Necessary and Sufficient Axioms for a Well-graded Knowledge Space. *Journal of Mathematical Psychology*, Vol.53, No.1, 40-42.

Dilworth, R.P. (1940). Lattices with Unique Irreducible Decomposition. *Ann. of Mathematics*, Vol.41, 771-777.

Djokovic, D.Z. (1973). Distance Preserving Subgraphs of Hypercubes. *J. Combin. Theory Ser.B*, Vol.14, 263-267.

Doignon, J.-P. & Falmagne, J.-Cl. (1997). Well-graded Families of Relations. *Discrete Mathematics*, Vol.173, 35–44.

Eppstein, D. (2005). The Lattice Dimension of a Graph. *European Journal of Combinatorics*, Vol.26, 585-592.

Eppstein, D. (2008). Upright-Quad Drawing of st-Planar Learning Spaces. *Journal of Graph Algorithms and Applications*, Vol.12, No.1, 51-72.

Eppstein, D.; Falmagne, J.-Cl. & Ovchinnikov, S. (2008). *Media Theory: Interdisciplinary Applied Mathematics*, Springer-Verlag, Berlin.

Falmagne, J.-Cl. & Doignon, J.-P., (2011). *Learning Spaces: Interdisciplinary Applied Mathematics*, Springer-Verlag, Berlin.

Felsner, S. & Knauer, K. (2009). ULD-Lattices and Δ-Bondes. *Combinatorics, Probability and Computing*, Vol.18, No.5, 707-724.

Glasserman, P. & Yao, D.D. (1994). *Monotone Structure in Discrete Event Systems*, Wiley Inter-Science, Series in Probability and Mathematical Statistics.

Grabisch, M. (2009). The Core of Games on Ordered Structures and Graphs. *4OR: A Quarterly Journal of Operations Research*, Vol.7, No.3, 207-238.

Graham, R.L. & Pollak, H.O. (1971). On the Addressing Problem for Loop Switching. *Bell System Technical Journal*, Vol.50, 2495-2519.

Honda, A. & Grabisch, M. (2008). An Axiomatization of Entropy of Capacities on Set Systems, *European Journal of Operational Research*, Vol.190 , 526-538.

Kempner, Y. & Levit, V.E. (2003). Correspondence between Two Antimatroid Algorithmic Characterizations. *The Electronic Journal of Combinatorics*, Vol.10.

Kempner, Y. & Levit, V.E. (2007). A Geometric Characterization of Poly-antimatroids, *ENDM (Electronic Notes in Discrete Mathematics)*, Vol.28, 357-364.

Kempner, Y. & Muchnik, I. (2003). Clustering on Antimatroids and Convex Geometries, *WSEAS Transactions on Mathematics*, Vol.2, No.1, 54-59.

Korte, B.; Lovász, L. & R. Schrader, R. (1991). *Greedoids*, Springer-Verlag, New York/Berlin.

Magnien, C.; Phan, H.D. & Vuillon, L. (2001). Characterization of Lattices Induced by (extended) Chip Firing Games, *Discrete Mathematics and Theoretical Computer Science Proceedings AA (DM-CCG)*, 229-244.

Monjardet, B. (2003). The Presence of Lattice Theory in Discrete Problems of Mathematical Social Sciences. Why, *Math. Social Science*, Vol.46, 103-144.

Nakamura, M. (2005). Characterization of Polygreedoids and Poly-antimatroids by Greedy Algorithms, *Oper.Res. Lett.*, Vol.33, No. 4, 389-394.

Ovchinnikov, S. (2008). Partial cubes: Structures, Characterizations, and Constructions, *Discrete Mathmatics*, Vol.308, 5597–5621.

4

Pure Links Between Graph Invariants and Large Cycle Structures

Zh.G. Nikoghosyan[*]

Institute for Informatics and Automation Problems,
National Academy of Sciences,
Armenia

1. Introduction

Hamiltonian graph theory is one of the oldest and attractive fields in discrete mathematics, concerning various path and cycle existence problems in graphs. These problems mainly are known to be NP-complete that force the graph theorists to direct efforts toward understanding the global and general relationship between various invariants of a graph and its path and cycle structure.

This chapter is devoted to large cycle substructures, perhaps the most important cycle structures in graphs: Hamilton, longest and dominating cycles and some generalized cycles including Hamilton and dominating cycles as special cases.

Graph invariants provide a powerful and maybe the single analytical tool for investigation of abstract structures of graphs. They, combined in convenient algebraic relations, carry global and general information about a graph and its particular substructures such as cycle structures, factors, matchings, colorings, coverings, and so on. The discovery of these relations is the primary problem of graph theory.

In the literature, eight basic (initial) invariants of a graph G are known having significant impact on large cycle structures, namely order n, size q, minimum degree δ, connectivity κ, independence number α, toughness τ and the lengths of a longest path and a longest cycle in $G \setminus C$ for a given longest cycle C in G, denoted by $\overline{p}$ and $\overline{c}$, respectively.

In this chapter we have collected 37 pure algebraic relations between $n, q, \delta, \kappa, \alpha, \tau, \overline{p}$ and $\overline{c}$ ensuring the existence of a certain type of large cycles. The majority of these results are sharp in all respects.

Focusing only on basic graph invariants, as well as on pure algebraic relations between these parameters, in fact, we present the simplest kind of relations for large cycles having no forerunners in the area. Actually they form a source from which nearly all possible hamiltonian results (including well-known Ore's theorem, Pósa's theorem and many other

---

[*] G.G. Nicoghossian (up to 1997)

generalizations) can be developed further by various additional new ideas, generalizations, extensions, restrictions and structural limitations:

- **generalized and extended graph invariants** - degree sequences (Pósa type, Chvátal type), degree sums (Ore type, Fun type), neighborhood unions, generalized degrees, local connectivity, and so on,
- **extended list of path and cycle structures** - Hamilton, longest and dominating cycles, generalized cycles including Hamilton and dominating cycles as special cases, 2-factor, multiple Hamilton cycles, edge disjoint Hamilton cycles, powers of Hamilton cycles, k-ordered Hamilton cycles, arbitrary cycles, cycle systems, pancyclic-type cycle systems, cycles containing specified sets of vertices or edges, shortest cycles, analogous path structures, and so on,
- **structural (descriptive) limitations** - regular, planar, bipartite, chordal and interval graphs, graphs with forbidden subgraphs, Boolean graphs, hypercubes, and so on,
- **graph extensions** - hypergraphs, digraphs and orgraphs, labeled and weighted graphs, infinite graphs, random graphs, and so on.

We refer to (Bermond, 1978) and (Gould, 1991, 2003) for more background and general surveys.

The order n, size q and minimum degree δ clearly are easy computable graph invariants. In (Even & Tarjan, 1975), it was proved that connectivity κ can be determined in polynomial time, as well. Determining the independence number α and toughness τ are shown in (Garey & Johnson, 1983) and (Bauer et al., 1990a) to be CD_λ-hard problems. Moreover, it was proved (Bauer et al., 1990a) that for any positive rational number t, recognizing t-tough graphs (in particular 1-tough graphs) is an NP-hard problem.

The order n and size q are neutral with respect to cycle structures. Meanwhile, they become more effective combined together (Theorem 1). The minimum degree δ having high frequency of occurrence in different relations is, in a sense, a more essential invariant than the order and size, providing some dispersion of the edges in a graph. The combinations between order n and minimum degree δ become much more fruitful especially under some additional connectivity conditions. The impact of some relations on cycle structures can be strengthened under additional conditions of the type $\delta \geq \alpha \pm i$ for appropriate integer i. By many graph theorists, the connectivity κ is at the heart of all path and cycle questions providing comparatively more uniform dispersion of the edges. An alternate connectedness measure is toughness τ - the most powerful and less investigated graph invariant introduced by Chvátal (Chvátal, 1973) as a means of studying the cycle structure of graphs. Chvátal (Chvátal, 1973) conjectured that there exists a finite constant t_0 such that every t_0-tough graph is hamiltonian. This conjecture is still open. We have omitted a number of results involving toughness τ as a parameter since they are far from being best possible.

Large cycle structures are centered around well-known Hamilton (spanning) cycles. Other types of large cycles were introduced for different situations when the graph contains no Hamilton cycles or it is difficult to find it. Generally, a cycle C in a graph G is a large cycle if it dominates some certain subgraph structures in G in a sense that every such structure has a vertex in common with C. When C dominates all vertices in G then C is a Hamilton cycle. When C dominates all edges in G then C is called a dominating cycle introduced by Nash-Williams (Nash-Williams, 1971). Further, if C dominates all paths in G of length at least some fixed integer λ then C is a PD_λ (path dominating)-cycle introduced by Bondy (Bondy, 1981). Finally, if C dominates all cycles in G of length at least λ then C is a CD_λ (cycle dominating)-cycle, introduced in (Zh.G. Nikoghosyan, 2009a). The existence problems of generalized PD_λ and CD_λ-cycles are studied in (Zh.G. Nikoghosyan, 2009a) including Hamilton and dominating cycles as special cases.

Section 2 is devoted to necessary notation and terminology. In Section 3, we discuss pure relations between various basic invariants of a graph and Hamilton cycles. Next sections are devoted to analogous pure relations concerning dominating cycles (Section 4), CD_λ-cycles (Section 5), long cycles (Section 6), long cycles with Hamilton cycles (Section 7), long cycles with dominating cycles (Section 8) and long cycles with CD_λ-cycles (Section 9). In Section 10 we present the proofs of Theorems 6, 21 and 27. Concluding remarks are given in Section 11.

2. Terminology

We consider only finite undirected graphs without loops or multiple edges. A good reference for any undefined terms is (Bondy & Murty, 1976). The set of vertices of a graph G is denoted by $V(G)$ and the set of edges by $E(G)$. For S a subset of $V(G)$, we denote by $G \setminus S$ the maximum subgraph of G with vertex set $V(G) \setminus S$. For a subgraph H of G we use $G \setminus H$ short for $G \setminus V(H)$. Denote by $N(x)$ the neighborhood of a vertex x in G. Put $d(x) = |N(x)|$

A simple cycle (or just a cycle) C of length t is a sequence $v_1 v_2 ... v_t v_1$ of distinct vertices $v_1, v_2, ..., v_t$ with $v_i v_{i+1} \in E(G)$ for each $i \in \{1, ..., t\}$, where $v_{t+1} = v_1$. When $t = 2$, the cycle $C = v_1 v_2 v_1$ on two vertices v_1, v_2 coincides with the edge $v_1 v_2$, and when $t = 1$, the cycle $C = v_1$ coincides with the vertex v_1. So, all vertices and edges in a graph can be considered as cycles of lengths 1 and 2, respectively. A graph G is hamiltonian if G contains a Hamilton cycle, i.e. a cycle containing all vertices of G. Let λ be an integer. A cycle C' in G is a PD_λ-cycle if $|P| \le \lambda - 1$ for each path P in $G \setminus C'$ and is a CD_λ-cycle if $|C''| \le \lambda - 1$ for each cycle C'' in $G \setminus C'$. In particular, PD_0-cycles and CD_1-cycles are well-known Hamilton cycles and PD_1-cycles and CD_2-cycles are often called dominating cycles.

We reserve n, q, δ, κ and α to denote the number of vertices (order), number of edges (size), minimum degree, connectivity and independence number of a graph, respectively. Let c denote the circumference - the length of a longest cycle in a graph. In general, $c \ge 1$. For C a longest cycle in G, denote by $\bar{p}$ and $\bar{c}$ the lengths of a longest path and a longest cycle in $G \setminus C$, respectively. Let $s(G)$ denote the number of components of a graph G. A graph G is t-tough if $|S| \ge t \cdot s(G \setminus S)$ for every subset $S \subseteq V(G)$ with $s(G \setminus S) > 1$. The toughness of G, denoted $\tau(G)$, is the maximum value of t for which G is t-tough (taking $\tau(K_n) = \infty$ for all $n \ge 1$).

An (x,y)-path is a path with end vertices x and y. Given an (x,y)-path L of G, we denote by $\overrightarrow{L}$ the path L with an orientation from x to y. If $u,v \in V(L)$ then $u\overrightarrow{L}v$ denotes the consecutive vertices on $\overrightarrow{L}$ from u to v in the direction specified by $\overrightarrow{L}$. The same vertices, in reverse order, are given by $v\overleftarrow{L}u$. For $\overrightarrow{L} = x\overrightarrow{L}y$ and $u \in V(L)$, let $u^+\left(\overrightarrow{L}\right)$ (or just u^+) denotes the successor of u $(u \neq y)$ on $\overrightarrow{L}$, and u^- denotes its predecessor $(u \neq x)$. If $A \subseteq V(L) \setminus \{y\}$ then we denote $A^+ = \{v^+ \mid v \in A\}$. Similar notation is used for cycles. If Q is a cycle and $u \in V(Q)$, then $u\overrightarrow{Q}u = u$.

Let a, b, t, k be integers with $k \leq t$. We use $H(a,b,t,k)$ to denote the graph obtained from $tK_a + \overline{K}_t$ by taking any k vertices in subgraph $\overline{K}_t$ and joining each of them to all vertices of K_b. Denote by L_δ the graph obtained from $3K_b + K_1$ by taking one vertex in each of three copies of K_b and joining them each to other. For odd n, where $n \geq 15$, construct the graph G_n from $\overline{K}_{(n-1)/2} + K_\delta + K_{(n+1)/2-\delta}$, where $n/3 \leq \delta \leq (n-5)/2$, by joining every vertex in K_δ to all other vertices and by adding a matching between all vertices in $K_{(n+1)/2-\delta}$ and $(n+1)/2-\delta$ vertices in $\overline{K}_{(n-1)/2}$. It is easily seen that G_n is 1-tough but not hamiltonian. A variation of the graph G_n, with K_δ replaced by $\overline{K}_\delta$ and $\delta = (n-5)/2$, will be denoted by G_n^*.

3. Pure relations for Hamilton cycles

We begin with a pure algebraic relation between order n and size q insuring the existence of a Hamilton cycle based on the natural idea that if a sufficient number of edges are present in the graph then a Hamilton cycle will exist.

Theorem 1 (Erdös & Gallai, 1959). Let G be an arbitrary graph. If

$$q \geq \frac{n^2 - 3n + 5}{2}$$

then G is hamiltonian.

Example for sharpness. To see that the size bound $(n^2 - 3n + 5)/2$ in Theorem 1 is best possible, note that the graph formed by joining one vertex of K_{n-1} to K_1, contains $(n^2 - 3n + 4)/2$ edges and is not hamiltonian.

The next pure algebraic relation links the size q and minimum degree δ insuring the existence of a Hamilton cycle. In view of Theorem 1, it seems a little surprising, providing, in fact, a contrary statement.

Theorem 2 (Zh.G. Nikoghosyan, 2011). Let G be an arbitrary graph. If

$$q \leq \delta^2 + \delta - 1$$

then G is hamiltonian.

Example for sharpness. The bound $\delta^2 + \delta - 1$ in Theorem 2 can not be relaxed to $\delta^2 + \delta$ since the graph $K_1 + 2K_\delta$ consisting of two copies of $K_{\delta+1}$ and having exactly one vertex in common, has $\delta^2 + \delta$ edges but is not hamiltonian.

The earliest sufficient condition for a graph to be hamiltonian is based on the order n and minimum degree δ ensuring the existence of a Hamilton cycle with sufficient number of edges by keeping the minimum degree at a fairly high level.

Theorem 3 (Dirac, 1952). Let G be an arbitrary graph. If

$$\delta \geq \frac{n}{2}$$

then G is hamiltonian.

Example for sharpness: $2K_\delta + K_1$.

The graph $2K_\delta + K_1$ shows that the bound $n/2$ in Theorem 3 can not be replaced by $(n-1)/2$.

The minimum degree bound $n/2$ in Theorem 3 can be slightly relaxed for graphs under additional 1-tough condition.

Theorem 4 (Jung, 1978). Let G be a graph with $n \geq 11$ and $\tau \geq 1$. If

$$\delta \geq \frac{n-4}{2}$$

then G is hamiltonian.

Examples for sharpness: Petersen graph; $K_{\delta,\delta+1}$; G_n^*.

This bound $(n-4)/2$ itself was lowered further to $(n-7)/2$ under stronger conditions $n \geq 30$ and $\tau > 1$.

Theorem 5 (Bauer et al., 1991a). Let G be a graph with $n \geq 30$ and $\tau > 1$. If

$$\delta \geq \frac{n-7}{2}$$

then G is hamiltonian.

Furthermore, the bound $n/2$ was essentially lowered to $(n+\kappa)/3$ (when $k < n/2$) by incorporating connectivity κ into the minimum degree bound.

Theorem 6 (Zh.G. Nikoghosyan, 1981). Let G be a graph with $\kappa \geq 2$. If

$$\delta \geq \frac{n + \kappa}{3}$$

then G is hamiltonian.

Examples for sharpness: $2K_\delta + K_1; H(1, \delta - \kappa + 1, \delta, \kappa)$ $(2 \leq \kappa < n/2)$.

A short proof of Theorem 6 was given by Häggkvist (Häggkvist & Nicoghossian, 1981).

The minimum degree bound $(n + \kappa)/3$ in Theorem 6 was slightly lowered to $(n + \kappa - 2)/3$ for 1-tough graphs.

Theorem 7 (Bauer & Schmeichel, 1991b). Let G be a graph with $\tau \geq 1$. If

$$\delta \geq \frac{n + \kappa - 2}{3}$$

then G is hamiltonian.

Examples for sharpness: $K_{\delta, \delta+1}; L_\delta$.

Another essential improvement of Dirac's bound $n/2$ was established for 2-connected graphs under additional strong condition $\delta \geq \alpha$.

Theorem 8 (Nash-Williams, 1971). Let G be a graph with $\kappa \geq 2$. If

$$\delta \geq \max\left\{\frac{n + 2}{3}, \alpha\right\}$$

then G is hamiltonian.

Examples for sharpness:
$(\lambda + 1)K_{\delta - \lambda + 1} + K_\lambda$ $(\delta \geq 2\lambda); (\lambda + 2)K_{\delta - \lambda} + K_{\lambda+1}$ $(\delta \geq 2\lambda + 1); H(\lambda, \lambda + 1, \lambda + 3, \lambda + 2)$.

Theorem 8 was slightly improved by replacing the condition $\kappa \geq 2$ with a stronger condition $\tau \geq 1$.

Theorem 9 (Bigalke & Jung, 1979). Let G be a graph with $\tau \geq 1$. If

$$\delta \geq \max\left\{\frac{n}{3}, \alpha - 1\right\}$$

then G is hamiltonian.

Examples for sharpness: $K_{\delta, \delta+1}$ $(n \geq 3); L_\delta$ $(n \geq 7); K_{\delta, \delta+1}$ $(n \geq 3)$.

For λ a positive integer, the bound $(n+2)/3$ in Theorem 8 was essentially lowered under additional condition of the type $\delta \geq \alpha + \lambda$, including Theorem 8 as a special case.

Theorem 10 (Fraisse, 1986). Let G be a graph, λ a positive integer and

$$\delta \geq \max\left\{\frac{n+2}{\lambda+2}+\lambda-1, \alpha+\lambda-1\right\}.$$

If $\kappa \geq \lambda+1$ then G is hamiltonian.

Examples for sharpness:
$(\lambda+1)K_{\delta-\lambda+1}+K_\lambda \ (\delta \geq 2\lambda); (\lambda+2)K_{\delta-\lambda}+K_{\lambda+1} \ (\delta \geq 2\lambda+1); H(\lambda, \lambda+1, \lambda+3, \lambda+2).$

Later, Theorem 8 was essentially improved for 3-connected graphs by incorporating the connectivity κ into the minimum degree bound.

Theorem 11 (Zh.G. Nikoghosyan, 1985a). Let G be a graph with $\kappa \geq 3$. If

$$\delta \geq \max\left\{\frac{n+2\kappa}{4}, \alpha\right\}$$

then G is hamiltonian.

Examples for sharpness: $3K_2+K_2; 4K_2+K_3; H(1,2,\kappa+1,\kappa).$

The graph $4K_2+K_3$ shows that for $\kappa=3$ the minimum degree bound $(n+2\kappa)/4$ in Theorem 11 can not be replaced by $(n+2\kappa-1)/4$.

Finally, the bound $(n+2\kappa)/4$ in Theorem 11 was reduced to $(n+\kappa+3)/4$ without any additional limitations providing a best possible result for each $\kappa \geq 3$.

Theorem 12 (Yamashita, 2008). Let G be a graph with $\kappa \geq 3$. If

$$\delta \geq \max\left\{\frac{n+\kappa+3}{4}, \alpha\right\}$$

then G is hamiltonian.

Examples for sharpness: $3K_{\delta-1}+K_2; H(2, n-3\delta+3, \delta-1, \kappa); H(1,2,\kappa+1,\kappa).$

The first pure relation between graph invariants involving connectivity κ as a parameter was developed in 1972.

Theorem 13 (Chvátal and Erdös, 1972). Let G be an arbitrary graph. If

$$\kappa \geq \alpha$$

then G is hamiltonian.

Example for sharpness: $K_{\delta, \delta+1}.$

4. Pure relations for dominating cycles

In view of Theorem 2, the following upper size bound is reasonable for dominating cycles.

Conjecture 1. Let G be a graph with $\kappa \geq 2$. If

$$q \leq \frac{3\left(\delta^2 + \delta - 2\right) - 1}{2}$$

then each longest cycle in G is a dominating cycle.

In 1971, it was proved that the minimum degree bound $(n+2)/3$ insures the existence of dominating cycles.

Theorem 14 (Nash-Williams, 1971). Let G be a graph with

$$\delta \geq \frac{n+2}{3}.$$

If $\kappa \geq 2$ then each longest cycle in G is a dominating cycle.

Examples for sharpness: $2K_3 + K_1$; $3K_{\delta-1} + K_2$; $H(1,2,4,3)$.

The graph $2K_3 + K_1$ shows that the connectivity condition $\kappa \geq 2$ in Theorem 14 can not be replaced by $\kappa \geq 1$. The second graph shows that the minimum degree condition $\delta \geq (n+2)/3$ can not be replaced by $\delta \geq (n+1)/3$ and the third graph shows that the conclusion "is a dominating cycle" can not be strengthened by replacing it with "is a Hamilton cycle".

The condition $\delta \geq (n+2)/3$ in Theorem 14 can be slightly relaxed under stronger 1-tough condition instead of $\kappa \geq 2$.

Theorem 15 (Bigalke & Jung, 1979). Let G be a graph with $\tau \geq 1$. If

$$\delta \geq \frac{n}{3}$$

then each longest cycle in G is a dominating cycle.

Examples for sharpness: $2(\kappa+1)K_2 + \kappa K_1$; L_3; G_n^*.

The bound $(n+2)/3$ in Theorem 14 can be lowered to $(n+2\kappa)/4$ by incorporating κ into the minimum degree bound.

Theorem 16 (Lu et al., 2005). Let G be graph with $\kappa \geq 3$. If

$$\delta \geq \frac{n+2\kappa}{4}$$

then each longest cycle in G is a dominating cycle.

Examples for sharpness: $3K_2 + K_2$; $4K_2 + K_3$; $H(1,2,\kappa+1,\kappa)$.

The graph $4K_2 + K_3$ shows that for $\kappa = 3$ the minimum degree bound $(n+2\kappa)/4$ in Theorem 16 can not be replaced by $(n+2\kappa-1)/4$.

In 2008, the bound $(n+2\kappa)/4$ itself was essentially reduced to $(n+\kappa+3)/4$ without any additional limitations, providing a best possible result for each $\kappa \geq 3$.

Theorem 17 (Yamashita, 2008). Let G be graph with $\kappa \geq 3$. If

$$\delta \geq \frac{n+\kappa+3}{4}$$

then each longest cycle in G is a dominating cycle.

Examples for sharpness: $3K_{\delta-1} + K_2$; $H(2,n-3\delta+3,\delta-1,\kappa)$; $H(1,2,\kappa+1,\kappa)$.

5. Pure relations for CD_λ-cycles

In 1990, the exact analog of Theorems 3 and 14 was established In terms of generalized CD_3-cycles.

Theorem 18 (Jung, 1990). Let G be a graph with

$$\delta \geq \frac{n+6}{4}.$$

If $\kappa \geq 3$ then each longest cycle in G is a CD_3-cycle.

Examples for sharpness:
$\lambda K_{\lambda+1} + K_{\lambda-1}$ $(\lambda \geq 2)$; $(\lambda+1)K_{\delta-\lambda+1} + K_\lambda$ $(\lambda \geq 1)$; $H(\lambda-1,\lambda,\lambda+2,\lambda+1)$ $(\lambda \geq 2)$.

In 2009, a common generalization of Theorems 3, 14 and 18 was proved by covering CD_λ-cycles for each integer $\lambda \geq 1$.

Theorem 19 (Zh.G. Nikoghosyan, 2009a). Let G be a graph, λ a positive integer and

$$\delta \geq \frac{n+2}{\lambda+1} + \lambda - 2.$$

Then each longest cycle in G is a $CD_{\min\{\lambda,\delta-\lambda+1\}}$-cycle.

Examples for sharpness:
$\lambda K_{\lambda+1} K_{\lambda-1}$ $(\lambda \geq 2)$; $(\lambda+1)K_{\delta-\lambda+1} + K_\lambda$ $(\lambda \geq 1)$; $H(\lambda-1,\lambda,\lambda+2,\lambda+1)$ $(\lambda \geq 2)$.

In (Zh.G. Nikoghosyan, 2009a), an analogous generalization has been conjectured in terms of PD_λ-cycles.

Conjecture 1 (Zh.G. Nikoghosyan, 2009a). Let G be a graph, λ a positive integer and $\kappa \geq \lambda$. If

$$\delta \geq \frac{n+2}{\lambda+1} + \lambda - 2$$

then each longest cycle in G is a $PD_{\min\{\lambda-1,\delta-\lambda\}}$ -cycle.

In view of Theorems 6 and 17, the next generalization seems reasonable.

Conjecture 2 (Yamashita, 2008). Let G be graph, λ an integer and $\kappa \geq \lambda \geq 2$. If

$$\delta \geq \frac{n+\kappa+\lambda(\lambda-2)}{\lambda+1}$$

then each longest cycle in G is a $PD_{\lambda-2}$ and $CD_{\lambda-1}$ -cycle.

6. Pure relations for long cycles

The earliest and simplest hamiltonian result links the circumference c and minimum degree δ.

Theorem 20 (Dirac, 1952). In every graph,

$$c \geq \delta + 1.$$

Example for sharpness: Join two copies of $K_{\delta+1}$ by an edge.

For C a longest cycle in a graph G, a lower bound for $|C|$ was developed based on the minimum degree δ and $\overline{p}$ - the length of a longest path in $G \setminus C$.

Theorem 21 (Zh.G. Nikoghosyan, 1998). Let G be a graph and C a longest cycle in G. Then

$$|C| \geq (\overline{p}+2)(\delta-\overline{p}).$$

Example for sharpness: $(\kappa+1)K_{\delta-\kappa+1} + K_{\kappa}$.

The next similar bound is based On the minimum degree δ and $\overline{c}$ - the length of a longest cycle in $G \setminus C$.

Theorem 22 (Zh.G. Nikoghosyan, 2000a). Let G be a graph and C a longest cycle in G. Then

$$|C| \geq (\overline{c}+1)(\delta-\overline{c}+1).$$

Example for sharpness: $(\kappa+1)K_{\delta-\kappa+1} + K_{\kappa}$.

In 2000, Theorem 22 was improved involving connectivity κ as a parameter combined with $\overline{c}$ and δ.

Theorem 23 (Zh.G. Nikoghosyan, 2000b). Let G be a graph with $\kappa \geq 2$ and C a longest cycle in G. If $\overline{c} \geq \kappa$ then

$$|C| \geq \frac{(\overline{c}+1)\kappa}{\overline{c}+\kappa+1}(\delta+2).$$

Otherwise,

$$|C| \geq \frac{(\overline{c}+1)\overline{c}}{2\overline{c}+1}(\delta+2).$$

Example for sharpness: $(\kappa+1)K_{\delta-\kappa+1} + K_{\kappa}$.

In view of Theorem 23, the following seems reasonable for PD_λ-cycles.

Conjecture 3 (Zh.G. Nikoghosyan, 2009a). Let G be a graph with $\kappa \geq 2$ and C a longest cycle in G. If $\overline{p} \geq \kappa - 1$ then

$$|C| \geq \frac{(\overline{p}+2)\kappa}{\overline{p}+\kappa+2}(\delta+2).$$

Otherwise,

$$|C| \geq \frac{(\overline{p}+2)\overline{p}}{2\overline{p}+2}(\delta+2).$$

7. Pure relations for Hamilton cycles and long cycles

The following direct generalization includes Theorem 3 as a special case.

Theorem 24 (Alon, 1986). Let G be a graph and λ a positive integer. If $\delta \geq n/(\lambda+1)$ then

$$c \geq \frac{n}{\lambda}.$$

Examples for sharpness: $(\lambda+1)K_\lambda + K_1$; $\lambda K_{\lambda+1}$.

In 1952, a relationship was established linking the minimum degree δ, circumference c and Hamilton cycles for 2-connexted graphs.

Theorem 25 (Dirac, 1952). Let G be a graph with $\kappa \geq 2$. Then

$$c \geq \min\{n, 2\delta\}.$$

Examples for sharpness:
$(\lambda+1)K_{\lambda+1} + K_\lambda \ (\lambda \geq 1); (\lambda+3)K_{\lambda-1} + K_{\lambda+2} \ (\lambda \geq 2); (\lambda+2)K_\lambda + K_{\lambda+1} \ (\lambda \geq 1).$

For 1-tough graphs the bound 2δ in Theorem 25 was slightly enlarged.

Theorem 26 (Bauer and Schmeichel, 1986). Let G be a graph with $\tau \geq 1$. Then

$$c \geq \min\{n, 2\delta + 2\}.$$

Examples for sharpness: $K_{\delta, \delta+1}$; L_2.

The first essential improvement of Theorem 25 was achieved by incorporating connectivity κ into the relation without any essential limitation.

Theorem 27 (Zh.G. Nikoghosyan, 1981). Let G be a graph with $\kappa \geq 3$. Then

$$c \geq \min\{n, 3\delta - \kappa\}.$$

Examples for sharpness: $3K_{\delta-1} + K_2$; $H(1, \delta - \kappa + 1, \delta, \kappa)$.

In (Voss & Zuluaga, 1977), it was proved that the bound $\min\{n, 2\delta\}$ in Theorem 25 can be essentially enlarged under additional condition $\delta \geq \alpha$ combined with $\kappa \geq 3$.

Theorem 28 (Voss and Zuluaga, 1977). Let G be a graph with $\kappa \geq 3$. If $\delta \geq \alpha$ then

$$c \geq \min\{n, 3\delta - 3\}.$$

Examples for sharpness: $(\lambda + 2)K_{\lambda+2} + K_{\lambda+1}$; $(\lambda + 4)K_\lambda + K_{\lambda+3}$; $(\lambda + 3)K_{\lambda+1} + K_{\lambda+2}$.

In 2009, a direct generalization was established for each positive integer λ, including Theorem 28 as a special case $(\lambda = 1)$.

Theorem 29 (Zh.G. Nikoghosyan, 2009a). Let G be a graph and λ a positive integer. If $\kappa \geq \lambda + 2$ and $\delta \geq \alpha + \lambda - 1$ then

$$c \geq \min\{n, (\lambda + 2)(\delta - \lambda)\}.$$

Examples for sharpness: $(\lambda + 2)K_{\lambda+2} + K_{\lambda+1}$; $(\lambda + 4)K_\lambda + K_{\lambda+3}$; $(\lambda + 3)K_{\lambda+1} + K_{\lambda+2}$.

In 1985, the bound $3\delta - \kappa$ in Theorem 27 was enlarged to $4\delta - 2\kappa$ under additional condition $\delta \geq \alpha$ combined with $\kappa \geq 4$.

Theorem 30 (Zh.G. Nikoghosyan, 1985b). Let G be a graph with $\kappa \geq 4$ and $\delta \geq \alpha$. Then

$$c \geq \min\{n, 4\delta - 2\kappa\}.$$

Examples for sharpness: $4K_2 + K_3$; $H(1, n - 2\delta, \delta, \kappa)$; $5K_2 + K_4$.

The bound $4\delta - 2\kappa$ in Theorem 30 is sharp for $\kappa = 4$.

Furthermore, the bound $4\delta - 2\kappa$ in Theorem 30 was essentially improved to $4\delta - \kappa - 4$ without any additional limitations providing a best possible result for each $\kappa \geq 4$.

Theorem 31 (M.Zh. Nikoghosyan & Zh.G. Nikoghosyan, 2011). Let G be a graph with $\kappa \geq 4$ and $\delta \geq \alpha$. Then

$$c \geq \min\{n, 4\delta - \kappa - 4\}.$$

Examples for sharpness: $4K_{\delta-2} + K_3$; $H(1, 2, \kappa+1, \kappa)$; $H(2, n-3\delta+3, \delta-1, \kappa)$.

The next theorem provides a lower bound for the circumference in terms of n, δ and α under the hypothesis of Theorem 14.

Theorem 32 (Bauer et al., 1990b). Let G be a graph with $\kappa \geq 2$. If $\delta \geq (n+2)/3$ then

$$c \geq \min\{n, n+\delta - \alpha\}.$$

Examples for sharpness: $2K_\delta + K_1$; $3K_{\delta-1} + K_2$; $K_{2\delta-2,\delta}$.

An analogous bound was established for 1-tough graphs.

Theorem 33 (Bauer et al., 1988). Let G be a graph with $\tau \geq 1$. If $\delta \geq n/3$ then

$$c \geq \min\{n, n+\delta - \alpha + 1\}.$$

Examples for sharpness: $K_{\delta,\delta+1}$; L_δ; G_n^*.

8. Pure relations for dominating cycles and long cycles

The exact analog of Theorem 25 for dominating cycles can be formulated as follows.

Theorem 34 (Voss & Zuluaga, 1977). Let G be a graph with $\kappa \geq 3$. Then either

$$c \geq 3\delta - 3$$

or each longest cycle in G is a dominating cycle.

Examples for sharpness:
$(\lambda+1)K_{\lambda+1} + K_\lambda \ (\lambda \geq 1)$; $(\lambda+3)K_{\lambda+1} + K_{\lambda+2} \ (\lambda \geq 2)$; $(\lambda+2)K_\lambda + K_{\lambda+1} \ (\lambda \geq 1)$.

The bound $3\delta - 3$ in Theorem 34 was enlarged to $4\delta - 2\kappa$ by incorporating connectivity κ into the bound.

Theorem 35 (Zh.G. Nikoghosyan, 2009b). Let G be a graph with $\kappa \geq 4$. Then either

$$c \geq 4\delta - 2\kappa$$

or G has a dominating cycle.

Examples for sharpness: $4K_2 + K_3$; $5K_2 + K_4$; $H(1, n-2\delta, \delta, \kappa)$.

Theorem 35 is sharp only for $\kappa = 4$ as can be seen from $5K_2 + K_4$. Further, the bound $4\delta - 2\kappa$ in Theorem 35 was essentially improved to $4\delta - \kappa - 4$ without any limitation providing a sharp bound for each $\kappa \geq 4$.

Theorem 36 (M.Zh. Nikoghosyan & Zh.G. Nikoghosyan, 2011). Let G be a graph with $\kappa \geq 4$. Then either

$$c \geq 4\delta - \kappa - 4$$

or each longest cycle in G is a dominating cycle.

Examples for sharpness: $4K_{\delta-2} + K_3; H(2, \delta - \kappa + 1, \delta - 1, \kappa); H(1, 2, \kappa + 1, \kappa)$.

9. Pure relations for CD_λ-cycles and long cycles

TThe following theorem can be considered as a common generalization of Theorems 25 and 34 by covering CD_λ-cycles for all $\lambda \geq 1$ including Hamilton and dominating cycles as special cases.

Theorem 37 (Zh.G. Nikoghosyan, 2009a). Let G be a graph and λ a positive integer. If $\kappa \geq \lambda + 1$ then either

$$c \geq (\lambda + 1)(\delta - \lambda + 1)$$

or each longest cycle in G is a $CD_{\min\{\lambda, \delta-\lambda\}}$-cycle.

Examples for sharpness:
$(\lambda + 1)K_{\lambda+1} + K_\lambda \ (\lambda \geq 1); (\lambda + 3)K_{\lambda-1} + K_{\lambda+2} \ (\lambda \geq 2); (\lambda + 2)K_\lambda + K_{\lambda+1} \ (\lambda \geq 1)$.

In (Zh.G. Nikoghosyan, 2009a), another version of Theorem 37 was conjectured in terms of PD_λ-cycles, instead of CD_λ-cycles.

Conjecture 4 (Zh.G. Nikoghosyan, 2009a). Let G be a graph and λ a positive integer. If $\kappa \geq \lambda + 1$ then either

$$c \geq (\lambda + 1)(\delta - \lambda + 1)$$

or each longest cycle in G is a $PD_{\min\{\lambda-1, \delta-\lambda-1\}}$-cycle.

In view of Theorems 27 and 36, the following common generalization seems quite reasonable.

Conjecture 5. Let G be a graph and $\lambda \geq 2$ an integer. If $\kappa \geq \lambda + 1$ then either

$$c \geq (\lambda + 1)\delta - \kappa - (\lambda + 1)(\lambda - 2)$$

or each longest cycle in G is a $PD_{\lambda-2}$ and $CD_{\lambda-1}$-cycle.

10. Proofs of theorems 6, 21 and 27

In proofs of theorems and lemmas, the end of the proof is marked by $\square$. In proofs of claims, the end of the proof is marked by Δ.

Proof of Theorem 27 (Mosesyan et al., 2009). Let G be a 3-connected graph and S a minimum cut-set in G. Choose a longest cycle C in G so as to maximize $|V(C) \cap S|$. The result holds immediately if $|C| \geq 3\delta - 3$, since $3\delta - 3 \geq 3\delta - \kappa$. Otherwise, by Theorem 34, C is a dominating cycle. Assume first that $S \not\subseteq V(C)$ and let $v \in S \setminus V(C)$. Since C is dominating, $N(v) \subseteq V(C)$. Let $\xi_1,...,\xi_t$ be the elements of $N(v)$, occurring on $\overrightarrow{C}$ in a consecutive order. Put

$$M_1 = \left\{ \xi_i \mid V\left(\xi_i^+ \overrightarrow{C} \xi_{i+1}^- \right) \cap S \neq \varnothing \right\}, \quad M_2 = N(v) \setminus M_1.$$

Since $v \in S$, we have

$$|M_1| \leq \kappa - 1, \quad |M_2| = |N(v)| - |M_1| \geq \delta - \kappa + 1.$$

Further, since C is extreme and $|V(C) \cap S|$ is maximum,

$$N(v) \cap N^+(v) \cap M_2^{++} = \varnothing.$$

Hence

$$|C| \geq |N(v)| + |N^+(v)| + |M_2^{++}|$$
$$= 2|N(v)| + |M_2| \geq 3\delta - \kappa + 1.$$

Now assume that $S \subseteq V(C)$. Let $H_1,...,H_h$ be the connected components of $G \setminus S$. If $V(G \setminus C) = \varnothing$ then $|C| = n$ and we are done. Let $x \in V(G \setminus C)$. Assume without loss of generality that $x \in V(H_1)$. Since C is dominating, $N(x) \subseteq V(C)$. Put $Y_1 = N(x) \cup N^+(x)$. Clearly $|Y_1| \geq 2\delta$ and to prove that $|C| \geq 3\delta - \kappa$, it remains to find a subset Y_2 in $V(C)$ such that $Y_1 \cap Y_2 = \varnothing$ and $|Y_2| \geq \delta - \kappa$. Abbreviate, $V_1 = V(H_1) \cup S$. Suppose first that $Y_1 \subseteq V_1$. If $V(H_2) \subseteq V(C)$ then $Y_2 = V(H_2)$ since $|V(H_2)| \geq \delta - \kappa + 1$. Otherwise, there exist $y \in V(H_2 \setminus C)$. Since C is dominating, $N(y) \subseteq V(C)$ and we can take $Y_2 = N(y) \setminus S$. Now let $Y_1 \not\subseteq V_1$. Assume without loss of generality that $Y_1 \cap V(H_2) \neq \varnothing$. Since $N(x) \subseteq V_1$, we have $N^+(x) \cap V(H_2) \neq \varnothing$. Let $z \in N^+(x) \cap V(H_2)$. If $N(z) \subseteq V(C)$ then take $Y_2 = N(z) \setminus S$, since $N^+(x)$ is an independent set of vertices (by standard arguments) and therefore, $N(z) \cap N^+(x) = \varnothing$. Otherwise, choose $w \in N(z) \setminus V(C)$. Clearly

$$N(w) \subseteq V(C), w \in V(H_2), N(w) \cap N^+(x) = \{z\}.$$

Then by taking

$$Y_2 = (N(w) \setminus \{z\}) \setminus (S \setminus \{z^-\})$$

we complete the proof of Theorem 27. □

Proof of Theorem 6 (Mosesyan et al., 2009). Let G be a 2-connected graph with $\delta \geq (n+\kappa)/3$ and let S be a minimum cut-set in G. Since $\delta \geq (n+\kappa)/3 \geq (n+2)/3$, by Theorem 14, every longest cycle in G is a dominating cycle. As in proof of Theorem 27, we can show that either G is hamiltonian or $c \geq 3\delta - \kappa$. Since $n \leq 3\delta - \kappa$ (by the hypothesis), it follows from $c \geq 3\delta - \kappa$ that $c = 3\delta - \kappa = n$. So, in any case, G is hamiltonian. □

To prove Theorem 21, we need some special definitions. Let G be a graph, C a longest cycle in G and $\overrightarrow{M}$ a longest path (or a cycle) in $G \backslash C$. Further, let $u_1, \ldots, u_m$ be the elements of $V(M)$ occurring on $\overrightarrow{M}$ in a consecutive order.

Definition 1 { M_C-spreading; $\overrightarrow{\Upsilon}(u)$; $\acute{u}$; $\ddot{u}$ }. An M_C-spreading Υ is a family of pairwise disjoint paths $\overrightarrow{\Upsilon}(u_1), \ldots, \overrightarrow{\Upsilon}(u_m)$ in $G \backslash C$ with $\overrightarrow{\Upsilon}(u_i) = u_i \overrightarrow{\Upsilon}(u_i) \ddot{u}_i$ $(i = 1, \ldots, m)$. If $u \neq \ddot{u}$ for some $\overrightarrow{\Upsilon}(u)$, then we use $\acute{u}$ to denote the successor of u along $\overrightarrow{\Upsilon}(u)$.

Definition 2 $\{\Phi_u; \varphi_u; \Psi_u; \psi_u\}$. Let Υ be any M_C-spreading. For each $u \in V(M)$, put

$$\Phi_u = N(\ddot{u}) \cap V(\Upsilon), \quad \varphi_u = |\Phi_u|,$$

$$\Psi_u = N(\ddot{u}) \cap V(H), \quad \psi_u = |\Psi_u|.$$

Definition 3 $\{U_0; \bar{U}_0; U_1; U^*\}$. For Υ an M_C-spreading, put

$$U_0 = \{u \in V(M) \mid u = \ddot{u}\}, \quad \bar{U}_0 = V(M) \backslash U_0,$$
$$U^* = \{u \in \bar{U}_0 \mid \Phi_u \subseteq V(\Upsilon(u))\}, \quad U_1 = V(M) \backslash (U_0 \cup U^*).$$

Definition 4 {(U_0)-minimal M_C-spreading}. An M_C-spreading Υ is said to be (U_0)-minimal, if it is chosen such that $|U_0|$ is minimum.

Definition 5 $\{B_u; B_u^*; b_u; b_u^*\}$. For Υ an M_C-spreading and $u \in V(M)$, set

$$B_u = \{v \in U_0 \mid v\acute{u} \in E\}, \quad b_u = |B_u|.$$

Further, for each $u \in U_0$, set

$$B_u^* = \{v \in \bar{U}_0 \mid u\acute{v} \in E\}, \quad b_u^* = |B_u^*|.$$

Lemma 1. Let C be a longest cycle in a graph G and M a path in $G \backslash C$. Let $\vec{L_1}, ..., \vec{L_r}$ be vertex disjoint paths in $G \backslash C$ with $\vec{L_i} = v_i \, \vec{L_i} \, w_i$ $(i = 1, ..., r)$ having only $v_1, ..., v_r$ in common with M and let $. Z_i = N(w_i) \cap V(C)$ $(i = 1, ..., r)$ .. Then

$$|C| \geq \sum_{i=1}^{r} |Z_i| + \left| \bigcup_{i=1}^{r} Z_i \right|.$$

Proof. We can assume without loss of generality that $v_i = w_i$ $(i = 1, ..., r)$, since otherwise, we can use the same arguments. If $\bigcup_{i=1}^{r} Z_i = \varnothing$, then there is nothing to prove. Let $\bigcup_{i=1}^{r} Z_i \neq \varnothing$ and let $\xi_1, ..., \xi_t$ be the elements of $\bigcup_{i=1}^{r} Z_i$ occurring on $\vec{C}$ in a consecutive order. Set

$$F_i = N(\xi_i) \cap \{w_1, ..., w_r\} \quad (i = 1, ..., t).$$

Suppose first that $t = 1$. If $|F_1| = 1$ then $\sum_{i=1}^{r} |Z_i| = \left| \bigcup_{i=1}^{r} Z_i \right| = 1$ and the result follows from $|C| \geq 2$ immediately. If $|F_1| \geq 2$, then choosing a largest segment $u \vec{M} v$ on M with $u, v \in F_1$, we get a cycle $C' = \xi_1 u \vec{M} v \xi_1$ satisfying

$$|C| \geq |C'| \geq \sum_{i=1}^{r} |Z_1| + 1 = \sum_{i=1}^{r} |Z_1| + \left| \bigcup_{i=1}^{r} Z_i \right|.$$

Now assume $t \geq 2$. Put

$$f(\xi_i) = \left| \xi_i \vec{C} \xi_{i+1} \right| \quad (i = 1, 2, ..., t),$$

where $\xi_{i+1} = \xi_1$. Then it is easy to see that

$$|C| = \sum_{i=1}^{t} f(\xi_i), \quad \sum_{i=1}^{t} |F_i| = \sum_{i=1}^{r} |Z_i|, \quad t = \left| \bigcup_{i=1}^{r} Z_i \right|. \tag{1}$$

For each $i \in \{1, ..., t\}$, let $x_i \vec{M} y_i$ be the largest segment on $\vec{M}$ with $x_i, y_i \in F_i \cup F_{i+1}$ (indices mod t). Now we need to show that

$$f(\xi_i) \geq \left(|F_i| + |F_{i+1}| + 2 \right) / 2. \text{ Indeed, if } x_i \in F_i \text{ and } y_i \in F_{i+1}, \text{ then } f(\xi_i) \geq \left| \xi_i x_i \vec{M} y_i \xi_{i+1} \right| \text{ since}$$

C is extreme. It means that

$$f(\xi_i) \geq \max\{|F_i|,|F_{i+1}|\} + 1 \geq \frac{1}{2}(|F_i| + |F_{i+1}| + 2).$$

The same inequality holds from $f(\xi_i) \geq \left|\xi_i y_i \overleftarrow{M} x_i \xi_{i+1}\right|$ if $x_i \in F_{i+1}$ and $y_i \in F_i$, by a similar argument. Now suppose that either $x_i, y_i \in F_i$ or $x_i, y_i \in F_{i+1}$. Assume without loss of generality that $x_i, y_i \in F_i$. In addition, we have $x_i, y_i \notin F_{i+1}$, since otherwise we are in the previous case. Let $x_i' \overrightarrow{M} y_i'$ be the largest segment on $\overrightarrow{M}$ with $x_i', y_i' \in F_{i+1}$. If $\left|x_i \overrightarrow{M} x_i'\right| \geq (|F_i| - |F_{i+1}|)/2$, then $f(\xi_i) \geq \left|\xi_i x_i \overrightarrow{M} y_i' \xi_{i+1}\right|$ and hence

$$f(\xi_i) \geq \frac{1}{2}(|F_i| - |F_{i+1}|) + |F_{i+1}| + 1 = \frac{1}{2}(|F_i| + |F_{i+1}| + 2).$$

Finally, if $\left|x_i \overrightarrow{M} x_i'\right| \leq (|F_i| - |F_{i+1}| - 1)/2$, then

$$f(\xi_i) \geq \left|\xi_i y_i \overleftarrow{M} x_i' \xi_{i+1}\right| = \left|x_i' \overrightarrow{M} y_i\right| + 2 = \left|x_i \overrightarrow{M} y_i\right| - \left|x_i \overrightarrow{M} x_i'\right| + 2$$

$$\geq |F_i| - 1 - \frac{1}{2}(|F_i| - |F_{i+1}| - 1) + 2 > \frac{1}{2}(|F_i| + |F_{i+1}| + 2).$$

So, $f(\xi_i) \geq (|F_i| + |F_{i+1}| + 2)/2$ $(i = 1,...,t)$ in any case, implying that

$$\sum_{i=1}^{t} f(\xi_i) \geq \sum_{i=1}^{t} \frac{1}{2}(|F_i| + |F_{i+1}| + 2) = \sum_{i=1}^{t} |F_i| + t$$

and the result follows from (1). $\quad\square$

Lemma 2. Let C be a longest cycle in a graph G and M a longest cycle in $G \backslash C$ with a U_0-minimal M_C-spreading Υ. Then for each $u \in U_1$, $|M| \geq \varphi_u + b_u + 1$.

Proof. Let $u \in U_1$. For each $x \in V(M)$, put

$A_u(x) = (\Phi_u \cup B_u) \cap V(\Upsilon(x))$. By the definition,

$$|\Phi_u \cup B_u| = \sum_{x \in V(M)} |A_u(x)|. \tag{2}$$

If $A_u(x) \neq \varnothing$ for some $x \in V(M)$, then we choose a vertex $\rho_u(x)$ in $A_u(x)$ such that $|x \overrightarrow{\Upsilon}(x) \rho_u(x)|$ is maximum. By the definition, $\rho_u(u) = (\ddot{u})^-$. Put $\overline{\rho}_u(x) = \ddot{u}$ if $\rho_u(x) \in \Phi_u$, and $\overline{\rho}_u(x) = \dot{u}$ if $\rho_u(x) \in B_u \backslash \Phi_u$. Clearly $\overline{\rho}_u(u) = \ddot{u}$.

Let $\Lambda_u = \{x \in V(M) \mid A_u(x) \neq \varnothing\}$. Further, for each distinct $x, y \in \Lambda_u$, put $\Lambda_u(x,y) = x \dot{u} y$ if either $x = u$, $y \in U_0$ or $y = u$, $x \in U_0$. Otherwise,

$$\Lambda_u(x,y) = x\overrightarrow{\Upsilon}(x)\rho_u(x)\overline{\rho}_u(x)\Upsilon(u)\overline{\rho}_u(y)\rho_u(y)\overleftarrow{\Upsilon}(y)y.$$

Let $\xi_1,...,\xi_f$ be the elements of Λ_u, occurring on $\overrightarrow{M}$ in a consecutive order with $\xi_1 = u$. For each integer i $(1 \le i \le f)$, set

$$M_i = \xi_i \overrightarrow{M}\xi_{i+1}, \quad \omega_i = |A_u(\xi_i)| + |A_u(\xi_{i+1})| \quad (indices \bmod f)$$

Claim 1. $\sum_{i=1}^{f}|M_i| \ge \sum_{i=1}^{f}\omega_i$.

Proof. Since M is extreme, for each $i \in \{2,...,f-1\}$,

$$|M_i| \ge |\Lambda_u(\xi_i,\xi_{i+1})| \ge |A_u(\xi_i)| + |A_u(\xi_{i+1})| = \omega_i.$$

If $\Phi_u \cap V(\Upsilon(\xi_2)) \ne \varnothing$ and $\Phi_u \cap V(\Upsilon(\xi_f)) \ne \varnothing$, then the inequality $|M_i| \ge \omega_i$ $(i = 1, f)$ holds as in previous case and we are done. Now let $\Phi_u \cap V(\Upsilon(\xi_2)) = \varnothing$ and $\Phi_u \cap V(\Upsilon(\xi_f)) = \varnothing$. It means that $A_u(\xi_2) = \{\xi_2\}$ and therefore, $|M_1| \ge 2 = |A_u(\xi_2)| + 1 = \omega_1 - |A_u(u)| + 1$. Analogously, $|M_f| \ge \omega_f - |A_u(u)| + 1$. By the definition of Λ_u, $\dot{u}\xi_2 \in E$ and $\dot{u}\xi_f \in E$. Since $u \in U_1$, we have $\Phi_u \cap V(\Upsilon(\xi_s)) \ne \varnothing$ for some $3 \le s \le f-1$. Then we can choose i,j such that $2 \le i \le s-1$ and $s \le j \le f-1$ with $|M_i| \ge \omega_i + |A_u(u)| - 1$ and $|M_j| \ge \omega_j + |A_u(u)| - 1$, and the result follows. Finally, because of the symmetry, we can suppose that $\Phi_u \cap V(\Upsilon(\xi_2)) = \varnothing$ and $\Phi_u \cap V(\Upsilon(\xi_f)) \ne \varnothing$. Clearly $|M_1| \ge \omega_1 - |A_u(u)| + 1$. By the definition of Λ_u, $\dot{u}\xi_2 \in E$. Then we can choose $i \in \{2,...,f-1\}$ such that $|M_i| \ge \omega_i + |A_u(u)| - 1$ and again the result follows. $\triangle$

Claim 2. If $|\Upsilon(u)| \ge 2$ then $\Phi_u \cap U_0 = \varnothing$.

Proof. Suppose to the contrary and let $v \in \Phi_u \cap U_0$. Then replacing $\Upsilon(u)$ and $\Upsilon(v)$ by $u\overrightarrow{\Upsilon}(u)(\ddot{u})^-$ and $v\ddot{u}$, respectively, we can form a new M_H-spreading, contradicting the (U_0)-minimality of Υ. $\triangle$

By (2) and Claim 1,

$$|M| = \sum_{i=1}^{f}|M_i| \ge \sum_{i=1}^{f}\omega_i = \sum_{i=1}^{f}\left(|A_u(\xi_i)| + |A_u(\xi_{i+1})|\right)$$

(3)

$$= 2\sum_{i=1}^{f}|A_u(\xi_i)| = 2\sum_{x\in V(M)}|A_u(x)| = 2|\Phi_u \cup B_u|.$$

If $|\Upsilon(u)| \geq 2$ then by Claim 2, $|\Phi_u \cup B_u| = \varphi_u + b_u$, which by (3) gives $|M| \geq 2(\varphi_u + b_u) \geq \varphi_u + b_u + 1$.

Finally, if $|\Upsilon(u)| = 1$, i.e. $\ddot{u} = \dot{u}$, then $|\Phi_u \cup B_u| = |B_u| + |\{u\}| = b_u + 1 = \varphi_u$ and again by (3)

$$|M| \geq 2|\Phi_u \cup B_u| \geq 2\varphi_u \geq \varphi_u + b_u + 1. \qquad \square$$

Lemma 3. Let C be a longest cycle in a graph G and L a longest path in $G \setminus C$ with a U_0-minimal L_C-spreading Υ. Then for each $u \in \bar{U}_0$, $|L| \geq \varphi_u + b_u$.

Proof. Put $L = u_1 ... u_m$. Let Λ_u, $\Lambda_u(x,y)$ and ω_i be as defined in proof of Lemma 2. Let $\xi_1, ..., \xi_f$ be the elements of Λ_u occurring on $\vec{L}$ in a consecutive order. Set

$$\vec{M'} = u_1 \vec{L} \xi_1, \quad \vec{M''} = \xi_f \vec{L} u_m, \quad \vec{M_i} = \xi_i \vec{L} \xi_{i+1} \quad (i = 1, ..., f-1).$$

Let G' be the graph obtained from G by adding an extra edge $u_m u_1$. Set $\vec{M} = u_1 ... u_m u_1$ and $M_f = \xi_f \vec{M} \xi_1$. Let $\Lambda'_u(\xi_f, \xi_1)$ and $\Lambda''_u(\xi_f, \xi_1)$ be the paths obtained from $\Lambda_u(\xi_f, \xi_1)$ by deleting the first and the last edges, respectively. Since L is extreme, $|M_i| \geq |\Lambda_u(\xi_i, \xi_{i+1})|$ $(i = 1, ..., f-1)$. As for M_f, observe that

$$|M'| \geq |\Lambda'_u(\xi_f, \xi_1)| = |\Lambda_u(\xi_f, \xi_1)| - 1,$$
$$|M''| \geq |\Lambda''_u(\xi_f, \xi_1)| = |\Lambda_u(\xi_f, \xi_1)| - 1,$$

Implying that

$$|M_f| = |M'| + |M''| + 1 \geq 2|\Lambda_u(\xi_f, \xi_1)| - 1 \geq |\Lambda_u(\xi_f, \xi_1)|.$$

So, $|M_i| \geq |\Lambda_u(\xi_i, \xi_{i+1})|$ for each $i \in \{1, ..., f\}$. Further, for each $u \in U_1$, we can argue exactly as in proof of Lemma 2 to get $|L| = |M| - 1 \geq \varphi_u + b_u$. Now let $u \in U^*$. By the definition, $\Phi_u \subseteq V(\Upsilon(u))$ and therefore, $|\Upsilon(u)| \geq |\Phi_u| = \varphi_u$. Since L is extreme, $|L| \geq 2(|B_u| + |\{u\}|) - 2 = 2b_u$. Hence,

$$|L| \geq |\Upsilon(u)| + \frac{1}{2}|\vec{L}| \geq \varphi_u + b_u.$$

Proof of Theorem 21 (Zh.G. Nikoghosyan, 1998). Let M be a longest path in $G \setminus C$ of length $\bar{p}$ with a (U_0)-minimal M_C-spreading Υ. If $\bar{p} = -1$, i.e. M is a Hamilton cycle, then $|C| \geq \delta + 1 = (\bar{p} + 2)(\delta - \bar{p})$. Let $\bar{p} \geq 0$. We claim that

(a1) if $u \in U_0$ and $v \in \bar{U}_0$ then $\Phi_u \cap V(\Upsilon(v)) \subseteq \{v, \dot{v}\}$,

(a2) if $u \in U_0$ then $\varphi_u \leq \bar{p} + b_u^*$,

(a3) if $v \in \bar{U}_0$ then $\varphi_u \leq \bar{p} - b_u$.

Let $u \in U_0$. If $v \in \bar{U}_0$ then to prove (a1) we can argue exactly as in proof of Claim 2 (see the proof of Lemma 2). The next claim follows immediately from (a1). To prove (a3), let $v \in \bar{U}_0$. Since M is extreme, by Lemma 3, $\bar{p} \geq \varphi_u + b_u$ for each $u \in \bar{U}_0$, and (a3) follows.

Observing that

$$\sum_{u \in U_0} b_u^* = \sum_{u \in \bar{U}_0} b_u$$

and using (a2) and (a3), we get

$$\sum_{u \in V(M)} \varphi_u \leq \bar{p}(\bar{p} + 1) + \sum_{u \in U_0} b_u^* - \sum_{u \in \bar{U}_0} b_u = \bar{p}(\bar{p} + 1).$$

Since Υ is extreme, we have

$$\psi_u = d(\ddot{u}) - \varphi_u \geq \delta - \varphi_u$$

for each $u \in V(M)$.

By summing, we get

$$\sum_{u \in V(M)} \psi_u = (\bar{p} + 1)\delta - \sum_{u \in V(M)} \varphi_u \geq (\bar{p} + 1)(\delta - \bar{p}).$$

In particular,

$$\max_u \psi_u \geq \delta - \bar{p}.$$

By Lemma 1,

$$|C| \geq \sum_{u \in V(M)} \psi_u + \max_u \psi_u$$
$$\geq (\bar{p} + 1)(\delta - \bar{p}) + \delta - \bar{p} = (\bar{p} + 2)(\delta - \bar{p}).$$

11. Conclusions

Graph invariants provide a powerful and maybe the single analytical tool for investigation of abstract structures of graphs. They, combined in convenient algebraic relations, carry global and general information about a graph and its particular substructures such as cycle structures, factors, matchings, colorings, coverings, and so on. The discovery of these relations is the primary problem of graph theory.

We focus on large cycle substructures, perhaps the most important cycle structures in graphs: Hamilton, longest and dominating cycles and some generalized cycles including Hamilton and dominating cycles as special cases.

In the literature, eight basic (initial) invariants of a graph G are known having significant impact on large cycle structures, namely order n, size q, minimum degree δ, connectivity κ, independence number α, toughness τ and the lengths of a longest path and a longest cycle in $G \setminus C$ for a given longest cycle C in G.

We have collected 37 pure algebraic relations between $n, q, \delta, \kappa, \alpha, \tau, \bar{p}$ and $\bar{c}$ ensuring the existence of a certain type of large cycles. The majority of these results are sharp in all respects.

Focusing only on basic graph invariants, as well as on pure algebraic relations between these parameters, in fact, we present the simplest kind of relations for large cycles having no forerunners in the area. Actually they form a source from which nearly all possible hamiltonian results (including well-known Ore's theorem, Pósa's theorem and many other generalizations) can be developed further by various additional new ideas, generalizations, extensions, restrictions and structural limitations.

12. References

Alon, N. (1986). The longest cycle of a graph with a large minimum degree, *Journal of Graph Theory*, Vol. 10, No. 1, pp. (123-127).

Bauer, D. & Schmeichel, E. (1986). Long cycles in tough graphs, *Technical Report 8612, Stevens Institute of Technology*, Hoboken.

Bauer, D., Schmeichel, E. & Veldman, H.J. (1988). A generalization of a Theorem of Bigalke and Jung, *Ars Combinatoria*, Vol. 26, pp. (53-58).

Bauer, D., Hakimi, S.L. & Schmeichel, E. (1990a). Recognizing tough graphs is NP-hard, *Discrete Applied Mathematics*, Vol. 28, No. 3, pp. (191-195).

Bauer, D., Morgana, A., Schmeichel, E. & Veldman, H.J. (1990b). Long cycles in graphs with large degree sums, *Discrete Mathematics*, Vol. 79, No. 1, pp. (59-70).

Bauer, D., Chen, G. & Lasser, L. (1991a). A degree condition for Hamilton cycles in t-tough graphs with $t > 1$, In: *Advances in graph theory*, V.R. Kulli, (Ed.), pp. (20-33), Vishwa International publications, Gulbarga, India.

Bauer, D. & Schmeichel, E. (1991b). On a Theorem of Häggkvist and Nicoghossian, *Graph Theory, Combinatorics, Algorithms and Applications*, pp. (20-25).

Bermond, J.C. (1978). Hamiltonian graphs, In: *Selected topics in graph theory*, L. Beineke and R.J. Wilson, (Eds.), Academic press, London.

Bigalke, A. & Jung, H.A. (1979). Über Hamiltonsche Kreise und unabhängige Ecken in Graphen, *Monatshefte für Mathematik*, Vol. 88, No. 3, pp. (195-210).

Bondy, J.A. & Murty, U.S.R. (1976). *Graph Theory with Applications*, Macmillan, London, ISBN 0-444-19451-7.

Bondy, J.A. (1981). Integrity in graph theory, *In: The Theory and Application of Graphs*, G. Chartrand, Y. Alavi, D.L. Goldsmith, L. Lesniak-Foster, D.R. Lick (Eds.), Wiley, New York, pp. (117-125). MR83e:05070.

Chvátal, V. & Erdös, P. (1972). A note on hamiltonian circuits, *Discrete Mathematics*, Vol. 2, No. 2, pp. (111-113).

Chvátal, V. (1973). Tough graphs and Hamiltonian circuits, *Discrete Mathematics*, Vol. 5, No. 3, pp. (215-228).

Dirac, G.A. (1952). Some theorems on abstract graphs, *Proceedings of the London Mathematical Society*, Vol. 2, No. 1, pp. (69-81).

Even, S. & Tarjan, R.E. (1975). Network flow and testing graph connectivity, *SIAM journal on computing*, Vol. 4, No. 4, pp. (507-518).

Erdös, P. & Gallai, T. (1959). On maximal paths and circuits of graphs, *Acta Mathematica Hungarica*, Vol. 10, No. 3-4, pp. (337-356).

Fraisse, P. (1986). D_λ-cycles and their applications for Hamiltonian graphs, Universite de Paris-sud, preprint.

Garey, M.R. & Johnson, D.S. (1983). *Computers and Intractability: A Guide to the Theory of NP-Completenes*, Freeman, New York.

Gould, R.J. (1991). Updating the Hamiltonian Problem - A survey, *Journal of Graph Theory*, Vol. 15, No. 2, pp. (121-157).

Gould, R.J. (2003). Advances on the Hamiltonian Problem - A survey, *Graphs and Combinatorics*,Vol. 19, No. 1, pp. (7-52).

Häggkvist, R. & Nicoghossian, G.G. (1981). A remark on hamiltonian cycles, *Journal of Combinatorial Theory*, Ser. B, Vol. 30, No. 1, pp. (118-120).

Jung, H.A. (1978). On maximal circuits in finite graphs, *Annals of Discrete Mathematics*, Vol. 3, pp. (129-144).

Jung, H.A. (1990). Long Cycles in Graphs with Moderate Connectivity, In: *Topics in combinatorics and graph theory*, R.Bodendieck and R.Henn (Eds), Phisika Verlag, Heidelberg, pp. (765-778).

Lu, M., Liu, H. & Tian, F. (2005). Two sufficient conditions for dominating cycles, *Journal of Graph Theory*, Vol. 49, No. 2, pp. (135-150).

Mosesyan, C.M., Nikoghosyan, M.Zh. & Nikoghosyan, Zh.G. (2009). Simple proofs of two Dirac-type theorems involving connectivity, *In: arXiv:0906.3630v2 [math.CO]*, 27 Jul 2009, Available from: http://arxiv.org/abs/0906.3630.

Nash-Williams, C.St.J.A. (1971). Edge-disjoint hamiltonian cycles in graphs with vertices of large valency, In: *Studies in Pure Mathematics*, L. Mirsky, (Ed.), pp. (157-183), Academic Press, San Diego/London.

Nikoghosyan, M.Zh. & Nikoghosyan, Zh.G. (2011). Large cycles in 4-connected graphs, *Discrete Mathematics*, Vol. 311, No. 4, pp. (302-306).

Nikoghosyan, Zh.G. (1981). On maximal cycle of a graph, *DAN Arm. SSR*, Vol. LXXII, No. 2, pp. (82-87) (in Russian).

Nikoghosyan, Zh.G. (1985a). A sufficient condition for a graph to be Hamiltonian, *Matematicheskie voprosy kibernetiki i vichislitelnoy tekhniki*, Vol. 14, pp. (34-54) (in Russian).

Nikoghosyan, Zh.G. (1985b). On maximal cycles in graphs, *DAN Arm. SSR*, Vol. LXXXI, No. 4, pp. (166-170) (in Russian).

Nikoghosyan, Zh.G. (1998). Path-Extensions and Long Cycles in Graphs, *Mathematical Problems of Computer Science*, Transactions of the Institute for Informatics and Automation Problems of the NAS (Republic of Armenia) and Yerevan State University, Vol. 19, pp. (25-31).

Nikoghosyan, Zh.G. (2000a). Cycle-Extensions and Long Cycles in Graphs, *Mathematical Problems of Computer Science*, Transactions of the Institute for Informatics and Automation Problems of the NAS (Republic of Armenia) and Yerevan State University, Vol. 21, pp. (121-128).

Nikoghosyan, Zh.G. (2000b). Cycle-Extensions and Long Cycles in κ -connected Graphs, *Mathematical Problems of Computer Science*, Transactions of the Institute for Informatics and Automation Problems of the NAS (Republic of Armenia) and Yerevan State University, Vol. 21, pp. (129-155).

Nikoghosyan, Zh.G. (2009a). Dirac-type generalizations concerning large cycles in graphs, *Discrete Mathematics*, Vol. 309, No. 8, pp. (1925-1930).

Nikoghosyan, Zh.G. (2009b). On the circumference, connectivity and dominating cycles, *In: arXiv:0906.1857v1 [math.CO]*, 10 Jun 2009, Available from: http://arxiv.org/abs/0906.1857.

Nikoghosyan, Zh.G. (2011). A Size Bound for Hamilton Cycles, *In: arXiv:1107.2201v1 [math.CO]* 12 Jul 2011, Available from: http://arxiv.org/abs/1107.2201.

Voss H.-J. & Zuluaga, C. (1977). Maximale gerade und ungerade Kreise in Graphen I, *Wiss. Z. Tech. Hochschule*, Ilmenau, Vol. 4, pp. (57-70).

Yamashita, T. (2009). A degree sum condition with connectivity for relative length of longest paths and cycles, *Discrete Mathematics*, Vol. 309, No. 23-24, pp. (6503-6507).

5

A Semi-Supervised Clustering Method Based on Graph Contraction and Spectral Graph Theory

Tetsuya Yoshida

Graduate School of Information Science and Technology, Hokkaido University
Japan

1. Introduction

Semi-supervised learning is a machine learning framework where learning from data is conducted by utilizing a small amount of labeled data as well as a large amount of unlabeled data (Chapelle et al., 2006). It has been intensively studied in data mining and machine learning communities recently. One of the reasons is that, it can alleviate the time-consuming effort to collect "ground truth" labeled data while sustaining relatively high performance by exploiting a large amount of unlabeled data. (Blum & Mitchell, 1998) showed the PAC learnability of semi-supervised learning, especially in classification problem.

On the other hand, data clustering, also called unsupervised learning, is a method of creating groups of objects, or clusters, in such a way that objects in one cluster are very similar and objects in different clusters are quite distinct. Clustering is one of the most frequently performed analysis (Jain et al., 1999). For example, in web activity logs, clusters can indicate navigation patterns of different user groups. Another direct application could be clustering of gene expression data so that genes within a same group evinces similar behavior.

Although labeled data is not required in clustering, sometimes constraints on data assignment might be available as domain knowledge about the data to be clustered. In such a situation, it is desirable to utilize the available constraints as semi-supervised information and to improve the performance of clustering (Basu et al., 2008). By regarding constraints on data assignment as supervised information, various research efforts have been conducted on semi-supervised clustering (Basu et al., 2004; 2008; Li et al., 2008; Tang et al., 2007; Xing et al., 2003). Although various forms of constraints can be considered, based on the previous work (Li et al., 2008; Tang et al., 2007; Wagstaff et al., 2001; Xing et al., 2003), we deal with the following two kinds of pairwise constraints in this paper: must-link constraints and cannot-link constraints. In this chapter, the former is also called as must-links, and the latter as cannot-links.

When similarities among data instances are specified, by connecting each pair of instances with an edge with the corresponding similarity, the entire data instances can be represented as an edge-weighted graph. In this chapter we present our semi-supervised clustering method based on graph contraction in general graph theory and graph Laplacian in spectral graph theory. Graph representation enables to deal with two kinds of pairwise constraints as well as pairwise similarities over a unified representation. Then, the graph is modified by contraction in graph theory (Diestel, 2006) and graph Laplacian in spectral graph theory (Chung, 1997; von Luxburg, 2007) to reflect the pairwise constraints.

Representing the relations (both pairwise constraints and similarities) among instances as an edge-weighted graph and modifying the graph structure based on the specified constraints enable to enhancing semi-supervised clustering. In our approach, the entire data instances are projected onto a subspace which is constructed with respect to the modified graph structure, and clustering is conducted over the projected data representation of instances. Although our approach utilizes graph Laplacian as in (Belkin & Niyogi, 2002), our approach differs from previous ones since pairwise constraints for semi-supervised clustering are also utilized in our approach for constructing the projected data representation (Yoshida, 2010; Yoshida & Okatani, 2010).

We report the performance evaluation of our approach, and compare it with other state-of-the-art semi-supervised clustering methods in terms of accuracy and running time. Extensive experiments are conducted over real-world datasets. The results are encouraging and indicate the effectiveness of our approach. Especially, our approach can leverage small amount of pairwise constraints to increase the performance. We believe that this is a good property in the semi-supervised learning setting.

The rest of this chapter is organized as follows. Section 2 explains the framework of semi-supervised clustering. Section 3 explains the details of our approach for clustering under pairwise constraints. Section 4 reports the performance evaluation over various document datasets. Section 5 discusses the effectiveness of our approach. Section 6 summarizes our contributions and suggests future directions.

2. Semi-supervised clustering

2.1 Preliminaries

Let X be a set of instances. For a set X, $|X|$ represents its cardinality.

A graph $G = (V, E)$ consists of a finite set of vertices V, a set of edges E over $V \times V$. The set E can be interpreted as representing a binary relation over V. A pair of vertices (v_i, v_j) is in the binary relation defined by a graph $G = (V, E)$ if and only if the pair $(v_i, v_j) \in E$.

An edge-weighted graph $G = (V, E, W)$ is defined as a graph $G = (V, E)$ with a weight on each edge in E. When $|V| = n$, i.e., the number of vertices in a graph is n, the weights in W can be represented as an n by n matrix $\mathbf{W}$ [1], where w_{ij} in $\mathbf{W}$ stands for the weight on the edge for the pair $(v_i, v_j) \in E$. $\mathbf{W}_{ij}$ also stands for the element w_{ij} in the matrix. We set $w_{ij} = 0$ for pairs $(v_i, v_j) \notin E$. In addition, we assume that $G = (V, E, W)$ is an undirected, simple graph without self-loops. Thus, the weight matrix $\mathbf{W}$ is symmetric and its diagonal elements are zeros.

2.2 Clustering

In general, clustering methods can be divided into two approaches: hierarchical methods and partitioning methods. (Jain et al., 1999). Hierarchical methods construct a cluster hierarchy, or a tree of clusters (called a dendrogram), whose leaves are the data points and whose internal nodes represent nested clusters of various sizes (Guha et al., 1998). Hierarchical methods can be further subdivided into *agglomerative* and *divisive* ones. On the other hand, partitioning methods return a single partition of the entire data under a fixed parameters (number of clusters, thresholds, etc.). Each cluster can be represented by its centroid

---

[1] A bold italic symbol W denotes a set, while a bold symbol $\mathbf{W}$ denotes a matrix.

(k-means algorithms (Hartigan & Wong, 1979)), or by one of its instances located near its center (k-medoid algorithms (Ng & Han, 2002)). For a recent overview of various clustering methods, please refer to (Jain et al., 1999).

When pairwise similarities among instances are specified, the entire data can be represented as an edge-weighted graph. Various graph-theoretic clustering approaches have been proposed to find subsets of vertices in a graph based on the edges among the vertices. Several methods utilizes graph coloring techniques (Guénoche et al., 1991; Yoshida & Ogino, 2011). Other methods are based on the flow or cut in graph, such as spectral clustering (von Luxburg, 2007). Graph-based spectral approach is also utilized in information-theoretic clustering (Yoshida, 2011).

2.3 Semi-supervised clustering

When the auxiliary or side information for data assignment in clustering is represented as a set of constraints, the *semi-supervised clustering* problem is (informally) described as follows.

Problem 1 (Semi-Supervised Clustering). *For a given set of data X and specified constraints, find a partition (a set of clusters) $T = \{t_1, \ldots, t_k\}$ which satisfies the specified constraints.*

There can be various forms of constraints. Based on the previous work (Li et al., 2008; Tang et al., 2007; Wagstaff et al., 2001; Xing et al., 2003), we consider the following two kinds of constraints defined in (Wagstaff et al., 2001):

Definition 1 (Pairwise Constraints). *For a given data instances X and a partition (a set of clusters) $C = \{c_1, \ldots, c_k\}$, must-link constraints C_{ML} and cannot-link constraints C_{CL} are sets of pairs such that:*

$$\exists(x_i, x_j) \in C_{ML} \Rightarrow \exists c \in C, (x_i \in c \wedge x_j \in c) \tag{1}$$

$$\exists(x_i, x_j) \in C_{ML} \Rightarrow \exists c_a, c_b \in C, c_a \neq c_b, (x_i \in c_a \wedge x_j \in c_b) \tag{2}$$

Intuitively, must-link constraints (also called must-links in this paper) specifies the pairs of instances in the same cluster, and cannot-link constraints (also called cannot-links) specifies the pairs of instances in different clusters.

3. Graph-based semi-supervised clustering

3.1 A graph-based approach

By assuming that some similarity measure for the pairs of instances is specified, we have proposed a graph-based approach for constrained clustering problem (Yoshida, 2010; Yoshida & Okatani, 2010). Based on the similarities, the entire data instances X can be represented as an edge-weighted graph $G = (V, E, W)$ where w_{ij} represents the similarity between a pair (x_i, x_j). In our approach, each data instance $x \in X$ corresponds to a vertex $v \in V$ in G. Thus, we abuse the symbol X to denote the set of vertices in G in the rest of the paper. Also, we assume that all w_{ij} is non-negative.

Definition 1 specifies two kinds of constraints. For must-link constraints, our approach utilizes a method based on graph contraction in general graph theory (Diestel, 2006) and treat it as hard constraints (Sections 3.2); for cannot-link constraints, our approach utilizes a method based on graph Laplacian in spectral graph theory (Chung, 1997; von Luxburg, 2007) and

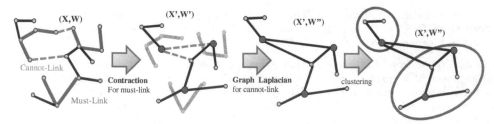

Fig. 1. Overview of our approach.

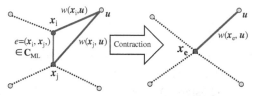

Fig. 2. Contraction for must-link constraints.

treat them as soft constraints under the optimization framework (Section 3.3). The overview of our approach is illustrated in Fig. 1.

3.2 Graph contraction for must-link constraints

When must-link constraints are treated as hard constraints, the transitive law holds among the constraints. This means that, for any two pairs (x_i, x_j) and $(x_j, x_l) \in C_{ML}$, x_i and x_l should also be in the same cluster (however, the cluster label is not known). In order to enforce the transitive law in must-links, we utilize graph contraction in general graph theory (Diestel, 2006) and modify the graph G for a data set X based on the specified must-links.

Definition 2 (Contraction). *Let $e=(x_i, x_j)$ be an edge of a graph $G = (X, E)$. define By G/e, we denote the graph (X', E') obtained from G by contracting the edge e into a new vertex x_e, where:*

$$X' = (X \backslash \{x_i, x_j\}) \cup \{x_e\} \tag{3}$$

$$E' = \{(u,v) \in E | \{u,v\} \cap \{x_i, x_j\} = \phi\}$$
$$\cup \{(x_e, u) | (x_i, u) \in E \backslash \{e\} \text{ or } (x_j, u) \in E \backslash \{e\}\} \tag{4}$$

G/e stands for the graph obtained from G by *contracting* an edge e into a new vertex x_e. The created vertex x_e becomes adjacent to all the former neighbors of x_i and x_j.

By contracting an edge e into a new vertex x_e, the newly created vertex x_e becomes adjacent to all the former neighbors of x_i and x_j. Repeated application of contraction for all the edges (pairs of instance) for must-links guarantees that the transitive law in must-links is sustained in the cluster assignment.

As described above, the entire dataset X is represented as an edge-weighted graph G in our approach. Thus, after contracting an edge $e=(x_i, x_j) \in C_{ML}$ into the newly created vertex x_e, it is necessary to define the weights in the contracted graph G/e. The weights in G represent the similarities among vertices. The original similarities should at least be sustained after contracting an edge in C_{ML}, since must-link constraints are for enforcing the similarities, not for reducing.

Based on the above observation, we define the weights in the contracted graph G/e as:

$$w(x_e, u)' = \max(w(x_i, u), w(x_j, u)) \quad if \ (x_i, u) \in E \ or (x_j, u) \in E \tag{5}$$

$$w(u, v)' = w(u, v) \quad otherwise \tag{6}$$

where $w(\cdot, \cdot)'$ stands for the weight in the contracted graph G/e. In eq.(5), the function max realizes the above requirement, and guarantees the non-decreasing properties of similarities (weights) after contraction of an edge. On the other hand, the original weight is preserved in eq.(6).

For each pair of edges in must-links, we apply graph contraction and define weights in the contracted graph based on eq.(5) and eq.(6). This results in modifying the original graph G into another graph $G' = (X', E', W')$ (as illustrated in Fig. 2). The number of vertices in the contracted graph G' is denoted as $n' = |X'|$. Note that the originally specified cannot-links also need to be modified during graph contraction with respect to must-links. The updated cannot-links over the created graph G' is denoted as C'_{CL}.

3.3 Graph Laplacian for cannot-link constraints

3.3.1 Spectral clustering

The objective of clustering is to assign similar instances to the same cluster and dissimilar ones to different clusters. To realize this, we utilize spectral clustering, which is based on the minimum cut of a graph. In spectral clustering (Ng et al., 2001; von Luxburg, 2007), data clustering is realized by seeking a function $f: X \to \mathcal{R}$ over the dataset X such that the learned function assigns similar values for similar instances and vice versa. The values assigned for the entire dataset can be represented as a vector. By denoting the assigned value for the i-th data instance as f_i, data clustering can be formalized as an optimization problem to find the vector f which minimizes the following objective function :

$$J_0 = f^t \mathbf{L} f \tag{7}$$

where f^t is a transpose of vector f, and the matrix $\mathbf{L}$ is defined as:

$$\mathbf{D} = diag(d_1, \ldots, d_n) \ \ (d_i = \sum_{j=1}^{n} w_{ij}) \tag{8}$$

$$\mathbf{L} = \mathbf{D} - \mathbf{W} \tag{9}$$

where $diag()$ in eq.(8) represents a diagonal matrix with the specified diagonal elements. The matrix $\mathbf{D}$ in eq.(8) is the degree matrix of a graph, and is calculated based on the weights in the graph. The matrix $\mathbf{L}$ in eq.(9) is called graph Laplacian (Chung, 1997; Ng et al., 2001; von Luxburg, 2007). Some clustering method, such as kmeans (Hartigan & Wong, 1979) or spherical kmeans (skmeans) (Dhillon & Modha, 2001)[2], is applied to the constructed data representation of instances (Ng et al., 2001; von Luxburg, 2007).

3.3.2 Graph Laplacian for cannot-link constraints

We utilized the framework of spectral clustering in Section 3.3.1. Furthermore, to reflect cannot-link constraints in the clustering process, we formalize the clustering under constraints

---

[2] skmeans is a standard clustering algorithm for high-dimensional sparse data.

as an optimization problem, and consider the minimization of the following objective function:

$$J = \frac{1}{2}\{\sum_{i,j} w'_{ij}||f_i - f_j||^2 - \lambda \sum_{u,v \in C'_{CL}} w'_{uv}||f_u - f_v||^2\} \tag{10}$$

where i and j sum over the vertices in the contracted graph G', and C'_{CL} stands for the cannot-link constraints over G'. $\lambda \in [0,1]$ is a hyper-parameter in our approach. The first term corresponds to the smoothness of the assigned values in spectral graph theory, and the second term represents the influence of cannot-links in optimization. Note that by setting $\lambda \in [0,1]$, the objective function in (10) is guaranteed to be a convex function.

From the above objective function in eq.(10), we can derive the following unnormalized graph Laplacian L'' which incorporates cannot-links as:

$$J = \frac{1}{2}\{\sum_{i,j} w'_{ij}||f_i - f_j||^2 - \lambda \sum_{u,v \in C'_{CL}} w'_{uv}||f_u - f_v||^2\} = f^t L'' f \tag{11}$$

The matrix L'' is defined based on the following matrices:

$$(C')_{uv} = \begin{cases} 1 & (x_u, x_v) \in C'_{CL} \\ 0 & \text{otherwise} \end{cases} \tag{12}$$

$$W^c = C' \odot W', \quad W'' = W' - \lambda W^c \tag{13}$$

$$d_i = \sum_{j=1}^{n'} w'_{ij}, \quad d_i^c = \sum_{j=1}^{n'} w_{ij}^c \tag{14}$$

$$D'' = diag(d''_1, \ldots, d''_{n'}), \quad d''_i = d_i - \lambda d_i^c \tag{15}$$

$$L'' = D'' - W'' \tag{16}$$

where $\odot$ stands for the Hadamard product (element-wise multiplication) of two matrices.

The above process amounts to modifying the representation of the contracted graph G' into another graph G'', with the modified weights W'' in eq.(13). Thus, as illustrated in Fig. 1, our approach modifies the original graph G into the contracted graph G' with must-link constraints, and then into another graph G'' with cannot-link constraints and similarities.

It is known that some form "balancing" among clusters needs to be considered for obtaining meaningful results (von Luxburg, 2007). Based on eq.(14) and eq.(16), we utilize the following normalized objective function:

$$J_{sym} = \sum_{i,j} w''_{ij}||\frac{f_i}{\sqrt{d''_i}} - \frac{f_j}{\sqrt{d''_j}}||^2 \tag{17}$$

over the graph G''. Minimizing J_{sym} in eq.(17) amounts to solving the generalized eigen-problem $L'' f = \alpha D'' f$, where α corresponds to an eigenvalue and f corresponds to the generalized eigenvector with the eigenvalue.

Algorithm 1 graph-based semi-supervised clustering (GBSSC)

Require: $G = (X, E, W)$; //an edge-weighted graph
Require: C_{ML}; //must-link constraints
Require: C_{CL}; //cannot-link constraints
Require: l; //the number of generalized eigenvectors
Require: k; //the number of clusters

1: **for** each $e \in C_{ML}$ **do**
2:     contract e and create the contracted graph G/e;
3: **end for**
    // Let $G' = (X', E', W')$ be the contracted graph.
4: create $\mathbf{C}'_{uv}, \mathbf{W}^c, \mathbf{W}'', \mathbf{D}''$ as eq.(12) $\sim$ eq.(15).
5: $\mathbf{L}''_{sym} = \mathbf{I} - \mathbf{D}''^{-\frac{1}{2}} \mathbf{W}'' \mathbf{D}''^{-\frac{1}{2}}$
6: Find l eigenvectors $\mathbf{F} = \{f^1, \ldots, f^l\}$ for $\mathbf{L}''_{sym}$, with the smallest non-zero eigenvalues.
7: Conduct clustering of data which are represented as $\mathbf{F}$ and construct clusters.
8: **return** clusters

Furthermore, the number of generalized eigenvectors can be extended to more than one. In that case, the generalized eigenvectors with positive eigenvalues are selected with ascending order of eigenvalues. The generalized eigenvectors with respect to the modified graph corresponds to the embeeded representation of the whole data instances.

3.4 Algorithm

The graph-based semi-supervised clustering method (called GBSSC) is summarized in Algorithm 1. The contracted graph G' is constructed from lines 1 to 3 based on the specified must-links. Lines 4 to 6 conduct the minimization of J_{sym} in eq.(17), which is represented as the normalized graph Laplacian $\mathbf{L}''_{sym}$ at line 5.

These correspond to the spectral embedding of the entire data instances X onto the subspace spanned by $\mathbf{F} = \{f^1, \ldots, f^l\}$ (Belkin & Niyogi, 2002). Note that pairwise constraints for semi-supervised clustering are also utilized on the construction of the embedded representation in our approach and thus differs from (Belkin & Niyogi, 2002). Some clustering method is applied to the data at line 7 and the constructed clusters are returned. Currently spherical kmeans (skmeans) (Dhillon & Modha, 2001) is utilized at line 7.

4. Evaluations

4.1 Experimental settings

4.1.1 Datasets

Based on the previous work (Dhillon et al., 2003; Tang et al., 2007), we evaluated our approach on 20 Newsgroup dataset (hereafter, called 20NG) [3] and TREC datasets [4]. Clustering of these datasets corresponds to document clustering, and each document is represented in the standard vector space model based on the occurrences of terms. Since the number of terms are

---

[3] http://people.csail.mit.edu/˜jrennie/20Newsgroups/.   (20news-18828 was utilized)
[4] http://glaros.dtc.umn.edu/gkhome/cluto/cluto/download

dataset	included groups
Multi5	comp.graphics, rec.motorcycles,rec.sport.baseball, sci.space talk.politics.mideast
Multi10	alt.atheism, comp.sys.mac.hardware,misc.forsale, rec.autos,rec.sport.hockey, sci.crypt,sci.med, sci.electronics,sci.space,talk.politics.guns
Multi15	alt.atheism, comp.graphics, comp.sys.mac.hardware, misc.forsale, rec.autos, rec.motorcycles, rec.sport.baseball, rec.sport.hockey, sci.crypt, sci.electronics, sci.med, sci.space, talk.politics.guns, talk.politics.mideast, talk.politics.misc

Table 1. Datasets from 20 Newsgroup dataset

dataset	#attributes	#classes	#data
hitech	126372	6	2301
reviews	126372	5	4069
sports	126372	7	8580
la1	31372	6	3204
la2	31372	6	3075
la2	31372	6	6279
k1b	21839	6	2340
ohscal	11465	10	11162
fbis	2000	17	2463

Table 2. TREC datasets (original representation)

huge in general, these are high-dimensional sparse datasets. Please note that our approach is generic and not specific to document clustering.

As in (Dhillon et al., 2003; Tang et al., 2007), 50 documents were sampled from each group (cluster) in order to create a sample for one dataset, and 10 samples were created for each dataset. For each sample, we conducted stemming using porter stemmer [5] and MontyTagger [6], removed stop words, and selected 2,000 words with descending order of mutual information (Cover & Thomas, 2006).

For TREC datasets, we utilized 9 datasets in Table 2. We followed the same procedure in 20NG and created 10 samples for each dataset[7]. Since these datasets are already preprocessed and represented as count data, we did not conduct stemming or tagging.

4.1.2 Evaluation measures

For each dataset, the cluster assignment was evaluated with respect to Normalized Mutual Information (NMI) (Strehl & Ghosh, 2002; Tang et al., 2007). Let C, $\hat{C}$ stand for random variables over the true and assigned clusters. NMI is defined as

$$NMI = \frac{I(\hat{C};C)}{(H(\hat{C}) + H(C))/2} \tag{18}$$

---

[5] http://www.tartarus.org/~martin/PorterStemmer
[6] http://web.media.mit.edu/~hugo/montytagger
[7] On fbis, 35 data were sampled for each class.

where $H(\cdot)$ is Shannon Entropy, and $I(\cdot;\cdot)$ is Mutual Information among the random variables C and $\hat{C}$. NMI corresponds to the accuracy of assignment. Thus, the larger NMI is, the better the cluster assignment is with respect to the "ground-truth" labels in each dataset.

All the compared methods first construct the representation for clustering and then apply some clustering method (e.g., skmeans). The running time (CPU time in second) for representation construction was measured on a computer with Debian/GNU Linux, Intel Xeon W5590, 36 GB memory. All the methods were implemented with R language and R packages.

4.1.3 Comparison

We compared our approach with SCREEN (Tang et al., 2007) and PCP (Li et al., 2008) (details are described in Section 5.2). Since all the compared methods are partitioning based clustering methods, we assume that the number of clusters k in each dataset is available.

SCREEN (Tang et al., 2007) conducts semi-supervised clustering by projecting the given data instances onto the subspace where the covariance with respect to the given data representation is maximized. To realize this, the covariance matrix with respect to the original data representation is constructed and their eigenvectors are utilized for projection. For high-dimensional data such as documents, this process is rather expensive, since the number of attributes (e.g., terms) gets large. To alleviate this problem, PCA (Principal Component Analysis) was first utilized as pre-processing to reduce the number of dimension in the data representation. We followed the same process in (Tang et al., 2007) and pre-processed data by PCA using 100 eigenvectors, and SCREEN was applied to the pre-processed data as in (Tang et al., 2007).

PCP (Li et al., 2008) first conducts metric learning based on the semi-definite programming, and then kernel k-means clustering is conducted over the learned metric. Some package (e.g. Csdp) is utilized to solve the semi-definite programming based on the specified pairwise constraints and similarities.

4.1.4 Parameters

The parameters under the pairwise constraints in Definition 1 are:

1) the number of constraints
2) the pairs of instances for constraints

As for **2)**, pairs of instances were randomly sampled from each dataset to generate the constraints. Thus, the main parameter is **1)**, the number of constraints, for must-links and cannot-links. We set the numbers of these two types of constrains to be the same, and varied the number of constraints.

Each data instance x in a dataset was normalized such that $x^t x = 1$, and Euclidian distance was utilized for SCREEN as in (Tang et al., 2007). With this normalization, cosine similarity, which is widely utilized as the standard similarity measure in document processing, was utilized for GBSSC and PCP, and the initial edge-weighted graph for each dataset was constructed with the similarities. The number of generalized eigenvectors l was set to the number of clusters k. In addition, following the procedure in (Li et al., 2008), m-nearest neighbor graph was constructed for PCP (m was set to 10 in the experiment). The hyper-parameter λ in eq.(10) was set to 0.5, since GBSSC is robust to this value as reported in Section 4.2.

4.1.5 Evaluation procedure

For each number of constraints, the pairwise constraints (must-links and cannot-links) were generated randomly based on the ground-truth labels in the datasets, and clustering was conducted with the generated constraints. Clustering with the same number of constraints was repeated 10 times with different initial configuration in clustering. In addition, the above process was also repeated 10 times for each number of constraints. Thus, for each dataset and the number of constraints, 100 runs were conducted. Furthermore, this process was repeated over 10 samples for each dataset. Thus, the average of 1,000 runs is reported for each dataset.

4.2 Evaluation of graph-based approach

Our approach modifies the data representation in a dataset according to the specified constraints. Especially, the similarities among instances (weights in a graph) are modified. The other possible approach would be to set the weights (similarities) as:

i) each pair $(x_i, x_j) \in C_{ML}$ to the maximum similarity
ii) each pair $(x_i, x_j) \in C_{CL}$ to the minimum similarity

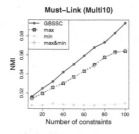
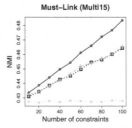

Fig. 3. Weight medication comparison.

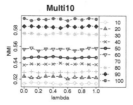
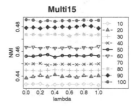

Fig. 4. Influence of λ.

First, we compared our approach for the handling of must-links in Section 3.2 with the above approaches on Multi10 and Multi15 datasets. The results are summarized in Fig. 3. In Fig. 3, horizontal axis corresponds to the number of constraints; vertical one corresponds to NMI. In the legend, max (black lines with boxes) stands for i), min (blue dotted lines with circles) stands for ii), and max&min (green dashed lines with crosses) stands for when both i) and ii) are employed. GBSSC (red solid lines with circles) stands for our approach.

The results in Fig. 3 show that GBSSC outperformed others and that it is effective in terms of the weight modification in a graph. One of the reasons for the results in Fig. 3 is that, when i) (max) is utilized, only the instances connected with must-links are affected, and thus they tend to be collected into a smaller "isolated" cluster. Creating rather small clusters makes the performance degraded. On the other hand, in our approach, instances adjacent to must-links are also affected via contraction.

As for ii) (min), the instances connected with cannot-links are by definition dissimilar with each other and their weights would be small in the original representation. Thus, setting the weights over must-links to the minimal value in the dataset does not affect the overall performance so much. These are illustrated in Fig. 5 and Fig. 6.

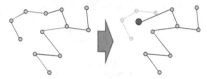

Fig. 5. Contraction of must-link constraints.

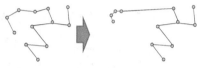

Fig. 6. Weight modification of must-link constraints.

Next, we evaluated the handling of cannot-links in Section 3.3. We varied the value of hyper-parameter λ in eq.(10) and analyzed its influence. The results are summarized in Fig. 4. In Fig. 4, horizontal axis corresponds to the value of λ, and the values in the legend corresponds to the number of pairwise constraints (e.g., 10 corresponds to the situation where the number of pairwise constraints are 10). The performance of GBSSC was not so much affected by the value of λ. Thus, our approach can be said as relatively robust with respect to this parameter. In addition, the accuracy (*NMI*) increased *monotonically* as the number of constraints increased. Thus, it can be concluded that GBSSC reflects the pairwise constraints and improves the performance based on semi-supervised information.

4.3 Evaluation on real world datasets

We report the comparison of our approach with other compared methods. In the reported figures, horizontal axis corresponds to the number of constraints; vertical one corresponds to either *NMI* or CPU time (in sec.).

In the legend in the figures, red lines correspond to our GBSSC, black dotted lines to SCREEN, green lines to PCP. Also, +PCA stands for the case where the dataset was first pre-processed by PCA (using 100 eigenvectors as in (Tang et al., 2007)) and then the corresponding method was applied. GBSSC+PCP (with purple lines) corresponds to the situation where must-links were handled by contraction in Section 3.2 and cannot-links by PCP.

4.3.1 20 Newsgroup datasets

The results for 20NG dataset are summarized in Figs. 7. These are the average of 10 datasets for each set of groups (i.e., average of 1000 runs). The results indicate that our approach outperformed other methods with respect to *NMI* (Fig. 7) when $l=k$ [8]. For Multi5, although the

---

[8] The number of generalized eigenvectors l was set to the number of clusters k. Note that we did not conduct any tuning for the value of l in these experiments. (Tang et al., 2007) reports that SCREEN could be improved by tuning the number of dimensions.

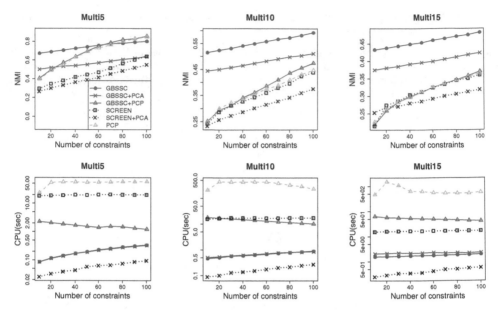

Fig. 7. Results on 20-Newsgroup

performance of PCP got close to that of GBSSC as the number of constraints increased, GBSSC was faster more than two orders of magnitude (100 times faster). Likewise, GBSSC+PCP and PCP were almost the same with respect to *NMI*, but the former was faster with more than one order (10 times faster). Although SCREEN+PCA was two to five times faster than GBSSC, it was inferior with respect to *NMI*. Utilization of PCA as the pre-processing enables this speed-up for SCREEN, in compensation for the accuracy (*NMI*).

Dimensionality reduction with PCA was effective for the speed-up of SCREEN, but it was not for GBSSC. On the other hand, it *deteriorated* their performance with respect to *NMI*. Thus, it is not necessary to utilize pre-processing such as PCA for GBSSC, and still our approach showed better performance.

4.3.2 TREC datasets

The results for TREC datasets are summarized in Fig. 8 and Fig. 9. As shown in Table 2, the number of dimensions (attributes) are huge in TREC datasets. Since calculating the eigenvalues of the covariance matrix with large number of attributes takes too much time, when SCREEN was applied to non-preprocessed data with PCA, it was too slow. Thus, SCREEN was applied only to the pre-processed data in TREC datasets. (shown as SCREEN+PCA).

On the whole, the results were quite similar to those in 20NG. Our approach outperformed SCREEN (in TREC datasets, SCREEN+PCA) with respect to *NMI*. It also outperformed PCP in most datasets, however, as the number of constraints increased, the latter showed better performance for review and sports datasets. In addition, PCP seems to improve the performance as the number of constraints increase. When GBSSC is utilized with PCP

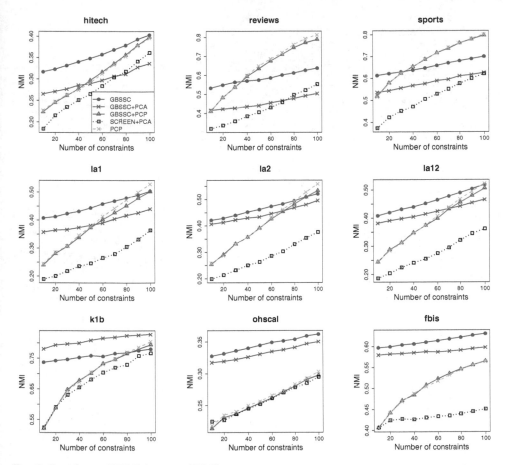

Fig. 8. Results on TREC datasets (*NMI*)

(denoted as GBSSC+PCP in the figure), it showed almost equivalent performance with respect to *NMI*, but the former was faster with more than one order.

5. Discussions

5.1 Effectiveness

The reported results show that our approach is effective in terms of the accuracy of cluster assignment (*NMI*). GBSSC outperformed SCREEN in all the datasets. Although it did not outperformed PCP in some TREC datasets with respect to *NMI*, but it was faster more than two orders of magnitude. Utilization of PCA as data pre-processing for dimensionality reduction enables the speed-up of SCREEN, in compensation for the accuracy of cluster assignment. On the other hand, PCP showed better performance in some datasets with respect to accuracy of cluster assignment, in compensation for the running time. Besides, since SCREEN originally conducts linear dimensionality reduction based on constraints, utilization

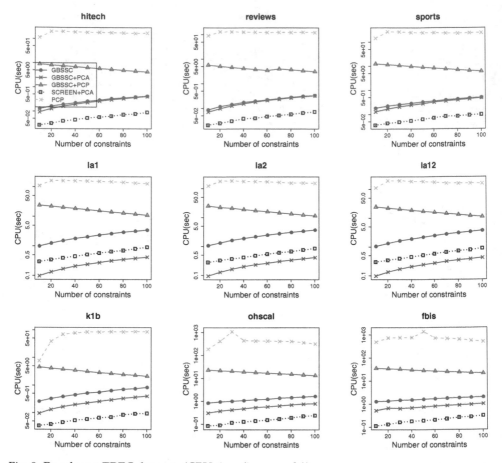

Fig. 9. Results on TREC datasets (CPU time (in seconds))

of *another* linear dimensionality reduction (such as PCA) as pre-processing might obscure its effect.

From these results, our approach can be said as effective in terms of the balance between the accuracy of cluster assignment and running time. Especially, it can leverage small amount of pairwise constraints to increase the performance. We believe that this is a good property in the semi-supervised learning setting.

5.2 Related work

Various approaches have been conducted on semi-supervised clustering. Among them are: constraint-based, distance-based, and hybrid approaches (Tang et al., 2007). The constraint-based approach tries to guide the clustering process with the specified pairwise instance constraints (Wagstaff et al., 2001). The distance-based approach utilizes metric learning techniques to acquire the distance measure during the clustering process based on the specified pairwise instance constraints (Li et al., 2008; Xing et al., 2003). The hybrid approach combines these two approaches under a probabilistic framework (Basu et al., 2004).

As for the semi-supervised clustering problem, (Wagstaff et al., 2001) proposed a clustering algorithm called COP-kmeans based on the famous kmeans algorithm. When assigning each data item to the cluster with minimum distance as in kmeans, COP-kmeans checks the constraint satisfaction and assigns each data item only to the admissible cluster (which does not violate the constraints).

SCREEN (Tang et al., 2007) first converts the data representation based on must-link constraints and removes the constraints. This process corresponds to contraction in our approach, but the weight definition is different. After that, based on cannot-link constraints, it finds out the linear mapping (linear projection) to a subspace where the variance among the data is maximized. Finally, clustering of the mapped data is conducted on the subspace.

PCP (Li et al., 2008) deals with the semi-supervised clustering problem by finding a mapping onto a space where the specified constraints are reflected. Using the specified constraints, it conducts metric learning based on the semi-definite programming and learn the kernel matrix on the mapped space. Although the explicit representation of the mapping or the data representation on the mapped space is not learned, kernel k-means clustering (Girolami, 2002) is conducted over the learned metric.

6. Conclusion

In this chapter we presented our semi-supervised clustering method based on graph contraction in general graph theory and graph Laplacian in spectral graph theory. Our approach can exploit a small amount of pairwise constraints as well as pairwise relations (similarities) among the data instances. Utilization of graph representation of instances enables to deal with the pairwise constraints as well as pairwise similarities over a unified representation. In order to reflect the pairwise constraints on the clustering process, the graph structure for the entire data instances is modified by graph contraction in general graph theory (Diestel, 2006) and graph Laplacian in spectral graph theory (Chung, 1997; von Luxburg, 2007).

We reported the performance of our approach over two real-world datasets with respect to the type of constraints as well as the number of constraints. We also compared with other state-of-the-art semi-supervised clustering methods in terms of accuracy of cluster assignment and running time. The experimental results indicate that our approach is effective in terms of the balance between the accuracy of cluster assignment and running time. Especially, it could leverage a small amount of pairwise constraints to improve the clustering performance. We plan to continue this line of research and to improve the presented approach in future.

7. Acknowledgments

The author is grateful to Mr. Okatani and Mr. Ogino for their help on implementation.

8. References

Basu, S., Bilenko, M. & Mooney, R. J. (2004). A probabilistic framework for semi-supervised clustering, *KDD-04*, pp. 59–68.
Basu, S., Davidson, I. & Wagstaff, K. (eds) (2008). *Constrained Clustering: Advances in Algorithms, Theory, and Applications*, Chapman & Hall/CRC Press.

Belkin, M. & Niyogi, P. (2002). Laplacian eigenmaps for dimensionality reduction and data representation, *Neural Computation* 15: 1373–1396.

Blum, A. & Mitchell, T. (1998). Combining labeled and unlabeled data with to-training, *Proc. 11th Computational Learning Theory*, pp. 92–100.

Chapelle, O., Schölkopf, B. & Zien, A. (eds) (2006). *Semi-Supervised Learning*, MIT Press.

Chung, F. (1997). *Spectral Graph Theory*, American Mathematical Society.

Cover, T. & Thomas, J. (2006). *Elements of Information Theory*, Wiley.

Dhillon, J., Mallela, S. & Modha, D. (2003). Information-theoretic co-clustering, *Proc. KDD'03*, pp. 89–98.

Dhillon, J. & Modha, D. (2001). Concept decompositions for large sparse text data using clustering, *Machine Learning* 42: 143–175.

Diestel, R. (2006). *Graph Theory*, Springer.

Girolami, M. (2002). Mercer kernel-based clustering in feature space, *IEEE Transactions on Neural Networks* 13(3): 780–784.

Guénoche, A., Hansen, P. & Jaumard, B. (1991). Efficient algorithms for divisive hierarchical clustering with the diameter criterion, *J. of Classification* 8: 5–30.

Guha, S., Rastogi, R. & Shim, K. (1998). Cure: An efficient clustering algorithm for large databases, *Proc. the ACM SIGMOD Conference*, pp. 73–84.

Hartigan, J. & Wong, M. (1979). Algorithm AS136: A k-means clustering algorithm, *Journal of Applied Statistics* 28: 100–108.

Jain, A., Murty, M. & P.J., F. (1999). Data clustering: A review, *ACM Computing Surveys* 31: 264–323.

Li, Z., Liu, J. & Tang, X. (2008). Pairwise constraint propagation by semidefinite programming for semi-supervised classification, *ICML-08*, pp. 576–583.

Ng, A. Y., Jordan, M. I. & Weiss, Y. (2001). On Spectral Clustering: Analysis and an algorithm, *Proc. NIPS 14*, pp. 849–856.

Ng, R. & Han, J. (2002). Clarans: a method for clustering objects for spatial data mining, *IEEE Transactions on Knowledge and Data Engineering* 14(5): 1003–1016.

Strehl, A. & Ghosh, J. (2002). Cluster Ensembles -A Knowledge Reuse Framework for Combining Multiple Partitions, *J. Machine Learning Research* 3(3): 583–617.

Tang, W., Xiong, H., Zhong, S. & Wu, J. (2007). Enhancing semi-supervised clustering : A feature projection perspective, *Proc. KDD'07*, pp. 707–716.

von Luxburg, U. (2007). A tutorial on spectral clustering, *Statistics and Computing* 17(4): 395–416.

Wagstaff, K., Cardie, C., Rogers, S. & Schroedl, S. (2001). Constrained k-means clustering with background knowledge, *In ICML01*, pp. 577–584.

Xing, E. P., Ng, A. Y., Jordan, M. I. & Russell, S. (2003). Distance metric learning, with application to clustering with side-information, *NIPS 15*, pp. 505–512.

Yoshida, T. (2010). Performance Evaluation of Constraints in Graph-based Semi-Supervised Clustering, *Proc. AMT-2010, LNAI 6335*, pp. 138–149.

Yoshida, T. (2011). A graph model for mutual information based clustering, *Journal of Intelligent Information Systems* 37(2): 187–216.

Yoshida, T. & Ogino, H. (2011). A re-coloring approach for graph b-coloring based clustering, *International Journal of Knowledge-Based & Intelligent Engineering Systems* . accepted.

Yoshida, T. & Okatani, K. (2010). A Graph-based projection approach for Semi-Supervised Clustering, *Proc. PKAW-2010, LNAI 6232*, pp. 1–13.

6

Visibility Algorithms: A Short Review

Angel M. Nuñez, Lucas Lacasa, Jose Patricio Gomez and Bartolo Luque
Universidad Politécnica de Madrid, Spain
Spain

1. Introduction

1.1 Motivation

Disregarding any underlying process (and therefore any physical, chemical, economical or whichever meaning of its mere numeric values), we can consider a time series just as an ordered set of values and play the naive mathematical game of turning this set into a different mathematical object with the aids of an abstract mapping, and see what happens: which properties of the original set are conserved, which are transformed and how, what can we say about one of the mathematical representations just by looking at the other... This exercise is of mathematical interest by itself. In addition, it turns out that time series or signals is a universal method of extracting information from dynamical systems in any field of science. Therefore, the preceding mathematical game gains some unexpected practical interest as it opens the possibility of analyzing a time series (i.e. the outcome of a dynamical process) from an alternative angle. Of course, the information stored in the original time series should be somehow conserved in the mapping. The motivation is completed when the new representation belongs to a relatively mature mathematical field, where information encoded in such a representation can be effectively disentangled and processed. This is, in a nutshell, a first motivation to map time series into networks.

This motivation is increased by two interconnected factors: first, although a mature field, time series analysis has some limitations, when it refers to study the so called complex signals. Beyond the linear regime, there exist a wide range of phenomena (not exclusive to physics) which are usually embraced in the field of the so called Complex Systems. Under this vague definition lies a common feature: the relevant effect of nonlinearities in their mathematical representation. This feature can be reflected in the temporal evolution of (at least one of) the variables describing the system and necessitates the use of specific tools for nonlinear analysis [1]. Dynamical phenomena such as chaos, long-range correlated stochastic processes, intermittency, multifractality, etc... are examples of complex phenomena where time series analysis is pushed to its own limits. Nonlinear time series analysis develops from techniques such as nonlinear correlation functions, embedding algorithms, multrifractal spectra, projection theorems... tools that increase in complexity parallel to the complexity of the process/series under study. New approaches, new paradigms to deal with complexity are not only welcome, but needed. Approaches that deal with the intrinsic nonlinearity

---

[1] We should note that nonlinearity is not the only feature that characterize a complex system; many interacting parts, randomness and emergence could also be cited but, as we are going to see later, nonlinearity will be sufficient for our purposes in this chapter

by being intrinsically nonlinear, that deal with the possible multiscale character of the underlying process by being designed to naturally incorporate multiple scales. And such is the framework of networks, of graph theory. Second, the technological era brings us the possibility of digitally analyze myriads of data in a glimpse. Massive data sets can nowadays be parsed, and with the aid of well suited algorithms, we can have access and filter data from many processes, let it be of physical, technological or even social garment. It is now time to develop new approaches to filter such plethora of information.

It is in this context that the network approach for time series analysis was born. The family of visibility algorithms constitute one of other possibilities to map a time series into a graph and subsequently analyze the structure of the series through the set of tools developed in the graph /complex network theory. In this chapter we will review some of its basic properties and show some of its first applications.

1.2 Different methods to map time series into graphs

The idea of mapping time series into graphs seems attractive because it lays a bridge between two prolific fields of modern science as Nonlinear Signal Analysis and Complex Networks Theory, so much so that it has attracted the attention of several research groups which have contributed to the topic with different strategies of mapping. While an exhaustive list of such strategies is beyond the scope of this work, we shall briefly outline some of them.

Zhang & Small (2006) developed a method that mapped each cycle of a pseudoperiodic time series into a node in a graph. The connection between nodes was established by a distance threshold in the reconstructed phase space when possible or by the linear correlation coefficient between cycles in the presence of noise. Noisy periodic time series mapped into random graphs while chaotic time series did it into scale-free, small-world networks due to the presence of unstable periodic orbits. This method was subsequently applied to characterize cardiac dynamics.

Xu et al. (2008) concentrated in the relative frequencies of appearance of four-node motifs inside a particular graph in order to classify it into a particular superfamily of networks which corresponded to specific underlying dynamics of the mapped time series. In this case, the method of mapping consisted in embedding the time series in an appropiated phase space where each point corresponded to a node in the network. A threshold was imposed not only in the minimum distance between two neighbours to be eligible (temporal separation should be greater than the mean period of the data) but also to the maximum number of neighbours a node could have. Different superfamilies were found for chaotic, hyperchaotic, random and noisy periodic underlying dynamics, unique fingerprints were also found for specific dynamical systems within a family.

Donner et al. (2010; 2011) presented a technique which was based on the properties of recurrence in the phase space of a dynamical system. More precisely, the recurrence matrix obtained by imposing a threshold in the minimum distance between two points in the phase space (as in Xu et al. (2008)) was interpreted as the adjacency matrix of an undirected, unweighted graph. Properties of such graphs at three different scales (local, intermediated and global) were presented and studied on several paradigmatic systems (Hénon map, Rossler system, Lorenz system, Bernoulli map). The variation of some of the properties of the graphs with the distance threshold was analyzed, the use of specific measures like the local clustering coefficient was proposed as a way for detecting dynamically invariant objects (saddle points or unstable periodic orbits) and studying the graph properties dependent on the embedding dimension was suggested as a means to distinguish between chaotic and stochastic systems.

Campanharo et al. (2011) contributed with an idea along the lines of Shirazi et al. (2009), Strozzi et al. (2009) and Haraguchi et al. (2009) of a surjective mapping which admits an inverse opperation. This approach opens the reciprocal possibility of benefiting from time series analysis to study the structure and properties of networks. Time series are treated as Markov processes, values are grouped in quantiles which will correspond to nodes in the associated graph. Weighted and directed connections are stablished between nodes as a function of the probability of transition between quantiles. An inverse operation can be defined without any a priori knowledge of the correspondance between nodes and quantiles just by imposing a continuity condition in the time series by means of a cost function defined on the weighted adjacency matrix of the graph. A random walk is performed on the network and a time series with properties equivalent to the original one is recovered. This method was applied to a battery of cases which included a periodic-to-random family of processes parametrized by the probability of transition p, a pair of chaotic systems (Lorentz and Rossler attractors) and two human heart rate time series. Reciprocally, the inverse map was applied to the metabolic network of Arabidopsis Thaliana and to the '97 year Internet Network. Time series obtained were demostrated to exhibit different dynamics.

Among all these methods of mapping, in this chapter we are going to concentrate our attention on the one developed in Lacasa et al. (2008) and subsequent works. To cite some of its most relevant features, we will stress its intrinsic nonlocality, its low computational cost, its straightforward implementation and its quite 'simple' way of inherit the time series properties in the structure of the associated graphs. These features are going to make it easier to find connections between the underlying processes and the networks obtained from them by a direct analysis of the latter. In what follows we will firstly present different versions of the algorithm along with its most notable properties, that in many cases can be derived analytically (theorems are reported when possible). Based on these latter properties, several applications are addressed.

2. Visibility algorithms: Theory

2.1 Natural visibility algorithm: definition

Let $\{x(t_i)\}_{i=1..N}$ be a time series of N data. The natural visibility algorithm (Lacasa et al., 2008) assigns each datum of the series to a node in the natural visibility graph (from now on NVg). Two nodes i and j in the graph are connected if one can draw a straight line in the time series joining $x(t_i)$ and $x(t_j)$ that does not intersect any intermediate data height $x(t_k)$ (see figure 1 for a graphical illustration). Hence, i and j are two connected nodes if the following geometrical criterion is fulfilled within the time series:

$$x(t_k) < x(t_i) + (x(t_j) - x(t_i))\frac{t_k - t_i}{t_j - t_k}. \tag{1}$$

It can easily checked that by means of the present algorithm, the associated graph extracted from a time series is always:

(*i*) connected: each node sees at least its nearest neighbors (left-hand side and right-hand side).
(*ii*) undirected: the way the algorithm is built up, there is no direction defined in the links.
(*iii*) invariant under affine transformations of the series data: the visibility criterium is invariant under rescaling of both horizontal and vertical axis, as well as under horizontal and vertical translations.
(*iv*) "lossy": some information regarding the time series is inevitably lost in the mapping from the fact that the network structure is completely determined in the (binary) adjacency matrix. For instance, two periodic series with the same period as $T1 = ..., 3, 1, 3, 1, ...$ and $T2 = ..., 3, 2, 3, 2, ...$ would have the same visibility graph, albeit being quantitatively different.

Fig. 1. Illustrative example of the visibility algorithm. In the upper part we plot a periodic time series and in the bottom part we represent the graph generated through the visibility algorithm. Each datum in the series corresponds to a node in the graph, such that two nodes are connected if their corresponding data heights fulfill the visibility criterion of equation 1. Note that the degree distribution of the visibility graph is composed by a finite number of peaks, much in the vein of the Discrete Fourier Transform of a periodic signal. We can thus interpret the visibility algorithm as a geometric transform.

One straightforward question is: what does the visibility algorithm stand for? In order to deepen on the geometric interpretation of the visibility graph, let us focus on a periodic series. It is straightforward that its visibility graph is a concatenation of a motif: a repetition of a pattern (see figure 1). Now, which is the degree distribution $P(k)$ of this visibility graph? Since the graph is just a motif's repetition, the degree distribution will be formed by a finite number of non-null values, this number being related to the period of the associated periodic series. This behavior reminds us the Discrete Fourier Transform (DFT), which for periodic series is formed by a finite number of peaks (vibration modes) related to the series period. Using this analogy, we can understand the visibility algorithm as a geometric (rather than integral) transform. Whereas a DFT decomposes a signal in a sum of (eventually infinite) modes, the visibility algorithm decomposes a signal in a concatenation of graph's motifs, and the degree distribution simply makes a histogram of such 'geometric modes'. While the time series is defined in the time domain and the DFT is defined on the frequency domain, the visibility graph is then defined on the 'visibility domain'. At this point we can mention that whereas a generic DFT fails to capture the presence of nonlinear correlations in time series (such as the presence of chaotic behavior), we will see that the visibility algorithm can distinguish between stochastic and chaotic series. Of course this analogy is, so far, a simple metaphor to help our intuition (this transform is not a reversible one for instance).

2.2 Horizontal visibility algorithm: definition

An alternative criterion for the construction of the visibility graph is defined as follows: let $\{x_i\}_{i=1..N}$ be a time series of N data. The so called horizontal visibility algorithm (Luque et al., 2009) assigns each datum of the series to a node in the horizontal visibility graph (from now on HVg). Two nodes i and j in the graph are connected if one can draw a horizontal line in the time series joining x_i and x_j that does not intersect any intermediate data height (see figure 2 for a graphical illustration). Hence, i and j are two connected nodes if the following geometrical criterion is fulfilled within the time series:

$$x_i, x_j > x_n \text{ for all } n \text{ such that } i < n < j \tag{2}$$

This algorithm is a simplification of the NVa. In fact, the HVg is always a subgraph of its associated NVg for the same time series (see figure 2). Beside this, the HVg graph will also be (*i*) connected, (*ii*) undirected, (*iii*) invariant under affine transformations of the series and (*iv*) "lossy". Some concrete properties of these graphs can be found in Gutin et al. (2011); Lacasa et al. (2010); Luque et al. (2009; 2011). In the next sections we are going to focus on properties of this particular method as it is a quite more analytically tractable version.

2.3 Topological properties of the HVg associated to periodic series: mean degree

Theorem 2.1. *The mean degree of an horizontal visibility graph associated to an infinite periodic series of period T (with no repeated values within a period) is*

$$\bar{k}(T) = 4\left(1 - \frac{1}{2T}\right) \tag{3}$$

A proof can be found in Núñez et al. (2010).

An interesting consequence of the previous result is that every time series extracted from a dynamical system has an associated HVG with a mean degree $2 \leq \bar{k} \leq 4$, where the lower bound is reached for constant series, whereas the upper bound is reached for aperiodic (random or chaotic) series (Luque et al., 2009).

2.4 Topological properties of the HVg associated to random time series

Let $\{x_i\}$ be a bi-infinite sequence of independent and identically distributed random variables extracted from a continous probability density $f(x)$, and consider its associated HVg. In the following sections we outline some theorems regarding the topological properties of these graphs.

2.4.1 Degree distribution of the visibility graph associated to a random time series

Theorem 2.2. *The degree distribution of its associated horizontal visibility graph is*

$$P(k) = \frac{1}{3}\left(\frac{2}{3}\right)^{k-2}, \, k = 2, 3, 4, ... \tag{4}$$

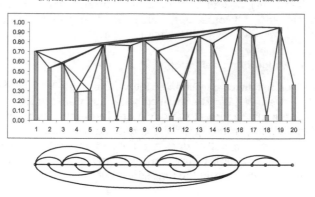

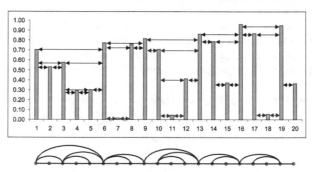

Fig. 2. Illustrative example of the natural and horizontal visibility algorithms. We plot the same time series and we represent the graphs generated through both visibility algorithms below. Each datum in the series corresponds to a node in the graph, such that two nodes are connected if their corresponding data heights fulfill respectively the visibility criteria of equations 1 and 2 respectively.

A lengthy constructive proof can be found in Luque et al. (2009) and alternative, shorter proofs can be found in Núñez et al. (2010).

Observe that the mean degree $\bar{k}$ of the horizontal visibility graph associated to an uncorrelated random process is then:

$$\bar{k} = \sum kP(k) = \sum_{k=2}^{\infty} \frac{k}{3}\left(\frac{2}{3}\right)^{k-2} = 4 \tag{5}$$

in good agreement with the prediction of eq. 3 in the limit $T \to \infty$, i.e. an aperiodic series.

2.4.2 Degree versus height

An interesting aspect worth exploring is the relation between data height and the node degree, that is, to study whether a functional relation between the height of a datum and the degree of its associated node holds. In this sense, let us define $P(k|x)$ as the conditional probability that a given node has degree k provided that it has height x. $P(k|x)$ is easily deduced in Luque et al. (2009), resulting in

$$P(k|x) = \sum_{j=0}^{k-2} \frac{(-1)^{k-2}}{j!(k-2-j)!}[1 - F(x)]^2 \cdot [\ln(1 - F(x))]^{k-2} \tag{6}$$

The average value of the degree of a node associated to a datum of height x, $K(x)$, in then

$$K(x) = \sum_{k-2}^{\infty} kP(k|x) = 2 - 2\ln(1 - F(x)) \tag{7}$$

where $F(x) = \int_{-\infty}^{x} f(x')dx'$.

Since $F(x) \in [0,1]$ and $\ln(x)$ are monotonically increasing functions, $K(x)$ will also be monotonically increasing. We can thus conclude that graph hubs (that is, the most connected nodes) are the data with largest values, that is, the extreme events of the series.

2.4.3 Local clustering coefficient distribution

The local clustering coefficient C (Boccaletti et al., 2006; Newmann, 2003) of an horizontal visibility graph associated to a random series can be easy deduced by means of geometrical arguments (Luque et al., 2009):

$$C(k) = \frac{k-1}{\binom{k}{2}} = \frac{2}{k} \tag{8}$$

what indicates a so called hierarchical structure (Ravasz et al., 2002). This relation between k and C allows us to deduce the local clustering coefficient distribution $P(C)$:

$$P(k) = \frac{1}{3}\left(\frac{2}{3}\right)^{k-2} = P(2/C)$$

$$P(C) = \frac{1}{3}\left(\frac{2}{3}\right)^{2/C-2} \tag{9}$$

2.4.4 Long distance visibility, mean degree and mean path length

The probability $P(n)$ that two data separated by n intermediate data be two connected nodes in the graph can be demostrated to be (see Luque et al. (2009))

$$P(n) = \left(\frac{1}{n} - 1\right) \int_0^1 f(x_0)F^n(x_0)dx_0 + \int_0^1 f(x_0)F^{n-1}(x_0)dx_0$$

$$= \frac{2}{n(n+1)} \tag{10}$$

where $P(n)$ is independent of the probability distribution f(x) of the random variable. Notice that the latter result can also be obtained, alternatively, with a purely combinatorial argument: take a random series with $n+1$ data and choose its two largest values. This latter pair can be placed with equiprobability in $n(n+1)$ positions, while only two of them are such that the largest values are placed at distance n, so we get $P(n) = \frac{2}{n(n+1)}$ on agreement with the previous development.

. 4.5 Small World property

If we looked the adjacency matrix (Newmann, 2003) of the horizontal visibility graph associated to a random series (Luque et al., 2009), we would see that every data x_i has visibility of its first neighbors x_{i-1}, x_{i+1}, every node i will be connected by construction to nodes $i - 1$ and $i + 1$: the graph is thus connected. The graph evidences a typical homogeneous structure: the adjacency matrix is predominantly filled around the main diagonal. Furthermore, the matrix evidences a superposed sparse structure, reminiscent of the visibility probability $P(n) = 2/(n(n + 1))$ that introduces some shortcuts in the horizontal visibility graph, much in the vein of the Small-World model (Strogatz, 2001). Here the probability of having these shortcuts is given by $P(n)$. Statistically speaking, we can interpret the graph's structure as quasi-homogeneous, where the size of the local neighborhood increases with the graph's size. Accordingly, we can approximate its mean path length $L(N)$ as:

$$L(N) \approx \sum_{n=1}^{N-1} nP(n) = \sum_{n=1}^{N-1} \frac{2}{n+1} = 2\log(N) + 2(\gamma - 1) + O(1/N) \tag{11}$$

where we have made use of the asymptotic expansion of the harmonic numbers and γ is the Euler-Mascheroni constant. As can be seen, the scaling is logarithmic, denoting that the horizontal visibility graph associated to a generic random series is Small-World (Newmann, 2003).

2.5 Topological properties of the HVg associated to other stochastic and chaotic processes

It was proved that $P(k) = (1/3)(2/3)^{k-2}$ for uncorrelated random series. To find out a similar closed expression in the case of generic chaotic or stochastic correlated processes is a very difficult task, since variables can be long-range correlated and hence the probabilities cannot be separated (lack of independence). This leads to a very involved calculation which is typically impossible to solve in the general case. However, some analytical developments can be made in order to compare them with our numerical results. Concretely, for Markovian systems global dependence is reduced to a one-step dependence. We will make use of such property to derive exact expressions for $P(2)$ and $P(3)$ in some Markovian systems (both deterministic and stochastic).

2.5.1 Ornstein-Uhlenbeck process: degree distribution

Suppose a short-range correlated series (exponentially decaying correlations) of infinite size generated through an Ornstein-Uhlenbeck process (Van Kampen, 2007), and generate its associated HVg. Let us consider the probability that a node chosen at random has degree $k = 2$. This node is associated to a datum labelled x_0 without lack of generality. Now, this node will have degree $k = 2$ if the datum first neighbors, x_1 and x_{-1} have values larger than x_0:

$$P(k = 2) = P(x_{-1} > x_0 \cap x_1 > x_0) \tag{12}$$

In this case the variables are correlated, so in general we should have

$$P(2) = \int_{-\infty}^{\infty} dx_0 \int_{x_0}^{\infty} dx_{-1} \int_{x_0}^{\infty} dx_1 \, f(x_{-1}, x_0, x_1) \tag{13}$$

We use the Markov property $f(x_{-1}, x_0, x_1) = f(x_{-1})f(x_0|x_{-1})f(x_1|x_0)$, that holds for an Ornstein-Uhlenbeck process with correlation function $C(t) \sim \exp(-t/\tau)$ (Van Kampen, 2007):

$$f(x) = \frac{\exp(-x^2/2)}{\sqrt{2\pi}} \qquad f(x_2|x_1) = \frac{\exp(-(x_2 - Kx_1)^2/2(1 - K^2))}{\sqrt{2\pi(1 - K^2)}}, \qquad (14)$$

where $K = \exp(-1/\tau)$.

Numerical integration allows us to calculate $P(2)$ for every given value of the correlation time τ. A procedure to compute $P(3)$ can also be found in Lacasa et al. (2010).

2.5.2 Logistic map: degree distribution

A chaotic map of the form $x_{n+1} = F(x_n)$ does also have the Markov property, and therefore a similar analysis can be applied (even if chaotic maps are deterministic). For chaotic dynamical systems whose trajectories belong to the attractor, there exists a probability measure that characterizes the long-run proportion of time spent by the system in the various regions of the attractor. In the case of the logistic map $F(x_n) = \mu x_n(1 - x_n)$ with parameter $\mu = 4$, the attractor is the whole interval $[0, 1]$ and the probability measure $f(x)$ corresponds to the *beta* distribution with parameters $a = 0.5$ and $b = 0.5$:

$$f(x) = \frac{x^{-0.5}(1 - x)^{-0.5}}{\mathrm{B}(0.5, 0.5)} \qquad (15)$$

Now, for a deterministic system, the transition probability is

$$f(x_{n+1}|x_n) = \delta[x_{n+1} - F(x_n)], \qquad (16)$$

where $\delta(x)$ is the Dirac delta distribution. Departing from equation 12, for the logistic map $F(x_n) = 4x_n(1 - x_n)$ and $x_n \in [0, 1]$, we have

$$P(2) = \int_0^1 dx_0 \int_{x_0}^1 f(x_{-1})f(x_0|x_{-1})dx_{-1} \int_{x_0}^1 f(x_1|x_0)dx_1 =$$
$$\int_0^1 dx_0 \int_{x_0}^1 f(x_{-1})\delta(x_0 - F(x_{-1}))dx_{-1} \int_{x_0}^1 \delta(x_1 - F(x_0))dx_1. \qquad (17)$$

Now, notice that, using the properties of the Dirac delta distribution, $\int_{x_0}^1 \delta(x_1 - F(x_0))dx_1$ is equal to one iff $F(x_0) \in [x_0, 1]$, what will happen iff $0 < x_0 < 3/4$, and zero otherwise. Therefore the only effect of this integral is to restrict the integration range of x_0 to be $[0, 3/4]$.

On the other hand,

$$\int_{x_0}^1 f(x_{-1})\delta[x_0 - F(x_{-1})]dx_{-1} = \sum_{x_k^*|F(x_k^*)=x_0} f(x_k^*)/|F'(x_k^*)|,$$

that is, the sum over the roots of the equation $F(x) = x_0$, iff $F(x_{-1}) > x_0$. But since $x_{-1} \in [x_0, 1]$ in the latter integral, it is easy to see that again, this is verified iff $0 < x_0 < 3/4$ (as a matter of fact, if $0 < x_0 < 3/4$ there is always a *single* value of $x_{-1} \in [x_0, 1]$ such that $F(x_{-1}) = x_0$, so the sum restricts to the adequate root). It is easy to see that the particular value is $x^* = (1 + \sqrt{1 - x_0})/2$. Making use of these piecewise solutions and equation 15, we finally have

$$P(2) = \int_0^{3/4} \frac{f(x^*)}{4\sqrt{1 - x_0}} dx_0 = 1/3, \tag{18}$$

Note that a similar development can be fruitfully applied to other chaotic maps, provided that they have a well defined natural measure. Analytical and numerical developments for $P(3)$ can be found in Lacasa et al. (2010).

2.6 Directed horizontal visibility graph

So far, undirected visibility graphs have been considered, as visibility did not have a predefined temporal arrow. However, such a directionality can be made explicit by making use of directed networks or digraphs (Newmann, 2003).

Let a *directed* horizontal visibility graph (DHVg, Lacasa et al. (2011)) be a horizontal visibility graph, where the degree $k(x_i)$ of the node x_i is now splitted in an *ingoing* degree $k_{in}(x_i)$, and an *outgoing* degree $k_{out}(x_i)$, such that $k(x_i) = k_{in}(x_i) + k_{out}(x_i)$. The ingoing degree $k_{in}(x_i)$ is defined as the number of links of node x_i with other *past* nodes associated with data in the series (that is, nodes with $j < i$). Conversely, the outgoing degree $k_{out}(x_i)$, is defined as the number of links with *future* nodes ($i < j$).

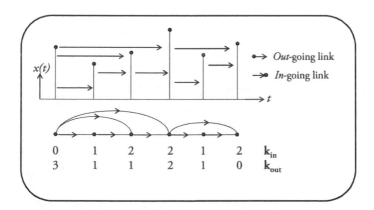

Fig. 3. Graphical illustration of the method. In the top we plot a sample time series $\{x(t)\}$. Each datum in the series is mapped to a node in the graph. Arrows, describing allowed directed visibility, link nodes. The associated directed horizontal visibility graph is plotted below. In this graph, each node has an ingoing degree k_{in}, which accounts for the number of links with *past* nodes, and an outgoing degree k_{out}, which in turn accounts for the number of links with *future* nodes. The asymmetry of the resulting graph can be captured in a first approximation through the invariance of the outgoing (or ingoing) degree series under time reversal.

For a graphical illustration of the method, see figure 3. The degree distribution of a graph describes the probability of an arbitrary node to have degree k (i.e. k links, Newmann (2003)). We define the *in* and *out* (or ingoing and outgoing) degree distributions of a DHVg as the probability distributions of k_{out} and k_{in} of the graph which we call $P_{out}(k) \equiv P(k_{out} = k)$ and $P_{in}(k) \equiv P(k_{in} = k)$, respectively.

2.6.1 Uncorrelated stochastic series: degree distribution

Theorem 2.3. *Let $\{x_t\}_{t=-\infty,...,\infty}$ be a bi-infinite sequence of independent and identically distributed random variables extracted from a continuous probability density $f(x)$. Then, both the in and out degree distributions of its associated directed horizontal visibility graph are*

$$P_{in}(k) = P_{out}(k) = \left(\frac{1}{2}\right)^k, \quad k = 1, 2, 3, ... \tag{19}$$

Proof. (out-distribution) Let x be an arbitrary datum of the aforementioned series. The probability that the horizontal visibility of x is interrupted by a datum x_r on its right is independent of $f(x)$,

$$\Phi_1 = \int_{-\infty}^{\infty} \int_{x}^{\infty} f(x)f(x_r)dx_r dx = \int_{-\infty}^{\infty} f(x)[1 - F(x)]dx = \frac{1}{2},$$

The probability $P(k)$ of the datum x being capable of exactly seeing k data may be expressed as

$$P(k) = Q(k)\Phi_1 = \frac{1}{2}Q(k), \tag{20}$$

where $Q(k)$ is the probability of x seeing at least k data. $Q(k)$ may be recurrently calculated via

$$Q(k) = Q(k-1)(1 - \Phi_1) = \frac{1}{2}Q(k-1), \tag{21}$$

from which, with $Q(1) = 1$, the following expression is obtained

$$Q(k) = \left(\frac{1}{2}\right)^{k-1}, \tag{22}$$

which together with equation (20) concludes the proof. □

An analogous derivation holds for the *in* case. This result is independent of the underlying probability density $f(x)$: it holds not only for Gaussian or uniformly distributed random series, but for any series of independent and identically distributed (i.i.d.) random variables extracted from a continuous distribution $f(x)$.

3. Towards a graph theory of time series?

In the preceding section, specific properties of the visibility graphs (either NVg, HVg or the directed version of HVg) associated to different time series have been considered. Relying on the aforementioned dualities between time series structure and network topological features, we proceed here to make the first steps for a graph theoretical analysis of time series and dynamical systems, addressing several nontrivial problems of time series analysis through the visibility algorithm apparatus.

3.1 Estimating the Hurst exponent with NVg

Self-similar processes such as fractional Brownian motion (fBm, Mandelbrot & Van Ness (1968)) are currently used to model fractal phenomena of different nature, ranging from Physics or Biology to Economics or Engineering (see Lacasa et al. (2009) and references therein). A fBm $B_H(t)$ is a non-stationary random process with stationary self-similar increments (fractional Gaussian noise) that can be characterized by the so called Hurst exponent, $0 < H < 1$. The one-step memory Brownian motion is obtained for $H = \frac{1}{2}$, whereas time series with $H > \frac{1}{2}$ shows persistence and anti-persistence if $H < \frac{1}{2}$. While different fBm generators and estimators have been introduced in the last years, the community lacks consensus on which method is best suited for each case. This drawback comes from the fact that fBm formalism is exact in the infinite limit, i.e. when the whole infinite series of data is considered. However, in practice, real time series are finite. Accordingly, long range correlations are partially broken in finite series, and local dynamics corresponding to a particular temporal window are overestimated. The practical simulation and the estimation from real (finite) time series is consequently a major issue that is, hitherto, still open. An overview of different methodologies and comparisons can be found in Carbone (2007); Kantelhardt (2008); Karagiannis et al. (2004); Mielniczuk & Wojdyllo (2007); Pilgram & Kaplan (1998); Podobnik & Stanley (2008); Simonsen et al. (1998); Weron (2002) and references therein.

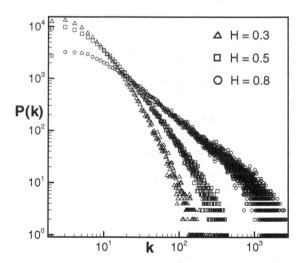

Fig. 4. Degree distribution of three visibility graphs, namely (i) triangles: extracted from a fBm series of 10^5 data with $H = 0.3$, (ii) squares: extracted from a fBm series of 10^5 data with $H = 0.5$, (iii) circles: extracted from a fBm series of 10^5 data with $H = 0.8$. Note that distributions are not normalized. The three visibility graphs are scale-free since their degree distributions follow a power law $P(k) \sim k^{-\gamma}$ with decreasing exponents $\gamma_{0.3} > \gamma_{0.5} > \gamma_{0.8}$.

Here we address the problem of estimating the Hurst exponent of a fBm series via the NVg. If we map a fBm time series by means of the NVa, what we get is a scale-free graph (Lacasa et al.,

2008; 2009), see figure 4. As a matter of fact, that fBm yields scale free visibility graphs is not that surprising; the most highly connected nodes (hubs) are the responsible for the heavy tailed degree distributions. Within fBm series, hubs are related to extreme values in the series, since a datum with a very large value has typically a large connectivity (a fact reminiscent of eq. 7). It can be proved (Lacasa et al., 2009) that the degree distribution of a NVg extracted from a fBm with Hurst exponent H shows a power law shape $P(k) \sim k^{-\gamma}$, such that

$$\gamma(H) = 3 - 2H. \tag{23}$$

Numerical analysis corroborated this theoretical relation in Lacasa et al. (2009).

It is well known that fBm has a power spectra that behaves as $1/f^{\beta}$, where the exponent β is related to the Hurst exponent of an fBm process through the well known relation

$$\beta(H) = 1 + 2H. \tag{24}$$

Now according to eqs. 23 and 24, the degree distribution of the visibility graph corresponding to a time series with $f^{-\beta}$ noise should be again power law $P(k) \sim k^{-\gamma}$ where

$$\gamma(\beta) = 4 - \beta. \tag{25}$$

The theoretical prediction eq. 25 was also corroborated numerically in Lacasa et al. (2009). Finally, eq. 24 holds for fBm processes, while for the increments of an fBm process, known as a fractional Gaussian noise (fGn), the relation between β and H turns to be

$$\beta(H) = -1 + 2H, \tag{26}$$

The relation between γ and H for a fGn (where fGn is a series composed by the increments of a fBm) can be deduced to be

$$\gamma(H) = 5 - 2H. \tag{27}$$

In order to illustrate this latter case, we address a realistic and striking dynamics where long range dependence has been recently described. Gait cycle (the stride interval in human walking rhythm) is a physiological signal that has been shown to display fractal dynamics and long range correlations in healthy young adults (Goldenberger et al., 2002; Hausdorff et al., 1996). In the upper part of fig. 5 we have plotted to series describing the fluctuations of walk rhythm of a young healthy person, for slow pace (bottom series of 3304 points) and fast pace (up series of 3595 points) respectively (data available in www.physionet.org/physiobank/database/umwdb/ (Goldberger et al., 2000)). In the bottom part we have represented the degree distribution of their visibility graphs. These ones are again power laws with exponents $\gamma = 3.03 \pm 0.05$ for fast pace and $\gamma = 3.19 \pm 0.05$ for slow pace (derived through MLE). According to eq. 25, the visibility algorithm predicts that gait dynamics evidence $f^{-\beta}$ behavior with $\beta = 1$ for fast pace, and $\beta = 0.8$ for slow pace, in perfect agreement with previous results based on a Detrended Fluctuation Analysis (Goldenberger et al., 2002; Hausdorff et al., 1996). These series record the fluctuations of walk rhythm (that is, the increments), so according to eq. 27, the Hurst exponent is $H = 1$ for fast pace and $H = 0.9$ for slow pace, that is to say, dynamics evidences long range dependence (persistence) (Goldenberger et al., 2002; Hausdorff et al., 1996).

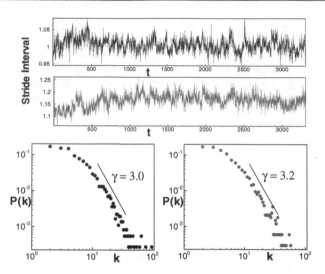

Fig. 5. Black signal: time series of 3595 points from the stride interval of a healthy person in fast pace. Red signal: time series of 3304 points from the stride interval of a healthy person in slow pace. Bottom: Degree distribution of the associated visibility graphs (the plot is in log-log). These are power laws where $\gamma = 3.03 \pm 0.05$ for the fast movement (black dots) and $\gamma = 3.19 \pm 0.05$ for the slow movement (red dots), what provides $\beta = 1$ and $\beta = 0.8$ for fast and slow pace respectively according to eq.25, in agreement with previous results (Goldenberger et al., 2002; Hausdorff et al., 1996).

3.2 Discriminating stochastic vs. chaotic series via HVg

Both stochastic and chaotic processes share many features, and the discrimination between them is indeed very subtle. The relevance of this problem is to determine whether the source of unpredictability (production of entropy) has its origin in a chaotic deterministic or stochastic dynamical system, a fundamental issue for modeling and forecasting purposes. Essentially, the majority of methods (Cecini et al., 2010; Kants H. & Schreiber, 2003) that have been introduced so far rely on two major differences between chaotic and stochastic dynamics. The first difference is that chaotic systems have a finite dimensional attractor, whereas stochastic processes arise from an infinite-dimensional one. Being able to reconstruct the attractor is thus a clear evidence showing that the time series has been generated by a deterministic system. The development of sophisticated embedding techniques (Kants H. & Schreiber, 2003) for attractor reconstruction is the most representative step forward in this direction. The second difference is that deterministic systems evidence, as opposed to random ones, short-time prediction: the time evolution of two nearby states will diverge exponentially fast for chaotic ones (finite and positive Lyapunov exponents) while in the case of a stochastic process such separation is randomly distributed. Whereas some algorithms relying on the preceding concepts are nowadays available, the great majority of them are purely phenomenological and often complicated to perform, computationally speaking. These drawbacks provide the motivation for a search for new methods that can directly distinguish, in a reliable way, stochastic from chaotic time series. We show here that the horizontal visibility algorithm offers a different, conceptually simple and computationally efficient method to distinguish between deterministic and stochastic dynamics, since the degree distribution of HVGs associated to stochastic and chaotic processes are exponential $P(k) \sim \exp(-\lambda k)$, where for stochastic dynamics $\lambda > \lambda_{un}$ and for chaotic dynamics $\lambda < \lambda_{un}$ (Lacasa et al., 2010), λ_{un} being the uncorrelated case, (theorem 2.2).

3.2.1 Correlated stochastic series

In order to analyze the effect of correlations between the data of the series, we focus on two generic and paradigmatic correlated stochastic processes, namely long-range (power-law decaying correlations) and Ornstein-Uhlenbeck (short-range exponentially decaying correlations) processes. We have computed the degree distribution of the HVg associated to different long-range and short-range correlated stochastic series (the method for generating the associated series is explained in Lacasa et al. (2010)) with correlation function $C(t) = t^{-\gamma}$ for different values of the correlation strength $\gamma \in [10^{-2} - 10^{1}]$ and with an exponentially decaying correlation function $C(t) = \exp(-t/\tau)$. In both cases the degree distribution of the associated HVG can be fitted for large k by an exponential function $\exp(-\lambda k)$. The parameter λ depends on γ or τ and is, in each case, a monotonic function that reaches the asymptotic value $\lambda = \lambda_{un} = \ln(3/2)$ in the uncorrelated limit $\gamma \to \infty$ or $\tau \to 0$, respectively. Detailed results of this phenomenology can be found in (Lacasa et al., 2010). In all cases, the limit is reached from above, i.e. $\lambda > \lambda_{un}$ (see figure 6). Interestingly enough, for the power-law correlations the convergence is slow, and there is still a noticeable deviation from the uncorrelated case even for weak correlations ($\gamma > 4.0$), whereas the convergence with τ is faster in the case of exponential correlations.

3.2.2 Chaotic maps

Poincaré recurrence theorem suggests that the degree distribution of HVgs associated to chaotic series should be asymptotically exponential (Luque et al., 2009). Several deterministic time series generated by chaotic maps have been analyzed:

(1) the α-map $f(x) = 1 - |2x - 1|^{\alpha}$, that reduces to the logistic and tent maps in their fully chaotic region for $\alpha = 2$ and $\alpha = 1$ respectively, for different values of α,
(2) the 2D Hénon map ($x_{t+1} = y_t + 1 - ax_t^2$, $y_{t+1} = bx_t$) in the fully chaotic region ($a = 1.4$, $b = 0.3$);
(3) a time-delayed variant of the Hénon map: $x_{t+1} = bx_{t-d} + 1 - ax_t^2$ in the region ($a = 1.6$, $b = 0.1$), where it shows chaotic behavior with an attractor dimension that increases linearly with the delay d (Sprott, 2006). This model has also been used for chaos control purposes (Buchner & Zebrowski, 2000), although here we set the parameters a and b to values for which we find high-dimensional chaos for almost every initial condition (Sprott, 2006);
(4) the Lozi map, a piecewise-linear variant of the Hénon map given by $x_{t+1} = 1 + y_n - a|x_t|$, $y_{t+1} = bx_t$ in the chaotic regime $a = 1.7$ and $b = 0.5$;
(5) the Kaplan-Yorke map $x_{t+1} = 2x_t \mod (1)$, $y_{t+1} = \lambda y_t + \cos(4\pi x_t) \mod (1)$; and
(6) the Arnold cat map $x_{t+1} = x_t + y_t \mod (1)$, $y_{t+1} = x_t + 2y_t \mod (1)$, a conservative system with integer Kaplan-Yorke dimension. References for these maps can be found in Sprott & Rowlands (2001).

We find that the tails of the degree distribution can be well approximated by an exponential function $P(k) \sim \exp(-\lambda k)$. Remarkably, we find that $\lambda < \lambda_{un}$ in every case, where λ seems to increase monotonically as a function of the chaos dimensionality [2], with an asymptotic value $\lambda \to \ln(3/2)$ for large values of the attractor dimension (see fig. 6 where we plot the specific values of λ as a function of the correlation dimension of the map (Sprott & Rowlands, 2001)). Again, we deduce that the degree distribution for uncorrelated series is a limiting case of the degree distribution for chaotic series but, as opposed to what we found for stochastic processes, the convergence flow towards λ_{un} is from below, and therefore $\lambda = \ln(3/2)$ plays the role of an effective frontier between correlated stochastic and chaotic processes.

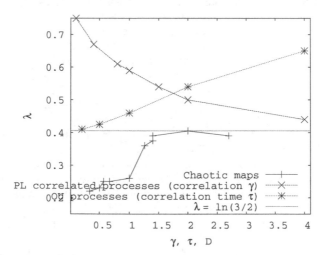

Fig. 6. Plot of the values of λ for several processes, namely: (i) for power-law correlated stochastic series with correlation function $C(t) = t^{-\gamma}$, as a function of the correlation γ, (ii) for Ornstein-Uhlenbeck series with correlation function $C(t) = \exp(-t/\tau)$, as a function of the correlation time τ, and (iii) for different chaotic maps, as a function of their correlation dimension D. Errors in the estimation of λ are incorporated in the size of the dots. Notice that stochastic processes cluster in the region $\lambda > \lambda_{un}$ whereas chaotic series belong to the opposite region $\lambda < \lambda_{un}$, evidencing a convergence towards the uncorrelated value $\lambda_{un} = \ln(3/2)$ (Luque et al., 2009) for decreasing correlations or increasing chaos dimensionality respectively.

In the following section we provide some heuristic arguments supporting our findings, for additional details, numerics and analytical developments we refer the reader to Lacasa et al. (2010).

3.2.3 Heuristics

We argue first that correlated series show lower data variability than uncorrelated ones, so decreasing the possibility of a node to reach far visibility and hence decreasing (statistically speaking) the probability of appearance of a large degree. Hence, the correlation tends to decrease the number of nodes with large degree as compared to the uncorrelated counterpart.

---

[2] This functional relation must nonetheless be taken in a cautious way, indeed, other chaos indicators (such as the Lyapunov spectra) may also play a relevant role in the final shape of $P(k)$ and such issues should be investigated in detail

Indeed, in the limit of infinitely large correlations ($\gamma \to 0$ or $\tau \to \infty$), the variability reduces to zero and the series become constant. The degree distribution in this limit case is, trivially,

$$P(k) = \delta(k-2) = \lim_{\lambda \to \infty} \frac{\lambda}{2} \exp(-\lambda|k-2|),$$

that is to say, infinitely large correlations would be associated to a diverging value of λ. This tendency is on agreement with the numerical simulations (figure 6) where we show that λ monotonically increases with decreasing values of γ or increasing values of τ respectively. Having in mind that in the limit of small correlations the theorem previously stated implies that $\lambda \to \lambda_{un} = \ln(3/2)$, we can therefore conclude that for a correlated stochastic process $\lambda_{stoch} > \lambda_{un}$.

Concerning chaotic series, remember that they are generated through a deterministic process whose orbit is continuous along the attractor. This continuity introduces a smoothing effect in the series that, statistically speaking, increases the probability of a given node to have a larger degree (uncorrelated series are rougher and hence it is more likely to have more nodes with smaller degree). Now, since in every case we have exponential degree distributions (this fact being related with the Poincaré recurrence theorem for chaotic series and with the return distribution in Poisson processes for stochastic series (Luque et al., 2009)), we conclude that the deviations must be encoded in the slope λ of the exponentials, such that $\lambda_{chaos} < \lambda_{un} < \lambda_{stoch}$, in good agreement with our numerical results.

3.3 Noise filtering using HVg: periodic series polluted with noise

In this section we address the task of filtering a noisy signal with a hidden periodic component within the horizontal visibility formalism, that is, we explore the possibility of using the method for noise filtering purposes (see (Núñez et al., 2010) for details). Periodicity detection algorithms (see for instance (Parthasarathy et al., 2006)) can be classified in essentially two categories, namely the time domain (autocorrelation based) and frequency domain (spectral) methods. Here we make use of the horizontal visibility algorithm to propose a third category: graph theoretical methods.

If we superpose a small amount of noise to a periodic series (a so-called *extrinsic* noise), while the degree of the nodes with associated small values will remain rather similar, the nodes associated to higher values will eventually increase their visibility and hence reach larger degrees. Accordingly, the delta-like structure of the degree distribution (associated with the periodic component of the series) will be perturbed, and an exponential tail will arise due to the presence of such noise (Lacasa et al., 2010; Luque et al., 2009). Can the algorithm characterize such kind of series? The answer is positive, since the degree distribution can be analytically calculated resulting in:

$$P(2) = 1/2,$$
$$P(3) = 0,$$
$$P(k+2) = \frac{1}{3}\left(\frac{2}{3}\right)^{k-2}, \, k \geq 2,$$
$$\text{or } P(k) = \frac{1}{4}\left(\frac{2}{3}\right)^{k-3}, \, k \geq 4, \tag{28}$$

that is to say, introducing a small amount of extrinsic uncorrelated noise in a periodic signal introduces an exponential tail in the HVG's degree distribution with the same slope as the one associated to a purely uncorrelated process. The mean degree $\bar{k}$ reads

$$\bar{k} = \sum_{k=2}^{\infty} kP(k) = 4,$$

which, according to equation 3, suggests aperiodicity, as expected.

3.3.1 A graph-theoretical noise filter

Let $S = \{x_i\}_{i=1,...,n}$ be a periodic series of period T (where $n >> T$) polluted by a certain amount of extrinsic noise (without loss of generality, suppose a white noise extracted from a uniform distribution $U[-0.5, 0.5]$), and define the filter f as a real valued scalar such that $f \in [\min x_i, \max x_i]$. The so called filtered Horizontal Visibility Graph (f-HVg) associated to S is constructed as it follows:
(i) each datum x_i in the time series is mapped to a node i in the f-HVg, (ii) two nodes i and j are connected in the f-HVg if the associated data fulfill

$$x_i, x_j > x_n + f, \forall n \mid i < n < j. \tag{29}$$

The procedure of filtering the noise from a noisy periodic signal goes as follows: one generates the f-HVg associated to S for increasing values of f, and in each case proceeds to calculate the mean degree $\bar{k}$. For the proper interval $f_{min} < f < f_{max}$, the f-HVg of the noisy periodic series S will be equivalent to the noise free HVg of the pure (periodic) signal, which has a well defined mean degree as a function of the series period. In this interval, the mean degree will therefore remain constant, and from equation 3 the period can be inferred. As an example, we

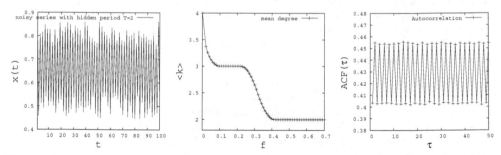

Fig. 7. *Left*: Periodic series of period $T = 2$ polluted with extrinsic noise extracted from a uniform distribution $U[-0.5, 0.5]$ of amplitude 0.1. *Middle*: Values of the HVg's mean degree $\bar{k}$ as a function of the amplitude of the graph theoretical filter. The first plateau is found for $\bar{k} = 3$, which renders a hidden period $T = (2 - \bar{k}/2)^{-1} = 2$. The second plateau corresponding to $\bar{k} = 2$ is found when the filter is large enough to screen each datum with its first neighbors, such that the mean degree reaches its lowest bound. *Right*: Autocorrelation function of the noisy periodic series, which is itself an almost periodic series with period $T = 2$, as it should.

have artificially generated a noisy periodic series of hidden period $T = 2$ (see figure 7). The results of the graph filtering technique yielded a net decreases of the mean degree, which has an initial value of 4 (as expected for the HVg ($f = 0$) of an aperiodic series such as a noisy periodic signal) and an asymptotic value of 2 (lower bound of the mean degree). The plateau is clearly found at $\bar{k} = 3$, which according to equation 3 yields a period

$$T = \left(2 - \frac{\bar{k}}{2}\right)^{-1} = 2,$$

as expected.

3.3.2 Noisy periodic versus chaotic

Let us now consider a simple case of chaotic map with disconnected attractors.The Logistic map

$$x_{t+1} = \mu x_t(1 - x_t),$$

with $x \in [0,1]$ and $\mu \in [3.6, 3.67]$ has an attractor that is partitioned in two disconnected chaotic bands, and the chaotic orbit makes an alternating journey between both bands (see fig. figintro). The map is ergodic, but the attractor is not the whole interval, as there is a gap between both chaotic bands. In this situation, the chaotic series is by definition not periodic, however, an autocorrelation function analysis indeed suggests the presence of periodicity, what is reminiscent of the disconnected two-band structure of the attractor. Interestingly enough, applying the aforementioned noise filter technique, at odds with the autocorrelation function, the results suggests that the method does not find any periodic structure, as it should (see Núñez et al. (2010) for details). Furthermore, information of both the phase space structure and the chaotic nature of the map becomes accessible from an analysis of the HVg's degree distribution. First, we find $P(2) = 1/2$, that indicates that half of the data are located in the bottom chaotic band, in agreement with the alternating nature of the chaotic orbit. This is reminiscent of the misleading result obtained from the autocorrelation function. Second, the tail of the degree distribution is exponential, with an asymptotic slope smaller than the one obtained rigorously (Luque et al., 2009) for a purely uncorrelated process. This is, according to Lacasa et al. (2010), characteristic of an underlying chaotic process.

3.4 The period-doubling route to chaos via HVg: Feigenbaum graphs

In low-dimensional dissipative systems chaotic motion develops out of regular motion in a small number of ways or routes, and amongst which the period-doubling bifurcation cascade or Feigenbaum scenario is perhaps the better known and most famous mechanism (Peitgen et al., 1992; Schuster, 1988). This route to chaos appears an infinite number of times amongst the family of attractors spawned by unimodal maps within the so-called periodic windows that interrupt stretches of chaotic attractors. In the opposite direction, a route out of chaos accompanies each period-doubling cascade by a chaotic band-splitting cascade, and their shared bifurcation accumulation points form transitions between order and chaos that are known to possess universal properties (Peitgen et al., 1992; Schuster, 1988; Strogatz, 1994). Low-dimensional maps have been extensively studied from a purely theoretical perspective, but systems with many degrees of freedom used to study diverse problems in physics, biology, chemistry, engineering, and social science, are known to display low-dimensional dynamics (Marvel et al., 2009).

In this section, we offer a distinct view of the Feigenbaum scenario through the specific HVg formalism, and provide a complete set of graphs, which we call Feigenbaum graphs, that encode the dynamics of all stationary trajectories of unimodal maps. We first characterize their topology via the order-of-visit and self-affinity properties of the maps. We will additionally define a renormalization group (RG) procedure that leads, via its flows, to or from network fixed-points to a comprehensive view of the entire family of attractors. Furthermore, the optimization of the entropy obtained from the degree distribution coincides with the RG fixed points and reproduces the essential features of the map's Lyapunov exponent independently of its sign. A general observation is that the HV algorithm extracts only universal elements of the dynamics, free of the peculiarities of the individual unimodal map, but also of universality classes characterized by the degree of nonlinearity. Therefore all the results presented in this section, while referring to the specific Logistic map for illustrative reasons apply to any unimodal map.

3.4.1 Feigenbaum graphs

According to the HV algorithm, a time series generated by the Logistic map for a specific value of μ (after an initial transient of approach to the attractor) is converted into a Feigenbaum graph (Luque et al., 2011). Notice that this is a well-defined subclass of HV graphs where consecutive nodes of degree $k = 2$, that is, consecutive data with the same value, do not appear, what is actually the case for series extracted from maps (besides the trivial case of a constant series). While for a period T there are in principle several possible periodic orbits, and therefore the set of associated Feigenbaum graphs is degenerate, it can be proved that the mean degree $\bar{k}(T)$ and normalized mean distance $\bar{d}(T)$ of all these Feigenbaum graphs fulfill $\bar{k}(T) = 4(1 - \frac{1}{2T})$ and $\bar{d}(T) = \frac{1}{3T}$ respectively, yielding a linear relation $\bar{d}(\bar{k}) = (4 - \bar{k})/6$ that is corroborated in the inset of figure 8. Aperiodic series ($T \to \infty$) reach the upper bound mean degree $\bar{k} = 4$.

3.4.2 Period-doubling cascade

A deep-seated feature of the period-doubling cascade is that the order in which the positions of a periodic attractor are visited is universal (Schroeder, 1991), the same for all unimodal maps. This ordering turns out to be a decisive property in the derivation of the structure of the Feigenbaum graphs. A plot the graphs for a family of attractors of increasing period $T = 2^n$, that is, for increasing values of $\mu < \mu_\infty$ can be found in (Luque et al., 2011). This basic pattern also leads to the expression for their associated degree distributions,

$$P(n,k) = \left(\tfrac{1}{2}\right)^{k/2}, \quad k = 2,4,6,...,2n, \tag{30}$$

$$P(n,k) = \left(\tfrac{1}{2}\right)^{n}, \quad k = 2(n+1),$$

and zero for k odd or $k > 2(n+1)$. At the accumulation point μ_∞ the period diverges ($n \to \infty$) and the distribution is exponential for all even values of the degree,

$$P(\infty,k) = \left(\frac{1}{2}\right)^{k/2}, \quad k = 2,4,6,..., \tag{31}$$

and zero for k odd. Observe that these relations are independent of the order of the map's nonlinearity: the HV algorithm sifts out every detail of the dynamics except for the basic storyline.

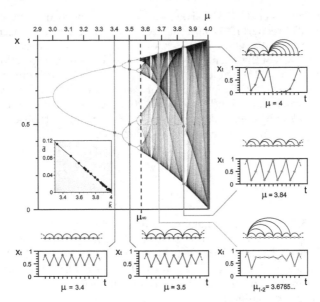

Fig. 8. Feigenbaum graphs from the Logistic map $x_{t+1} = f(x_t) = \mu x_t(1 - x_t)$. The main figure portrays the family of attractors of the Logistic map and indicates a transition from periodic to chaotic behavior at $\mu_\infty = 3.569946...$ through period-doubling bifurcations. For $\mu \geq \mu_\infty$ the figure shows merging of chaotic-band attractors where aperiodic behavior appears interrupted by windows that, when entered from their left-hand side, display periodic motion of period $T = m \cdot 2^0$ with $m > 1$ (for $\mu < \mu_\infty$, $m = 1$) that subsequently develops into m period-doubling cascades with new accumulation points $\mu_\infty(m)$. Each accumulation point $\mu_\infty(m)$ is in turn the limit of a chaotic-band reverse bifurcation cascade with m initial chaotic bands, reminiscent of the self-affine structure of the entire diagram. All unimodal maps exhibit a period-doubling route to chaos with universal asymptotic scaling ratios between successive bifurcations that depend only on the order of the nonlinearity of the map, the Logistic map belongs to the quadratic case. Adjoining the main figure, we show time series and their associated Feigenbaum graphs according to the HV mapping criterion for several values of μ where the map evidences both regular and chaotic behavior (see the text). Inset: numerical values of the mean normalized distance $\bar{d}$ as a function of mean degree $\bar{k}$ of the Feigenbaum graphs for $3 < \mu < 4$ (associated to time series of 1500 data after a transient and a step $\delta\mu = 0.05$), in good agreement with the theoretical linear relation (see the text).

3.4.3 Period-doubling bifurcation cascade of chaotic bands

We turn next to the period-doubling bifurcation cascade of chaotic bands that takes place as μ decreases from $\mu = 4$ towards μ_∞. For the largest value of the control parameter, at $\mu = 4$, the attractor is fully chaotic and occupies the entire interval $[0, 1]$ (see figure 8). This is the first chaotic band $n = 0$ at its maximum amplitude. As μ decreases in value within $\mu_\infty < \mu < 4$ band-narrowing and successive band-splittings (Peitgen et al., 1992; Schroeder, 1991; Schuster, 1988; Strogatz, 1994) occur. In general, after n reverse bifurcations the phase space is partitioned in 2^n disconnected chaotic bands, which are self-affine copies of the first chaotic band (Crutchfield et al., 1982). The values of μ at which the bands split are called Misiurewicz points (Schroeder, 1991), and their location converges to the accumulation point μ_∞ for $n \to \infty$. Significantly, while in the chaotic zone orbits are aperiodic, for reasons of continuity they visit each of the 2^n chaotic bands in the same order as positions are visited in the attractors of period $T = 2^n$ (Schroeder, 1991). A plot of the Feigenbaum graphs generated through chaotic time series at different values of μ that correspond to an increasing number of reverse bifurcations can be found in(Luque et al., 2011). Since chaotic bands do not overlap, one can derive the following degree distribution for a Feigenbaum graph after n chaotic-band reverse bifurcations by using only the universal order of visits

$$P_\mu(n,k) = \left(\tfrac{1}{2}\right)^{k/2}, \quad k = 2,4,6,...,2n,$$

$$P_\mu(n,k \geq 2(n+1)) = \left(\tfrac{1}{2}\right)^n, \tag{32}$$

and zero for $k = 3,5,7,...,2n+1$. We note that this time the degree distribution retains some dependence on the specific value of μ, concretely, for those nodes with degree $k \geq 2(n+1)$, all of which belong to the top chaotic band. The HV algorithm filters out chaotic motion within all bands except for that taking place in the top band whose contribution decreases as $n \to \infty$ and appears coarse-grained in the cumulative distribution $P_\mu(n,k \geq 2(n+1))$. As would be expected, at the accumulation point μ_∞ we recover the exponential degree distribution (equation 31), i.e. $\lim_{n\to\infty} P_\mu(n,k) = P(\infty,k)$.

3.4.4 Renormalization group

Before proceeding to interpret these findings via the consideration of renormalization group (RG) arguments, we recall that the Feigenbaum tree shows a rich self-affine structure: for $\mu > \mu_\infty$ periodic windows of initial period m undergo successive period-doubling bifurcations with new accumulation points $\mu_\infty(m)$ that appear interwoven with chaotic attractors. These cascades are self-affine copies of the fundamental one. The process of reverse bifurcations also evidences this self-affine structure, such that each accumulation point is the limit of a chaotic-band reverse bifurcation cascade. Accordingly, we label $G(m,n)$ the Feigenbaum graph associated with a periodic series of period $T = m \cdot 2^n$, that is, a graph obtained from an attractor within window of initial period m after n period-doubling bifurcations. In the same fashion, $G_\mu(n,m)$ is associated with a chaotic attractor composed by $m \cdot 2^n$ bands (that is, after n chaotic band reverse bifurcations of m initial chaotic bands). Graphs corresponding to $G(1,n)$ and $G_\mu(1,n)$ respectively can be found in (Luque et al., 2011). For the first accumulation point $G(1,\infty) = G_\mu(1,\infty) \equiv G_\infty$. Similarly, in each accumulation point $\mu_\infty(m)$, the identity $G(m,\infty) = G_\mu(m,\infty)$ is fulfilled.

In order to recast previous findings in the context of the renormalization group, let us define an RG operation $\mathcal{R}$ on a graph as the coarse-graining of every couple of adjacent nodes where one of them has degree $k = 2$ into a block node that inherits the links of the previous two nodes. This is a real-space RG transformation on the Feigenbaum graph (Newmann & Watts, 1999), dissimilar from recently suggested box-covering complex network renormalization schemes (Radicchi et al., 2008; Song et al., 2005; 2006). This scheme turns out to be equivalent for $\mu < \mu_\infty$ to the construction of an HV graph from the composed map $f^{(2)}$ instead of the original f, in correspondence to the original Feigenbaum renormalization procedure (Strogatz, 1994). We first note that $\mathcal{R}\{G(1,n)\} = G(1,n-1)$, thus, an iteration of this process yields an RG flow that converges to the (1st) trivial fixed point $\mathcal{R}^{(n)}\{G(1,n)\} = G(1,0) \equiv G_0 = \mathcal{R}\{G_0\}$. This is the stable fixed point of the RG flow $\forall \mu < \mu_\infty$. We note that there is only one relevant variable in our RG scheme, represented by the reduced control parameter $\Delta\mu = \mu_\infty - \mu$, hence, to identify a nontrivial fixed point we set $\Delta\mu = 0$ or equivalently $n \to \infty$, where the structure of the Feigenbaum graph turns to be completely self-similar under $\mathcal{R}$. Therefore we conclude that $G(1,\infty) \equiv G_\infty$ is the nontrivial fixed point of the RG flow, $\mathcal{R}\{G_\infty\} = G_\infty$. In connection with this, let $P_t(k)$ be the degree distribution of a generic Feigenbaum graph G_t in the period-doubling cascade after t iterations of $\mathcal{R}$, and point out that the RG operation, $\mathcal{R}\{G_t\} = G_{t+1}$, implies a recurrence relation $(1 - P_t(2))P_{t+1}(k) = P_t(k+2)$, whose fixed point coincides with the degree distribution found in equation 31. This confirms that the nontrivial fixed point of the flow is indeed G_∞.

Next, under the same RG transformation, the self-affine structure of the family of attractors yields $\mathcal{R}\{G_\mu(1,n)\} = G_\mu(1,n-1)$, generating a RG flow that converges to the Feigenbaum graph associated to the 1st chaotic band, $\mathcal{R}^{(n)}\{G_\mu(1,n)\} = G_\mu(1,0)$. Repeated application of $\mathcal{R}$ breaks temporal correlations in the series, and the RG flow leads to a 2nd trivial fixed point $\mathcal{R}^{(\infty)}\{G_\mu(1,0)\} = G_{\text{rand}} = \mathcal{R}\{G_{\text{rand}}\}$, where G_{rand} is the HV graph generated by a purely uncorrelated random process. This graph has a universal degree distribution $P(k) = (1/3)(2/3)^{k-2}$, independent of the random process underlying probability density (see (Lacasa et al., 2010; Luque et al., 2009)).

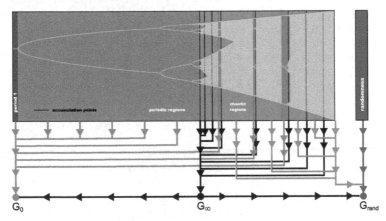

Fig. 9. Illustrative cartoon incorporating the RG flow of Feigenbaum graphs in the whole Feigenbaum diagram: aperiodic (chaotic or random) series generate graphs whose RG flow converge to the trivial fixed point G_{rand}, whereas periodic series (both in the region $\mu < \mu_\infty$ and inside windows of stability) generate graphs whose RG flow converges to the trivial fixed point $G(0,1)$. The nontrivial fixed point of the RG flow $G(\infty,1)$ is only reached through the critical manifold of graphs at the accumulation points $\mu_\infty(m)$.

Finally, let us consider the RG flow inside a given periodic window of initial period m. As the renormalization process addresses nodes with degree $k = 2$, the initial applications of $\mathcal{R}$ only change the core structure of the graph associated with the specific value m. The RG flow will therefore converge to the 1st trivial fixed point via the initial path $\mathcal{R}^{(p)}\{G(m,n)\} = G(1,n)$, with $p \leq m$, whereas it converges to the 2nd trivial fixed point for $G_\mu(m,n)$ via $\mathcal{R}^{(p)}\{G_\mu(m,n)\} = G_\mu(1,n)$. In the limit of $n \to \infty$ the RG flow proceeds towards the nontrivial fixed point via the path $\mathcal{R}^{(p)}\{G(m,\infty)\} = G(1,\infty)$. Incidentally, extending the definition of the reduced control parameter to $\Delta\mu(m) = \mu_\infty(m) - \mu$, the family of accumulation points is found at $\Delta\mu(m) = 0$. A complete schematic representation of the RG flows can be seen in figure 9.

Interestingly, and at odds with standard RG applications to (asymptotically) scale-invariant systems, we find that invariance at $\Delta\mu = 0$ is associated in this instance to an exponential (rather than power-law) function of the observables, concretely, that for the degree distribution. The reason is straightforward: $\mathcal{R}$ is not a conformal transformation (*i.e.* a scale operation) as in the typical RG, but rather, a translation procedure. The associated invariant functions are therefore non homogeneous (with the property $g(ax) = bg(x)$), but exponential (with the property $g(x+a) = cg(x)$).

3.4.5 Network entropy

Finally, we derive, via optimization of an entropic functional for the Feigenbaum graphs, all the RG flow directions and fixed points directly from the information contained in the degree distribution. Amongst the graph theoretical entropies that have been proposed we employ here the Shannon entropy of the degree distribution $P(k)$, that is $h = -\sum_{k=2}^{\infty} P(k) \log P(k)$. By making use of the Maximum Entropy formalism, it is easy to prove that the degree distribution $P(k)$ that maximizes h is exactly $P(k) = (1/3)(2/3)^{k-2}$, which corresponds to the distribution for the 2nd trivial fixed point of the RG flow G_{rand}. Alternatively, with the incorporation of the additional constraint that allows only even values for the degree (the topological restriction for Feigenbaum graphs $G(1,n)$), entropy maximization yields a degree distribution that coincides with equation 31, which corresponds to the nontrivial fixed point of the RG flow G_∞. Lastly, the degree distribution that minimizes h trivially corresponds to G_0, *i.e.* the 1st trivial fixed point of the RG flow. Remarkably, these results indicate that the fixed-point structure of the RG flow are obtained via optimization of the entropy for the entire family of networks, supporting a suggested connection between RG theory and the principle of Maximum Entropy (Robledo, 1999).

The network entropy h can be calculated exactly for $G(1,n)$ ($\mu < \mu_\infty$ or $T = 2^n$), yielding $h(n) = \log 4 \cdot (1 - 2^{-n})$. Because increments of entropy are only due to the occurrence of bifurcations h increases with μ in a step-wise way, and reaches asymptotically the value $h(\infty) = \log 4$ at the accumulation point μ_∞. For Feigenbaum graphs $G_\mu(1,n)$ (in the chaotic region), in general h cannot be derived exactly since the precise shape of $P(k)$ is unknown (albeit the asymptotic shape is also exponential (Luque et al., 2011)). Yet, the main feature of h can be determined along the chaotic-band splitting process, as each reverse bifurcation generates two self-affine copies of each chaotic band. Accordingly, the decrease of entropy associated with this reverse bifurcation process can be described as $h_\mu(n) = \log 4 + h_\mu(0)/2^n$, where the entropy $h_\mu(n)$ after n reverse bifurcations can be described in terms of the entropy associated with the first chaotic band $h_\mu(0)$. The chaotic-band reverse bifurcation process takes place in the chaotic region in the direction of decreasing μ's, and therefore leads in this case to a decrease of entropy with an asymptotic value of $\log 4$ for $n \to \infty$ at the accumulation point. These results suggest that the graph entropy behaves qualitatively as the map's Lyapunov exponent λ, with the peculiarity of having a shift of $\log 4$, as confirmed in figure 10. This unexpected qualitative agreement is reasonable in the chaotic region in view of the Pesin theorem (Peitgen et al., 1992), that relates the positive Lyapunov exponents of a map with its Kolmogorov-Sinai entropy (akin to a topological entropy) that for unimodal

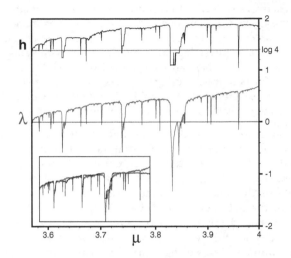

Fig. 10. Horizontal visibility network entropy h and Lyapunov exponent λ for the Logistic map. We plot the numerical values of h and λ for $3.5 < \mu < 4$ (the numerical step is $\delta\mu = 5 \cdot 10^{-4}$ and in each case the processed time series have a size of 2^{12} data). The inset reproduces the same data but with a rescaled entropy $h - \log(4)$. The surprisingly good match between both quantities is reminiscent of the Pesin identity (see text). Unexpectedly, the Lyapunov exponent within the periodic windows ($\lambda < 0$ inside the chaotic region) is also well captured by h.

maps reads $h_{KS} = \lambda$, $\forall\lambda > 0$, since h can be understood as a proxy for h_{KS}. Unexpectedly, this qualitative agreement seems also valid in the periodic windows ($\lambda < 0$), since the graph entropy is positive and varies with the value of the associated (negative) Lyapunov exponent even though $h_{KS} = 0$, hinting at a Pesin-like relation valid also out of chaos which deserves further investigation. The agreement between both quantities lead us to conclude that the Feigenbaum graphs capture not only the period-doubling route to chaos in a universal way, but also inherits the main feature of chaos, *i.e.* sensitivity to initial conditions.

3.5 Measuring irreversibility via HVg

A stationary process x_t is said to be statistically time reversible (hereafter time reversible) if for every n, the series $\{x_1, \cdots, x_n\}$ and $\{x_n, \cdots, x_1\}$ have the same joint probability distributions (Weiss, 1975). Roughly, this means that a reversible time series and its time reversed are, statistically speaking, equally probable. Reversible processes include the family of Gaussian linear processes (as well as Fourier-transform surrogates and nonlinear static transformations of them), and are associated with processes at thermal equilibrium in statistical physics. Conversely, time series irreversibility is indicative of the presence of nonlinearities in the underlying dynamics, including non-Gaussian stochastic processes and dissipative chaos, and are associated with systems driven out-of-equilibrium in the realm of thermodynamics (Kawai et al., 2007; Parrondo et al., 2009). Time series irreversibility is an important topic in basic and applied science. From a physical perspective, and based on the relation between statistical reversibility and physical dissipation (Kawai et al., 2007; Parrondo et al., 2009), the concept of time series irreversibility has been used to derive information about the entropy production of the physical mechanism generating the series, even if one ignores any detail of such mechanism (Roldan & Parrondo, 2011). In a more applied context, it has been suggested that irreversibility in complex physiological series decreases with aging or pathology, being maximal in young and healthy subjects (Costa et al., 2005; 2008; Yang et al., 2003), rendering this feature important for noninvasive diagnosis. As complex signals pervade natural and social sciences, the topic of time series reversibility is indeed relevant for scientists aiming to understand and model the dynamics behind complex signals.

The definition of time series reversibility is formal and therefore there is not an *a priori* optimal algorithm to quantify it in practice. Several methods to measure time irreversibility have been proposed (Andrieux et al., 2007; Cammarota & Rogora, 2007; Costa et al., 2005; Daw et al., 2000; Diks et al., 1995; Gaspard, 2004; Kennel, 2004; Wang et al., 2005; Yang et al., 2003). The majority of them perform a time series symbolization, typically making an empirical partition of the data range (Daw et al., 2000) (note that such a transformation does not alter the reversible character of the output series (Kennel, 2004)) and subsequently analyze the symbolized series, through statistical comparison of symbol strings occurrence in the forward and backwards series or using compression algorithms (Cover & Thomas, 2006; Kennel, 2004; Roldan & Parrondo, 2011). The first step requires an extra amount of *ad hoc* information (such as range partitioning or size of the symbol alphabet) and therefore the output of these methods eventually depend on these extra parameters. A second issue is that since typical symbolization is local, the presence of multiple scales (a signature of complex signals) could be swept away by this coarse-graining: in this sense multi-scale algorithms have been proposed recently (Costa et al., 2005; 2008). The *time directed* version of the horizontal visibility algorithm is proposed in this section as a simple and well defined tool for measuring time series irreversibility (see Lacasa et al. (2011) for details).

3.5.1 Quantifying irreversibility: DHVg and Kullback-Leibler divergence

The main conjecture of this application is that the information stored in the *in* and *out* distributions take into account the amount of time irreversibility of the associated series. More precisely, we claim that this can be measured, in a first approximation, as the distance (in a distributional sense) between the *in* and *out* degree distributions ($P_{in}(k)$ and $P_{out}(k)$). If needed, higher order measures can be used, such as the corresponding distance between the *in* and *out* degree-degree distributions ($P_{in}(k, k')$ and $P_{out}(k, k')$). These are defined as the *in* and *out* joint degree distributions of a node and its first neighbors (Newmann, 2003), describing the probability of an arbitrary node whose neighbor has degree k' to have degree k.

The Kullback-Leibler divergence (Cover & Thomas, 2006) is used as the distance between the *in* and *out* degree distributions. Relative entropy or Kullback-Leibler divergence (KLD) is introduced in information theory as a measure of distinguishability between two probability distributions. Given a random variable x and two probability distributions $p(x)$ and $q(x)$, KLD between p and q is defined as follows:

$$D(p||q) \equiv \sum_{x \in \mathcal{X}} p(x) \log \frac{p(x)}{q(x)}, \tag{33}$$

which vanishes if and only if both probability distributions are equal $p = q$ and it is bigger than zero otherwise.

We compare the outgoing degree distribution in the actual (forward) series $P_{k_{\text{out}}}(k|\{x(t)\}_{t=1,\ldots,N}) = P_{\text{out}}(k)$ with the corresponding probability in the time-reversed (or backward) time series, which is equal to the probability distribution of the ingoing degree in the actual process $P_{k_{\text{out}}}(k|\{x(t)\}_{t=N,\ldots,1}) = P_{\text{in}}(k)$. The KLD between these two distributions is

$$D[P_{\text{out}}(k)||P_{\text{in}}(k)] = \sum_k P_{\text{out}}(k) \log \frac{P_{\text{out}}(k)}{P_{\text{in}}(k)}. \tag{34}$$

This measure vanishes if and only if the outgoing and ingoing degree probability distributions of a time series are identical, $P_{\text{out}}(k) = P_{\text{in}}(k)$, and it is positive otherwise. We will apply it to several examples as a measure of irreversibility.

Notice that previous methods to estimate time series irreversibility generally proceed by first making a (somewhat *ad hoc*) local symbolization of the series, coarse-graining each of the series data into a symbol (typically, an integer) from an ordered set. Then, they subsequently perform a statistical analysis of word occurrences (where a word of length n is simply a concatenation of n symbols) from the forward and backwards symbolized series (Andrieux et al., 2007; Wang et al., 2005). Time series irreversibility is therefore linked to the difference between the word statistics of the forward and backwards symbolized series. The method presented here can also be considered as a symbolization if we restrict ourselves to the information stored in the series $\{k_{\text{out}}(t)\}_{t=1,\ldots,N}$ and $\{k_{\text{in}}(t)\}_{t=1,\ldots,N}$. However, at odds with other methods, here the symbolization process (i) lacks *ad hoc* parameters (such as number of symbols in the set or partition definition), and (ii) it takes into account *global* information: each coarse-graining $x_t \rightarrow (k_{\text{in}}(t), k_{\text{out}}(t))$ is performed using information from the whole series, according to the mapping criterion of fig. 3. Hence, this symbolization naturally takes into account multiple scales, which is desirable if we want to tackle complex signals (Costa et al., 2005; 2008).

3.5.2 Results for correlated stochastic series

The first example of a reversible series with $D[P_{\text{out}}(k)||P_{\text{in}}(k)] = 0$ are uncorrelated stochastic series which were considered in 2.6.1. As a further validation, linearly correlated stochastic processes have also been considered as additional examples of reversible dynamics (Weiss, 1975). An explanation of the method employed to generate the series can be consulted in (Lacasa et al., 2010), and results are summarized in table 1.

3.5.3 Results for a discrete flashing ratchet

A discrete flashing ratchet is an example of thermodynamic system which can be smoothly driven out of equilibrium by modifying the value of a physical parameter (the peak value V of an asymmetric potencial). We make use of a time series generated by a discrete flashing ratchet model introduced in (Roldan & Parrondo, 2010). For $V = 0$ detailed balance condition is satisfied, the system is in equilibrium and trajectories are statistically reversible. In this case both $D[P_{\text{out}}(k)||P_{\text{in}}(k)]$ and $D[P_{\text{out}}(k,k')||P_{\text{in}}(k,k')]$ using degree distributions and degree-degree distributions vanish. On the other hand, if V is increased, the system is driven out of equilibrium, what introduces a net statistical irreversibility which increases with V (Roldan & Parrondo, 2010). The amount of irreversibility estimated with KLD increases with

V for both measures, therefore the results produced by the method are qualitatively correct (see (Lacasa et al., 2011) for details). Interestingly enough, the tendency holds even for high values of the potential, where the statistics are poor and the KLD of sequences of symbols usually fail when estimating irreversibility (Roldan & Parrondo, 2010). However the values of the KLD obtained are far below the KLD per step between the forward and backward trajectories, which is equal to the dissipation as reported in (Roldan & Parrondo, 2010).

The degree distributions capture the irreversibility of the original series but it is difficult to establish a quantitative relationship between eq. (34) and the KLD between trajectories. The

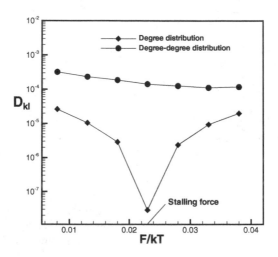

Fig. 11. Irreversibility measures $D[P_{\mathrm{out}}(k)||P_{\mathrm{in}}(k)]$ and $D[P_{\mathrm{out}}(k,k')||P_{\mathrm{in}}(k,k')]$ in the flashing ratchet ($r = 2, V = 2kT$) as a function of FL/kT. Here, F is the applied force and L is the spatial period of the ratchet, which in this case is equal to 1. For each value of the force, we make use of a single stationary series of size $N = 10^6$ containing partial information (the state information is removed).

measure based on the degree-degree distribution $D[P_{\mathrm{out}}(k,k')||P_{\mathrm{in}}(k,k')]$ takes into account more information of the visibility graph structure than the KLD using degree distributions, providing a closer bound to the physical dissipation as it is expected by the chain rule (Cover & Thomas, 2006), $D[P_{\mathrm{out}}(k,k')||P_{\mathrm{in}}(k,k')] \geq D[P_{\mathrm{out}}(k)||P_{\mathrm{in}}(k)]$. The improvement is even qualitatively significant in some situations. For instance, when a force opposite to the net current on the system is present (Roldan & Parrondo, 2010), the current vanishes for a given value of the force usually termed as *stalling force*. When the force reaches this value, the system is still out of equilibrium ($V > 0$) and it is therefore time irreversible, but no current of particles is observed if we describe the dynamics of the ratchet with partial information given by the position x. $D[P_{\mathrm{out}}(k)||P_{\mathrm{in}}(k)]$ tends to zero when the force approaches to the stalling value (see figure 11). Therefore, our measure of irreversibility (34) fails in this case, as do other KLD estimators based on local flows or currents (Roldan & Parrondo, 2010). However, $D[P_{\mathrm{out}}(k,k')||P_{\mathrm{in}}(k,k')]$ captures the irreversibility of the time series, and yields a positive value at the stalling force(Roldan & Parrondo, 2011).

Series description	$D[P_{out}(k)\|P_{in}(k)]$	$D[P_{out}(k,k')\|P_{in}(k,k')]$
Reversible Stochastic Processes		
$U[0,1]$ uncorrelated	$3.88 \cdot 10^{-6}$	$2.85 \cdot 10^{-4}$
Ornstein-Uhlenbeck ($\tau = 1.0$)	$7.82 \cdot 10^{-6}$	$1.52 \cdot 10^{-4}$
Long-range correlated stationary process ($\gamma = 2.0$)	$1.28 \cdot 10^{-5}$	$2.0 \cdot 10^{-4}$
Dissipative Chaos		
Logistic map ($\mu = 4$)	0.377	2.978
α map ($\alpha = 3$)	0.455	3.005
α map ($\alpha = 4$)	0.522	3.518
Henon map ($a = 1.4, b = 0.3$)	0.178	1.707
Lozi map	0.114	1.265
Kaplan Yorke map	0.164	0.390
Conservative Chaos		
Arnold Cat map	$1.77 \cdot 10^{-5}$	$4.05 \cdot 10^{-4}$

Table 1. Values of the irreversibility measure associated to the degree distribution $D[P_{out}(k)\|P_{in}(k)]$ and the degree-degree distribution $D[P_{out}(k,k')\|P_{in}(k,k')]$ respectively, for the visibility graphs associated to series of 10^6 data generated from reversible and irreversible processes. In every case chain rule is satisfied, since $D[P_{out}(k,k')\|P_{in}(k,k')] \geq D[P_{out}(k)\|P_{in}(k)]$. Note that that the method correctly distinguishes between reversible and irreversible processes, as KLD vanishes for the former and it is positive for the latter.

3.5.4 Results for chaotic series

This method was applied to several chaotic series and found that it is able to distinguish between dissipative and conservative chaotic systems. Dissipative chaotic systems are those that do not preserve the volume of the phase space, and they produce irreversible time series. This is the case of chaotic maps in which entropy production via instabilities in the forward time direction is quantitatively different to the amount of past information lost. In other words, those whose positive Lyapunov exponents, which characterize chaos in the forward process, differ in magnitude with negative ones, which characterize chaos in the backward process (Kennel, 2004). Several chaotic maps have been analyzed and the degree of reversibility of their associated time series has been estimated using using KLD, showing that for dissipative chaotic series it is positive while it vanishes for an example of conservative chaos. A summary of results cann be checked in table 1. In every case, we find an asymptotic positive value, in agreement with the conjecture that dissipative chaos is indeed time irreversible.

Finally, we also consider the *Arnold cat map*: $x_{t+1} = x_t + y_t \mod(1), y_{t+1} = x_t + 2y_t \mod(1)$. At odds with previous dissipative maps, this is an example of a *conservative* (measure-preserving) chaotic system with integer Kaplan-Yorke dimension (Sprott & Rowlands, 2001). The map has two Lyapunov exponents which coincide in magnitude $\lambda_1 = \ln(3 + \sqrt{5})/2 = 0.9624$ and $\lambda_2 = \ln(3 - \sqrt{5})/2 = -0.9624$. This implies that the amount of information created in the forward process (λ_1) is equal to the amount of information created in the backwards process $(-\lambda_2)$, therefore the process is time reversible. $D[P_{\text{out}}(k)||P_{\text{in}}(k)]$ for a time series of this map asymptotically tends to zero with series size, and the same happens with the degree-degree distributions (see table 1). This correctly suggests that albeit chaotic, the map is statistically time reversible.

3.5.5 Robustness: Irreversible chaotic series polluted with noise

Standard time series analysis methods evidence problems when noise is present in chaotic series. Even a small amount of noise can destroy the fractal structure of a chaotic attractor and mislead the calculation of chaos indicators such as the correlation dimension or the Lyapunov exponents (Kostelich & Schreiber, 1993). In order to check if our method is robust, we add an amount of white noise (measurement noise) to a signal extracted from a fully chaotic Logistic map ($\mu = 4.0$). The results for the KLD of the signal polluted with noise is significantly greater than zero, as it exceeds the one associated to the noise in four orders of magnitude, even when the noise reaches the 100% of the signal amplitude (Lacasa et al., 2011). Therefore our method correctly predicts that the signal is irreversible even when adding a large amount of noise.

4. Summary, perspectives and open problems

In this chapter a review on the state of the art of visibility algorithms as a method to make time series analysis through network theory has been presented. We have reported the properties of natural and horizontal visibility algorithms, and have explored their ability in several problems such as the estimation of Hurst exponent in self-similar (fractal series), the discrimination between uncorrelated, correlated stochastic and chaotic processes, the problem of noise filtering, the problem of determining the amount of irreversibility (i.e. entropy production) of a system, or the generic study of nonlinear systems as they undergo a period-doubling route to chaos.

Before commenting on the plethora of applications and challenging open problems to be faced, a few words on how to be cautious and make good science should be stated. The simplicity and straightforwardness of a method can be tricky, since they could convey the wrong impression to directly produce results when applied to concrete problems. From a physical point of view, the practical interest of this method lies in its ability to reveal properties of the system under study, i.e. to reveal hidden structures in a given series. But this capacity is intimately linked to the strength and extent of the theory behind the method. That is why, before venturing to study complex systems in nature, a method should provide a sufficient theoretical support. In the case under study, it should be clearly stated what information and which properties we are mapping into what and how, before attempting to measure all kind of features in a visibility graph.

According to this, the first general open problem lies just there: to generate a mathematically sound, rigorous theory that explains and shows how time series/dynamical systems properties are mapped into the associated visibility graph. In this review we have outlined the first steps in this direction, but a broad and general theory is still to be completely developed. This theory should deal with questions such as (i) what concrete information are the algorithms mapping? and (ii) how they do so? Once we know this, we can understand what network features are behind multifractality, spatio-temporal chaos, intermittency, quasi-periodicity, and many other complex dynamical processes.

Only when these questions have been rigorously responded, this tool could be ready to be unambiguously used by practitioners, since visibility algorithms will be a new and universal method to extract information from complex signals. Moreover, the possibility of defining mesoscale measures, which are typically network-based (for instance, modularity, community structure, etc), could be of interest to analyze non-local / multiscale dynamics. The potentials of the method could then apply to study long standing problems in Physics and Society, such as turbulence, stock market dynamics, or physiological signals such as electro-encephalogram, electro-cardiograms, and so on. On this respect, the initial naive approaches in those directions (turbulence (Liu et al., 2009), financial series (Liu et al., 2009; Yang et al., 2009), cardiac series (Shao, 2010)) are nowadays inconclusive because the theory behind the method is not fully developed. Eventually.

5. References

D. Andrieux, P. Gaspard, S. Ciliberto, N. Garnier, S. Joubaud, and A. Petrosyan, Entropy production and time asymmetry in nonequilibrium fluctuations, *Phys. Rev. Lett.* 98, 150601 (2007).

Boccaletti S, Latora V, Moreno Y, Chavez M, & Hwang DU (2006) Complex networks: Structure and dynamics. Phys. Rep. 424, 175.

Buchner T. and Zebrowski J.J., Logistic map with a delayed feedback: Stability of a discrete time-delay control of chaos. *Phys. Rev. E* 63, 016210 (2000).

C. Cammarota and E. Rogora, Time reversal, symbolic series and irreversibility of human heartbeat *Chaos, Solitons and Fractals* 32 (2007) 1649-1654.

Campanharo, Andriana S. L. O., Sirer, M. Irmak, Malmgren, R. Dean, Ramos, Fernando M. & Amaral, Luís A. Nunes (2011). Duality between Time Series and Networks. *PLoS ONE*, Vol. 6, No. 8, e23378.

A. Carbone, *Phys. Rev. E* 76, 056703 (2007).

Cencini M., Cecconi F., and Vulpiani A., *Chaos: From Simple Models to Complex Systems*, World Scientific (2010).

M. Costa, A.L. Goldberger, and C.-K. Peng, Broken Asymmetry of the Human Heartbeat: Loss of Time Irreversibility in Aging and Disease, *Phys. Rev. Lett.* 95, 198102 (2005).

M.D. Costa, C.K. Peng and A.L. Goldberger, Multiscale Analysis of Heart Rate Dynamics: Entropy and Time Irreversibility Measures, *Cardiovasc Eng* 8, (2008)

T.M. Cover and J.A. Thomas, *Elements of Information Theory* (Wiley, New Jersey, 2006).

Crutchfield JP, Farmer JD & Huberman BA (1982) Fluctuations and simple chaotic dynamics. Phys. Rep. 92, 2.

C.S. Daw, C.E.A. Finney, and M.B. Kennel, Symbolic approach for measuring temporal 'irreversibility', *Phys. Rev. E* 62, 2 (2000).

C. Diks, J.C. van Houwelingen, F. Takens, and J. DeGoede, Reversibility as a criterion for discriminating time series, *Phys. Lett. A* 201 (1995), 221-228.

Donner, R. V., Zou, Y., Donges, J. F., Marwan, N. & Juergen Kurths (2010). Recurrence networks - A novel paradigm for nonlinear time series analysis. *New Journal of Physics*, 12, 033025.

Donner, R. V., Small, M., Donges, J. F., Marwan, N., Zou, Y., Xiang, R. & Juergen Kurths (2010). Recurrence-based time series analysis by means of complex network methods. *International Journal of Bifurcation and Chaos*, Vol. 21, No. 4, 1019–1046.

P. Gaspard, Time-reversed dynamical entropy and irreversibility in markovian random processes, *J. Stat. Phys.* 117 (2004).

A.L. Goldberger *et al.*, *Circulation* 101, 23 (2000) 215-220.

A.R. Goldenberger *et al.*, *Proc. Natl. Acad. Sci. USA* 99, 1 (2002), 2466-2472.

Gutin, G., Mansour, T. & Severini, S. (2011). A characterization of horizontal visibility graphs and combinatorics on words. *PHYSICA A*, Vol. 390 (12), 2421–2428.

Haraguchi, Y., Shimada, Y., Ikeguchi, T. & Aihara, K. (2009). Transformation from complex networks to time series using classical multidimensional scaling, *Proceedings of the 19th International Conference on Artificial Neural Networks*, Heidelberg, Berlin, ICANN 2009, Springer-Verlag.

J.M. Hausdorff *et al., J. App. Physiol.* 80 (1996) 1448-1457.

J. W. Kantelhardt, Fractal and multifractal time series, in: *Springer encyclopaedia of complexity and system science* (in press, 2008) preprint arXiv:0804.0747.

Kants H. and Schreiber T. *Nonlinear Time Series Analysis*, (Camdrige University Press, 2003).

T. Karagiannis, M. Molle and M. Faloutsos, *IEEE internet computing* 8, 5 (2004) 57-64.

R. Kawai, JMR Parrondo and C Van den Broeck, Dissipation: the phase-space perspective, *Phys. Rev. Lett.* 98, 080602 (2007).

MB Kennel, Testing time symmetry in time series using data compression techniques, *Phys. Rev. E* 69, 056208 (2004).

E.J. Kostelich and T. Schreiber, *Phys. Rev. E* 48, 1752 (1993).

Lacasa L., Luque B., Ballesteros F., Luque J. & Nuño J.C. (2008). From time series to complex networks: the visibility graph. *Proc. Natl. Acad. Sci. USA* 105, 13, 4972-4975.

Lacasa L., Luque B., Nuño J.C. & Luque J. (2009). The Visibility Graph: a new method for estimating the Hurst exponent of fractional Brownian motion. *EPL* 86, 30001 (2009).

Lacasa, L. & Toral, R. (2010). Description of stochastic and chaotic series using visibility graphs. *Phys. Rev. E* 82, 036120 (2010).

Lacasa, L., Núñez, A. M., Roldan, E., Parrondo, J.M.R., & Luque, B. (2011).Time series irreversibility: a visibility graph approach. *arXiv:1108.1691* 2011.

Liu, C., Zhou, W-T. & Yuan, W-K. (2009). Statistical properties of visibility graph of energy dissipation rates in three-dimensional fully developed turbulence. *Physics Letters A*, Vol. 389 (13), 7.

Luque B., Lacasa L., Balleteros F., & Luque J. (2009). Horizontal visibility graphs: exact results for random time series. *Phys Rev E* 80, 046103 (2009).

B. Luque, L. Lacasa, F.J. Ballesteros & A. Robledo (2011). Feigenbaum graphs: a complex network perspective of chaos *PLoS ONE* 6, 9 (2011).

B.B Mandelbrot & J.W Van Ness (1968). Fractional Brownian Motions, Fractional Noises and Applications. *SIAM Review* 10, 4 (1968) 422-437.

Marvel, S.A., Mirollo, R.E. & Strogatz, S. H. (2009) Identical phase oscillators with global sinusoidal coupling evolve by Möbius group action. Chaos 19, 043104.

J. Mielniczuk and P. Wojdyllo, *Comput. Statist. Data Anal.* 51 (2007) 4510-4525.

Núñez, A. M., Lacasa, L., Valero, E., Gómez, J.P. & Luque, B. (2010). Detecting series periodicity with horizontal visibility graphs. *International Journal of Bifurcation and Chaos*, in press, 2010.

Newmann MEJ & Watts DJ (1999) Renormalization group analysis of the small-world network model. Phys. Lett. A 263:341-346.

M.E.J. Newmann, The structure and function of complex networks, *SIAM Review* 45, 167-256 (2003).

JMR Parrondo, C Van den Broeck and R Kawai, Entropy production and the arrow of time, *New. J. Phys.* 11 (2009) 073008.

Parthasarathy S, Mehta S., and Srinivasan S. (2006). Robust Periodicity Detection Algorithms, Proceedings of the 15th ACM international conference on Information and knowledge management.

Peitgen H.O., Jurgens H., and Saupe D. *Chaos and Fractals: New Frontiers of Science*, Springer-Verlag, New York.(1992).

B. Pilgram and D.T. Kaplan, *Physica D* 114 (1998) 108-112.

B. Podobnik, H.E. Stanley, *Phys. Rev.Lett.* 100, 084102 (2008).

Radicchi F, Ramasco JJ, Barrat A, & Fortunato S (2008) Complex Networks Renormalization: Flows and Fixed Points. Phys. Rev. Lett. 101, 148701.

E. Ravasz, A.L. Somera, D.A. Mongru, Z.N. Oltvai, A.-L. Barabasi, *Science* 297, 1551 (2002).

Robledo, A. (1999) Renormalization group, entropy optimization, and nonextensivity at criticality. Phys. Rev. Lett. 83, 12.

E. Roldan and JMR Parrondo, Estimating dissipation from single stationary trajectories, *Phys. Rev. Lett.* 105, 15 (2010).

E. Roldan and JMR Parrondo, Dissipation and relative entropy in discrete random stationary states, *in preparation*.

Schroeder M (1991) Fractals, chaos, power laws: minutes from an infinite paradise. Freeman and Co., New York.

Schuster, H.G. (1988) Deterministic Chaos. An Introduction. 2nd revised ed, Weinheim: VCH.

Shao, Z-G. (2010). Network analysis of human heartbeat dynamics. *Applied Physics Letters*, Vol. 96 073703 (2010).

A. H. Shirazi, G. Reza Jafari, J. Davoudi, J. Peinke, M. Reza Rahimi Tabar & Muhammad Sahimi (2011).Mapping stochastic processes onto complex networks. *Journal of Statistical Mechanics: Theory and Experiment*, Vol. 2009, No. 07, P07046.

I. Simonsen, A. Hansen and O.M. Nes, *Phys. Rev.E* 58, 3 (1998).

Song C, Havlin S, & Makse HA (2005) Self-similarity of complex networks. Nature 433, 392.

Song C, Havlin S, & Makse HA (2006) Origins of fractality in the growth of complex networks. Nat. Phys. 2.

Sprott J.C., and Rowlands G., Improved correlation dimension calculation, *International Journal of Bifurcation and Chaos* 11, 7 (2001) 1865-1880.

Sprott J.C., High-dimensional dynamics in the delayed Hénon map. *EJTP* 12 (2006) 19-35.

Strogatz, S.H. (1994) Nonlinear dynamics and chaos. Perseus Books Publishing, LLC.

Strogatz, S.H. (2001) Exploring complex networks. Nature 410:268-276.

Strozzi, F., Zaldívar, J. M., Poljansek, K., Bono, F. & Gutiérrez, E. (2009). From complex networks to time series analysis and viceversa: Application to metabolic networks. *JRC Scientific and Technical Reports*, EUR 23947, JRC52892.

Van Kampen N.G., *Stochastic processes in Physics and Chemistry*, Elsevier, The Netherlands (2007).

Q. Wang, S. R. Kulkarni and S. Verdú, Divergence estimation of continuous distributions based on data-dependent partitions, *IEEE Transactions on Information Theory*, 51, 9 (2005).

G. Weiss, Time-reversibility of linear stochastic processes, *J. Appl. Prob.* 12, 831-836 (1975).

R. Weron, *Physica A* 312 (2002) 285-299.

Xu, X., Zhang, J. & Small, M. (2008). Superfamily phenomena and motifs of networks induced from time series. *PNAS*, Vol. 105, No. 50, 19601–19605.

A.C. Yang, S.S. Hseu, H.W. Yien, A.L. Goldberger, and C.-K. Peng, Linguistic analysis of the humean heartbeat using frequency and rank order statistics, *Phys. Rev. Lett.* 90, 10 (2003).

Yang, Y., Jianbo, W., Yang, H. & Mang, J. (2009). Visibility graph approach to exchange rate
    series. *PHYSICA A*, Vol. 388 (20), 4431–4437.
Zhang, J. & Small, M. (2006).Complex Network from Pseudoperiodic Time Series: Topology
    versus Dynamics. *PRL*, Vol. 96, 238701.

7

A Review on Node-Matching Between Networks

Qi Xuan[1], Li Yu[1], Fang Du[2] and Tie-Jun Wu[3]
[1]Zhejiang University of Technology
[2]Johns Hopkins University
[3]Zhejiang University
[1,3]China
[2]USA

1. Introduction

The relationships between individuals in various systems are always described by networks. Recently, the quick development of computer science makes it possible to study the structures of those super-complex networks in many areas including sociology (Xuan et al., 2009; Xuan, Du & Wu, 2010a), biology (Barabási & Oltvai, 2004; Eguíluz et al., 2005), physics (Dorogovtsev et al., 2008; Rozenfeld et al., 2010), etc., by the tools in graph theory. Interestingly, it was revealed that many of these complex networks in various areas present several similar topological properties, such as small-world (Watts & Strogatz, 1998), scale-free (Barabási & Albert, 1999), self-similarity (Motter et al., 2003), symmetry (Xiao et al., 2008), etc. In order to explain these properties, a large number of models have been proposed (Barabási & Albert, 1999; Li & Chen, 2003; Mossa et al., 2002; Watts & Strogatz, 1998; Xiao et al., 2008; Xuan, Du, Wu & Chen, 2010; Xuan et al., 2006; 2007; 2008). However, most of current researches still focus on understanding the relationships between individuals in a single system, while the inter-system relationships are always ignored.

One of such inter-system relationships may be caused by the fact that an individual may be active in different systems with different identities (Xuan & Wu, 2009), and this type inter-system relationships may further lead to the similar structures of different complex networks. For instance, an ancient protein may evolve into various homologous proteins in different species, a concept may be expressed by different words in different languages, and a person may be active in different communication networks with different identities represented by telephone numbers (Onnela et al., 2007) and email addresses (Newman et al., 2002), etc. Therefore, revealing the different identities of an individual in several different systems has practical significance in many areas (Xuan, Du & Wu, 2010b), e.g., revealing homogeneous proteins, auto-translating languages, inter-network filtrating information, and so on. Through describing complex systems by networks, these different tasks can be transferred to a common node-matching problem between different complex networks, and thus can be solved in the same framework.

However, since many real-world complex networks are always highly symmetric (Xiao et al., 2008), i.e., there are always large numbers of nodes sharing the same neighbors in a network, it seems quite difficult to distinguish them in one network only by comparing their topological properties (Costa et al., 2007), such as degrees, clustering coefficient and so on, not to mention matching them between different complex networks. Fortunately, the researchers of different areas can use their own dedicated methods, such as chemical (Cootes et al., 2007; Kelley et al., 2003), semantic (Giunchiglia & Shvaiko, 2004) and others, to reveal a part of matched nodes, although their high economical or computational cost makes it almost impossible to examine and compare each pair of nodes between different large-scale networks. Such extra information are certainly very useful in solving the node-matching problem between complex networks. Based on these findings, we first introduced two kinds of co-evolving models (Xuan, Du & Wu, 2010b; Xuan & Wu, 2009) to create interacting networks, which can help better understand the co-evolution of different systems. Such co-evolution results in some structural similarity between complex networks, which made it possible to design node-matching algorithms by adopting the structural information. With the reason that the selection of the pairwise matched nodes revealed a priori by the dedicated methods is somewhat controllable, we then proposed several revealed matched nodes selecting strategies to improve the performances of node-matching algorithms. Finally, based on the similarities between nodes of different networks calculated by their connections to several pairs of preliminarily revealed matched nodes, we provided three different node-matching algorithms, including the classical optimal matching algorithm (Kuhn, 2005; Munkres, 1957; Xuan & Wu, 2009) in graph theory, one-to-one and one-to-many iterative node-matching algorithms (Du et al., 2010; Xuan, Du & Wu, 2010b) to solve artificial and real-world node-matching problems.

This chapter will review the overall process that we defined and solved the node-matching problems between different networks. In the next section, the node-matching problem is defined, and two co-evolving network models as well as a real-world node-matching data set are introduced. In Section 3, several revealed matched nodes selecting strategies are provided in order to improve the performances of the subsequent node-matching algorithms. Then in Section 4, the similarities between nodes of different networks are defined and several node-matching algorithms are introduced and the experiments are implemented. Finally, the chapter is concluded in Section 5.

2. Definitions and data sets

2.1 Definitions

The node matching problem between two different networks are described as follows (Xuan, Du & Wu, 2010b; Xuan & Wu, 2009): the two networks under study are denoted by $G_1 = (V_1, E_1)$ and $G_2 = (V_2, E_2)$, where $V_i = v_1^i, \ldots, v_M^i$ and E_i represent the node set and the link set of network $i(i = 1, 2)$, respectively. Assume that there are $M(M \leqslant \min\{N_1, N_2\})$ pairs of matched nodes $v_j^1 \leftrightarrow v_j^2$ defined by $\{v_1^i, \ldots, v_M^i\} \subseteq V_i(i = 1, 2)$, while $P_r(P_r < M)$ pairs of them have been already revealed, named as revealed matched nodes and denoted by $\{v_1^i, \ldots, v_{P_r}^i\} \subset V_i(i = 1, 2)$. Then the problem is: can we design a method to find the other $M - P_r$ pairs of matched nodes in these two distinct networks by using the structural information of G_1 and G_2 and the revealed matched nodes? If we can design such a method and finally $P_c(Pc \leqslant M - Pr)$ pairs of them are revealed correctly, the matching precision ϕ then can be calculated by

$$\phi = \frac{P_c}{M - P_r}. \tag{1}$$

2.2 Co-evolution network models

In order to better understand the interactions between different systems and test the subsequent node matching algorithms, two co-evolution network models need to be first introduced, where the parameters are set to be $N_1 = N_2 = M = N$ for convenience. Generally, there are two ways to create a pair of interactional networks, as is shown in Fig. 1 (a) and (b), respectively, both of which may work in reality. Inspired by the evolution of organisms, the first way is that the pair of interactional networks G_1 and G_2 are evolved from a common original network; in other words, they are derived from the same network (obtained by some model) through random rewiring processes. And the other way is that the pair of interactional networks are derived from two independent networks by a random interacting process composed of the following two steps (Xuan & Wu, 2009):

- **Networks initialization:** Two networks G_1 and G_2 with N nodes respectively are created by the same rule, where all the nodes are randomly matched, i.e., N pairs of randomly matched nodes $v_i^1 \leftrightarrow v_i^2$ are provided.

- **Interaction:** if v_i^1 (or v_i^2) and v_j^1 (or v_j^2) is connected in G_1 (or G_2) while v_i^2 (or v_i^1) and v_j^2 (or v_j^1) is not connected in G_2 (or G_1), then connect v_i^2 (or v_i^1) and v_j^2 (or v_j^1) with probability η_1 (or η_2).

Here, the second way will be adopted to create pairs of tested artificial interactional networks.

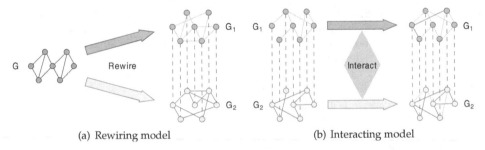

(a) Rewiring model   (b) Interacting model

Fig. 1. Two ways to create a pair of interactional networks (Xuan & Wu, 2009). (a) The pair of interactional networks G_1 and G_2 are derived from the same original network through random rewiring. The corresponding nodes are matched and connected by brown dashed lines. (b) The pair of interactional networks G_1 and G_2 are derived from a pair of independent networks by random interacting, i.e., two non-linked nodes in the network G_1 are connected by a green line with probability η_2 if their corresponding matched nodes in G_2 were linked while two non-linked nodes in G_2 are connected by a red line with probability η_1 if their corresponding matched nodes in G_1 were linked. η_1 and η_2 are named as interactional degree.

2.3 Real-world interactional networks

In reality, when two strangers chat with each other for some reason, e.g., demand of business, common interests, curiosity, warmheart, etc., they may be friends one day in the future if they enjoy with each other, in other words, the chat network may influence the evolution of the friendship network. On the other hand, there is also a natural trend that one prefers to chat with his friends or acquaintances rather than strangers, i.e., the friendship network determines the chat network to a certain extent. Therefore, chat network and friend network can be considered as a pair of real-world interactional networks, which can be figured on a quite large scale by advanced communication technologies and thus used to test the subsequent node matching algorithms.

As an example, we collected the communication records and the contact lists in a week from the database of *Alibaba trademanager* (an instant messenger (IM) mainly used for electronic commerce). We mainly focus on 14,800 employees of the *Alibaba* company and construct the chat network G_1 and the friendship network G_2 among them by these records. The two networks were then preprocessed by the following two steps (Du et al., 2010; Xuan, Du & Wu, 2010b):

- **Extract the giant cluster (GC):** Extract the GCs of G_1 and G_2, denoted by $G_1^g = (V_1^g, E_1^g)$ and $G_2^g = (V_2^g, E_2^g)$ where V_i^g and E_i^g represent the node set and the link set of the GC G_i^g respectively.

- **Calculate the intersection:** A pair of matched nodes in the networks correspond to the same *Alibaba* user. Select those users appearing in both the G_1^g and G_2^g, denoted by $V^c = V_1^g \cap V_2^g$, and get the sub-networks $G_1^c = (V^c, E_1^c)$ and $G_2^c = (V^c, E_2^c)$ where $E_i^c \subseteq E_i^g$ represents the set of links between nodes in V^c. Set $G_1 = G_1^c$ and $G_2 = G_2^c$, and terminate the preprocessing if both the networks G_1^c and G_2^c are connected, otherwise, turn to the first step.

After the preprocessing, both the networks G_1 and G_2 have 9859 nodes and are one-to-one matched, i.e., each node in G_1 has a matched node in G_2 and vice versa. Moreover, if there is a link between two nodes in G_1, we can find a link between their matched nodes in G_2 with probability 80.8%, and the probability is 18.4% from G_2 to G_1. Their basic topological properties, such as the number of nodes N, the average degree $\langle k \rangle$, the average clustering coefficient $\langle C \rangle$, and the average shortest path length $\langle L \rangle$ are presented in Table 1.

Networks	N	$\langle k \rangle$	$\langle C \rangle$	$\langle L \rangle$
Chat	9859	39.4	0.218	3.37
Friendship	9859	172	0.313	2.55

Table 1. The basic properties, i.e., the number of vertices N, the average degree $\langle k \rangle$, the average clustering coefficient $\langle C \rangle$, and the average shortest path length $\langle L \rangle$ for the chat network and the friendship network derived from *Alibaba trademanager* database (Du et al., 2010; Xuan, Du & Wu, 2010b).

3. Revealed matched nodes selecting strategies

Since the interactional networks under study are usually not completely identical (Xuan & Wu, 2009), it seems unpractical to match nodes between different networks just by their local structural properties. As a result, a few pairs of matched nodes would be better revealed as references before the node-matching algorithms are implemented.

Recent studies on real-world networks reveals that many of them have similar heterogeneous structure characterized by a power-law degree distribution (Barabási, 2009; Barrat et al., 2004; Eguíluz et al., 2005; Xuan et al., 2009). This property, first modeled by Barabási and Albert (BA) (Barabási & Albert, 1999), indicates that the connection of a heterogeneous network highly depends on hub nodes with quite large degrees, i.e., once these hub nodes are attacked, the average shortest path length of the network will increase quickly (Albert et al., 2000; Crucitti et al., 2004; Motter & Lai, 2002), as a result, the communication efficiency of the network will be largely weakened. For the node matching problem introduced here, we proved that (Xuan & Wu, 2009) such hub nodes can provide more structural information than those normal nodes and thus are more suitable to be revealed matched nodes. Based on the interactional model introduced in Fig. 1 (b), denoting the degree of v_i^1 by d_i^1 and the degree of v_j^2 by d_j^2, if they are randomly selected as a pair of matched nodes, then, averagely speaking, there are $d_i^1 d_j^2 / N$ other pairs of matched nodes around them before the interaction. And after the interaction, the degree of v_i^1 and that of v_j^2 can be calculated by Eq. (2) and Eq. (3) respectively,

$$\tilde{d}_i^1 = d_i^1 + d_j^2(1 - \frac{d_i^1}{N})\eta_2, \tag{2}$$

$$\tilde{d}_j^2 = d_j^2 + d_i^1(1 - \frac{d_j^2}{N})\eta_1. \tag{3}$$

And the number of pairs of other matched nodes around the matched nodes v_i^1 and v_j^2 after the interaction can be calculated by Eq. (4),

$$F_{ij} = d_j^2(1 - \frac{d_i^1}{N})\eta_2 + d_i^1(1 - \frac{d_j^2}{N})\eta_1 + \frac{d_i^1 d_j^2}{N}. \tag{4}$$

Since real-world complex networks always have a very huge number of nodes and a relatively small average degree, Eq. (2)-Eq. (4) can be further simplified to Eq. (5)-Eq. (7) respectively,

$$\tilde{d}_i^1 \approx d_i^1 + \eta_2 d_j^2, \tag{5}$$

$$\tilde{d}_j^2 \approx d_j^2 + \eta_1 d_i^1, \tag{6}$$

$$F_{ij} \approx \eta_1 d_i^1 + \eta_2 d_j^2. \tag{7}$$

Then we get Eq. (8) as

$$F_{ij} \approx \begin{cases} \frac{\eta_1(1-\eta_2)}{1-\eta_1\eta_2}\tilde{d}_i^1 + \frac{\eta_2(1-\eta_1)}{1-\eta_1\eta_2}\tilde{d}_j^2, & \eta_1\eta_2 < 1; \\ \frac{1}{2}\tilde{d}_i^1 + \frac{1}{2}\tilde{d}_j^2, & \eta_1\eta_2 = 1. \end{cases} \tag{8}$$

With the reason that the matched nodes are supposed unknown beforehand in reality, it seems unpractical to sort all the pairs of matched nodes by F_{ij} in descending order in order to improve the final matching precision ϕ, although larger F_{ij} corresponds to more pairs of unrevealed matched nodes around a pair of revealed matched nodes v_i^1 and v_j^2. Fortunately, Eq. (8) suggests a substitute way, i.e., selecting nodes with larger degree in the reference network, revealing their matched nodes in the other network by some dedicated methods, then these pairs of matched nodes are set to the revealed matched nodes.

3.1 Large degree priority strategies

Based on this principle, we proposed large degree priority strategies (Xuan & Wu, 2009) for the optimal node matching algorithm, as described by

- **Large Degree Priority in G_1 (LDP1)**: G_1 is selected as the reference network, where the nodes are sorted by their degrees in descending order, and the top P_r of them as well as their matched nodes in G_2 are selected as the revealed matched nodes.
- **Large Degree Priority in G_2 (LDP2)**: G_2 is selected as the reference network, where the nodes are sorted by their degree in descending order, and the top P_r of them as well as their matched nodes in G_1 are selected as the revealed matched nodes.

But which of them can bring higher matching precision? Can we answer this question just by comparing the structural properties (in particular, the degree sequences) of the two interactional networks? Without loss of generality, for a pair of interactional networks, suppose G_1 has larger average degree than G_2, i.e., $\langle \widetilde{d^1} \rangle > \langle \widetilde{d^2} \rangle$. Multiply Eq. (5) by η_1 and minus Eq. (6), we get

$$\eta_1 \langle \widetilde{d^1} \rangle - \langle \widetilde{d^2} \rangle = (\eta_1 \eta_2 - 1)\langle d^2 \rangle. \tag{9}$$

Since $\eta_1 \eta_2 \leqslant 1$, the value of η_1 can be roughly estimated by

$$\eta_1 \leqslant \frac{\langle \widetilde{d^2} \rangle}{\langle \widetilde{d^1} \rangle}, \tag{10}$$

while the value of η_2 cannot be estimated just by comparing the structural properties of the interactional networks. Suppose that the nodes are sorted by their degrees in descending order, denote by $R^i (i = 1, 2)$ the set of top P_r nodes in G_i, then from Eq. (8), we can see that more structural information may be provided when G_2 is selected as the reference network, if it is satisfied that

$$\eta_1 \sum_{v_i^1 \in R^1} \widetilde{d_i^1} + (1 - \eta_1)P_r \langle \widetilde{d^2} \rangle < \eta_1 P_r \langle \widetilde{d^1} \rangle + (1 - \eta_1) \sum_{v_i^2 \in R^2} \widetilde{d_i^2}, \tag{11}$$

which is equivalent to

$$\eta_1 \Big(\sum_{v_i^1 \in R^1} \widetilde{d_i^1} - P_r \langle \widetilde{d^1} \rangle \Big) + (1 - \eta_1)\Big(P_r \langle \widetilde{d^2} \rangle - \sum_{v_i^2 \in R^2} \widetilde{d_i^2} \Big) < 0. \tag{12}$$

Because it is always satisfied that

$$\sum_{v_i^1 \in R^1} \widetilde{d_i^1} \geqslant P_r \langle \widetilde{d^1} \rangle, \quad \sum_{v_i^2 \in R^2} \widetilde{d_i^2} \geqslant P_r \langle \widetilde{d^2} \rangle, \tag{13}$$

Eq. (12) must be satisfied if we have

$$\frac{\langle \widetilde{d^2} \rangle}{\langle \widetilde{d^1} \rangle}\Big(\sum_{v_i^1 \in R^1} \widetilde{d_i^1} - P_r \langle \widetilde{d^1} \rangle \Big) + \Big(1 - \frac{\langle \widetilde{d^2} \rangle}{\langle \widetilde{d^1} \rangle}\Big)\Big(P_r \langle \widetilde{d^2} \rangle - \sum_{v_i^2 \in R^2} \widetilde{d_i^2} \Big) < 0. \tag{14}$$

where all the parameters are known when two interactional networks are provided. That is, only when Eq. (14) is satisfied, we can say that LDP2 may be superior to LDP1.

3.2 Centralized large degree priority strategies

The above LDP strategies are designed for optimal node-matching algorithms, while for iterative node-matching algorithms, these strategies need to be further modified. Because in this case, the revealed pairwise matched nodes would better be centralized to a local world in the networks so as to improve the matching precision in the first round, then the second round and so on. Correspondingly, we propose two centralized large degree priority strategies specially for iterative node-matching algorithms (Xuan, Du & Wu, 2010b):

- **Centralized Large Degree Priority in G_1 (CLDP1).** G_1 is selected as the reference network, where a set R_1 ($|R_1| = P_r$) of nodes are picked up according to their degrees by following process. The node of the largest degree in G_1 is firstly selected as the only member of R_1. Denoting the neighbor set of R_1 as U_1 ($U_1 \cap R_1 = \varnothing$), i.e., each node in U_1 (but none of the nodes in $V_1 \setminus (U_1 \cup R_1)$) is at least connected to one node in R_1, at each time the nodes in $V_1 \setminus R_1$ are sorted by the number of neighbors belonging to U_1 in descending order and the top one is selected to join in R_1. Update R_1 and U_1 and repeat the selecting process until the set R_1 contains exactly P_r nodes. Then the set R_1 of nodes in G_1 as well as their matched nodes in G_2 are selected as the revealed pairwise matched nodes.

- **Centralized Large Degree Priority in G_2 (CLDP2).** G_2 is selected as the reference network, where a set R_2 ($|R_2| = P_r$) of nodes are picked up according to their degrees by following process. The node of the largest degree in G_2 is firstly selected as the only member of R_2. Denoting the neighbor set of R_2 as U_2 ($U_2 \cap R_2 = \varnothing$), i.e., each node in U_2 (but none of the nodes in $V_2 \setminus (U_2 \cup R_2)$) is at least connected to one node in R_2, at each time the nodes in $V_2 \setminus R_2$ are sorted by the number of neighbors belonging to U_2 in descending order and the top one is selected to join in R_2. Update R_2 and U_2 and repeat the selecting process until the set R_2 contains exactly P_r nodes. Then the set R_2 of nodes in G_2 as well as their matched nodes in G_1 are selected as the revealed pairwise matched nodes.

4. Node-matching algorithms

4.1 Similarities between nodes of interactional networks

Name	Definition
Common Neighbors (Newman, 2001)	$S_{ij} = n_M(v_i^1, v_j^2)$
Salton Index (Salton & McGill, 1983)	$S_{ij} = \dfrac{n_M(v_i^1, v_j^2)}{\sqrt{n_L(v_i^1) \times n_L(v_j^2)}}$
Jaccard Index (Jaccard, 1901)	$S_{ij} = \dfrac{n_M(v_i^1, v_j^2)}{n_L(v_i^1) + n_L(v_j^2) - n_M(v_i^1, v_j^2)}$
Sørensen Index (Sørensen, 1948)	$S_{ij} = \dfrac{2n_M(v_i^1, v_j^2)}{n_L(v_i^1) + n_L(v_j^2)}$
Hub Promoted Index (Ravasz et al., 2002)	$S_{ij} = \dfrac{n_M(v_i^1, v_j^2)}{\min\{n_L(v_i^1), n_L(v_j^2)\}}$
Hub Depressed Index (Lü & Zhou, 2011)	$S_{ij} = \dfrac{n_M(v_i^1, v_j^2)}{\max\{n_L(v_i^1), n_L(v_j^2)\}}$

Table 2. Several definitions of similarities between nodes of interactional networks based on their local structural information.

The similarity between two nodes belonging to different networks can be measured by the number of pairs of revealed matched nodes around them, e.g., the number of common friends they contact with in different communication networks, where a common friend is denoted by a pair of revealed matched nodes in corresponding communication networks. Denote by $n_L(v_i^1)$ and $n_L(v_j^2)$ the numbers of links connected to the node v_i^1 and v_j^2 in the networks G_1 and G_2, respectively, and by $n_M(v_i^1, v_j^2)$ the number of pairs of revealed matched nodes (v_k^1, v_k^2) where v_i^1 and v_j^2 are mutually connected, i.e., v_i^1 is connected to v_k^1 and v_j^2 is connected to v_k^2, in the corresponding networks. Then the similarity between v_i^1 and v_j^2 can be calculated by a number of methods (Jaccard, 1901; Lü & Zhou, 2011; Newman, 2001; Ravasz et al., 2002; Salton & McGill, 1983; Sørensen, 1948), as presented in Table. 2. Here, we adopt Jaccard Index to calculate the similarities between nodes of interactional networks.

4.2 Optimal node-matching algorithm

When revealed pairwise matched nodes are selected by LDP strategies, the similarity of each pair of the remaining nodes from different interactional networks can be calculated by Jaccard Index. Then, reviewing the definitions in Section 2.1, the node-matching problem between G_1 and G_2 can be transferred to a maximum matching problem for the bipartite graph $G_b = (U_1, U_2, W)$ where $U_i = \{v_{P_r+1}^i, v_{P_r+2}^i, \dots, v_N^i\}$ $(i = 1, 2)$, and W denotes the set of links weighted by the similarities between these two groups of nodes. Without loss of generality, under the assumption $N_1 \leqslant N_2$, the task is to find a set of nonadjacent weighted links $\{w_1, w_2, \dots, w_{N_1-P_r}\}$ to maximize the sum of their weights $\sum_{i=1}^{N_1-P_r} s_i$, which can be solved by the classical KM algorithm (Kuhn, 2005; Munkres, 1957). Note that, although the KM algorithm was developed for the case $N_1 = N_2$, it could be also feasible in the case $N_1 < N_2$ through factitiously adding $N_2 - N_1$ isolated nodes in G_1. For this reason we supposed $N_1 = N_2 = N$ for simplicity.

Since the KM algorithm has relatively high complexity $O(N^3)$, the sizes of the test networks cannot be very large. Here the two interactional networks G_1 and G_2 are both created by the BA model with $N = 100$ nodes and average degree $\langle k \rangle = 8$. Then they interact with each other with different interactional degrees $\eta_1 = 0.9$ and $\eta_2 = 0.1$ by the model shown in Fig. 1 (b). Denote the sample ratio by $\gamma = P_r / N$, the matching results are shown in Fig. 2, where we can see that, in most cases, LDP1 is prior to LDP2. This result is reasonable because when $\eta_1 \gg \eta_2$, Eq. (8) suggests that larger F_{ij} can be expected when select those nodes with large degrees in G_1 and their correspondences in G_2 as the revealed matched nodes. Note that, in this experiment, we set $M = N$ for simplicity, that is, every node in one network has its correspondence in the other network. In reality, M may be smaller than N, i.e., some individuals may be active in only one of the interactional networks. In this case, we need further select $M - P_r$ pairs of matched nodes from $N - P_r$ pairs of matched nodes obtained by the node-matching algorithm. If the value of M is known a priori, we can simply sort $N - P_r$ pairs of matched nodes by their attached similarities, then select the top $M - P_r$ pairs with larger similarities as the final pairs of matched nodes. However, if M is unknown, we have to set a threshold $\theta \in [0, 1)$ beforehand, and those pairs of matched nodes with similarities larger than θ then are selected as the final pairs of matched nodes, which will not be further discussed here. That is, in the following studies, we always set $M = N_1 = N_2 = N$ for simplicity.

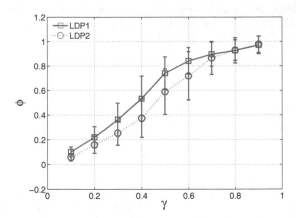

Fig. 2. The matching precision ϕ as the function of the sample ratio γ by adopting the two revealed matched nodes selection strategies, i.e., LDP1 and LDP2, for scale-free networks created by the BA model with $N = 100$ and $\langle k \rangle = 8$ and different interactional degrees $\eta_1 = 0.9$ and $\eta_2 = 0.1$ (Xuan & Wu, 2009). For each γ and each selection strategy, the experiment is implemented on 100 different pairs of scale-free networks, then the average matching precision as well as the error bar is recorded.

4.3 Iterative node-matching algorithm

As we can see in Fig. 2, the optimal node-matching algorithm fails to achieve acceptable results when there are only a relatively small number of pairwise matched nodes revealed beforehand, e.g., in order to achieve a matching precision of 80%, we have to reveal as many as 60% correspondences between nodes of the two networks in advance, which, as well as its long running time, hinders its efficient application in node-matching between real-world networks of quite large size. Based on the CDLP revealed matched nodes selecting strategies and Jaccard similarities between nodes of different networks, the iterative node-matching algorithm is simply composed of the following two steps (Xuan, Du & Wu, 2010b):

- **Node matching.** At each time, a pair of unmatched nodes belonging to different networks with the largest similarity are selected as a pair of matched nodes. Then this pair of matched nodes are considered as a pair of newly revealed matched nodes, then recalculate the similarities between the remaining nodes, and so forth.

- **Termination.** The iterative process is terminated when all of the nodes in the interactional networks have been matched.

The time complexity of the above node-matching algorithm mainly depends on the recalculation of the similarities. Generally, once a pair of nodes from different networks are matched at $(\tau - 1)$th round, we need to recalculated the similarities of about $k_\tau^1 k_\tau^2$ pairs of nodes mutually connected to that pair of matched nodes at τth round, where k_τ^i $(i = 1, 2)$ represents the degree of the matched node in G_i at $(\tau - 1)$th round. Provided $N_1 = N_2 = M = N$, the running time of the algorithm, denoted by Γ, can be calculated by Eq. (15) statistically,

$$\Gamma \sim E(\sum_{\tau=1}^{N} k_\tau^1 k_\tau^2).$$

(15)

If the two networks under study are strongly dependent each other, i.e., extremely G_1 and G_2 are identical and a node in one network only can be matched to the node of equal degree in the other network, Eq. (15) can be replaced by Eq. (16),

$$\Gamma \sim \sum_{\tau=1}^{N} E((k_\tau^1)^2). \tag{16}$$

For scale-free networks generated by the BA model, the degree distribution follows $p(k) \sim k^{-3}$, thus the running time can be simplified by Eq. (17),

$$\Gamma \sim N \int_1^N k^2 k^{-3} dk \sim N \ln N. \tag{17}$$

However, if the two target networks are relatively independent from each other, i.e., a node with large degree in one network can be matched to a node with small degree in the other network, which is more common in reality, Eq. (15) can be approximatively transferred to Eq. (18),

$$\Gamma \sim \sum_{\tau=1}^{N} E(k_\tau^1)E(k_\tau^2) \sim N\langle k^1 \rangle \langle k^2 \rangle, \tag{18}$$

where $\langle k^i \rangle$ represents the average degree of the network G_i. In most cases, $\langle k^i \rangle$ can be considered as a constant, therefore, Eq. (18) suggests a linear time complexity $O(N)$ of the algorithm (Xuan, Du & Wu, 2010b). Eqs. (17) and (18) mean that the iterative node-matching algorithm has much lower complexity than the optimal node-matching algorithm.

In order to compare to the optimal node-matching algorithm, here we take the same example to test the iterative node-matching algorithm. Since the iterative algorithm is able to solve node-matching problems between networks of quite large size, the two interactional networks G_1 and G_2 here are also created by the BA model with same average degree $\langle k \rangle = 8$, but much larger network size $N = 500$. Then these two networks interact with each other with different interactional degrees $\eta_1 = 0.9$ and $\eta_2 = 0.1$ by the same model shown in Fig. 1 (b). The matching results are show in Fig. 3 (a). At this time, in order to correctly reveal most of matched nodes in the networks (e.g., $\phi \geq 80\%$), we only need to have a very small percentage of matched nodes revealed beforehand (1% for CLDP1 and 1.6% for CLDP2), i.e., the iterative node-matching algorithm is far more efficient than the optimal node-matching algorithm on interactional artificial scale-free networks.

However, when we test this iterative node-matching algorithm on the real-world interactional chat network and friendship network introduced in Section 2.3, the matching results, as shown in Fig. 3 (b), are not that satisfactory, i.e. the final matching precision between the pair of real-world networks is much lower than that between the artificial networks generated by the BA model when adopting the same proportion of pairwise revealed matched nodes. For example, only about 40% matched nodes are revealed correctly, even though there are as many as 10% matched nodes are revealed beforehand. This phenomenon may be caused by the relatively high symmetry of the chat network and the friendship network. Generally, the local symmetry between the two non-linked nodes v_i and v_j in a network is defined by (Xuan, Du & Wu, 2010b)

$$\omega_{ij} = \frac{\chi_{ij}^c}{\chi_{ij}^t}, \tag{19}$$

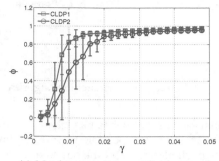
(a) Matching results on artificial networks

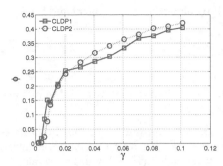
(b) Matching results on real-world networks

Fig. 3. The matching precision ϕ as the function of the sample ratio γ by adopting the two revealed matched nodes selection strategies, i.e., CLDP1 and CLDP2, for (a) the interactional scale-free networks created by the BA model with $N = 500$ and $\langle k \rangle = 8$ and different interactional degrees $\eta_1 = 0.9$ and $\eta_2 = 0.1$, and (b) the interactional real-world chat network and friendship network (Xuan, Du & Wu, 2010b). For artificial networks, the experiment is implemented on 100 different pairs of scale-free networks for each γ and each selection strategy, then the average matching precision as well as the error bar is recorded.

where χ_{ij}^c and χ_{ij}^t are the numbers of their common and total neighbors, respectively. If nodes v_i and v_j are connected, release the link and then calculate the symmetry between them following Eq. (8). Since it is impossible to distinguish two nodes v_i and v_j in a network with the symmetry $\chi_{ij} = 1$ (i.e. they share the same neighbors excluding themselves) just by adopting their topological information, those highly symmetric nodes in one network may be wrongly matched to the nodes in the other network with quite a high probability, and thus one-to-one node-matching algorithms may produce poor results in such situations.

4.4 One-to-many iterative node-matching algorithms

In order to overcome the above limitation of one-to-one node-matching algorithms, we proposed one-to-many node matching (Du et al., 2010) through expanding the number of nodes in each matching step. In fact, one-to-many node matching has its practical significance because it can help to quickly narrow down the searching range of a target individual in different complex systems. Particularly, a 1-to-M algorithm should output $N - P_r$ correspondences as defined by Eq. (20),

$$v_i^1 \to Q_i^2 = \{v_{i_1}^2, v_{i_2}^2, \ldots, v_{i_M}^2\}, \tag{20}$$

where v_i^1 $(i = P_r + 1, P_r + 2, \ldots, N)$ is a node in G_1, and Q_i^2 is a node set including the top M most likely matched nodes of v_i^1 in G_2. It should be noted that here 1-to-M match is just a natural generalization of 1-to-1 match, therefore, Eq. (20) also provides a consistent 1-to-1 match, i.e., $v_i^1 \leftrightarrow v_{i_1}^2$, satisfying that $v_{i_1}^2$ and $v_{j_1}^2$ represent two different nodes in G_2 if $i \neq j$. For a 1-to-M matching algorithm, a node v_i^1 in G_1 is considered correctly matched if its real matched node v_i^2 is contained in Q_i^2, i.e., $v_i^2 \in Q_i^2$. Denoting P_M $(P_M \leqslant N - P_r)$ as the number of nodes in G_1 that are correctly matched, the matching precision ϕ_M for the 1-to-M node matching algorithm can be calculated by Eq. (21), and naturally Eq. (22) is always satisfied.

$$\phi_M = \frac{P_M}{N - P_r},\tag{21}$$

$$\phi_M \geq \phi_{M-1}.\tag{22}$$

Next, we will introduce two different one-to-many iterative node-matching algorithms (Du et al., 2010).

1) **A1: Local mapping**. Since the similarity between each pair of nodes may change as the one-to-one iterative algorithm is implemented step by step, it is possible to correct some initially wrongly matched nodes by recalculating their similarities after the one-to-one node matching algorithm is terminated. This fact leads to the first one-to-many node matching algorithm based on local mapping. In particular, the Algorithm A1 is defined by the following two steps (Du et al., 2010):

- **Iterative 1-to-1 node matching**. $N - P_r$ pairs of nodes, i.e., $v_i^1 \leftrightarrow Q_i^2 = \{v_{i_1}^2\}$ ($i = P_r + 1, P_r + 2, \ldots, N$), are firstly matched by the iterative 1-to-1 node matching algorithm.
- **Candidate nodes selection**. Denote by X_i^1 the neighbor set of node v_i^1 in G_1, which has a matched node set X_i^2 in G_2 where the nodes are 1-to-1 matched to those in X_i^1, then denote by Y_i^2 the neighbor set of X_i^2, including all the nodes directly connected to those in X_i^2. Based on the definition of similarity, only the similarities between node v_i^1 ($i = P_r + 1, P_r + 2, \ldots, N$) and the nodes in Y_i^2 can be larger than 0 and thus are recalculated. Then the top $M - 1$ nodes with largest similarities are selected as the candidate corresponding nodes of v_i^1. It should be noted that $v_{i_1}^2$ is not reconsidered here, and if Y_i^2 only contains fewer than $M - 1$ nodes, other $M - 1 - |Y_i^2|$ nodes can be randomly selected from G_2 to be consistent with Eq. (20).

2) **A2: Ensembling**. In the area of machine learning, it is a common way to improve the generalization performance of an algorithm by combining the results of many different predictors (Breiman, 1996; Freund & Schapire, 1997; Krogh & Sollich, 1997; Miyoshi et al., 2005). However, the above iterative one-to-one node matching algorithm is totally deterministic, i.e., for a given pair of target networks and certain revealed matched nodes, the algorithm must produce the same matching result. Therefore, it cannot be directly used for ensemble, and thus a new statistical iterative one-to-one node matching algorithm have to be introduced first, where a pair of newly revealed matched nodes is adopted only with probability $p (p < 1)$ to calculate the similarities between those unrevealed nodes of different networks in the succeeding iterative process. Then a group of different one-to-one matching results can be obtained by implementing such a statistical iterative one-to-one node matching algorithm for several rounds, and the obtained results can be merged into a unique one-to-many matching result by a voting strategy. In particular, the algorithm A2 is defined by the following three steps (Du et al., 2010):

- **Iterative 1-to-1 node matching**. $N - P_r$ pairs of nodes, i.e., $v_i^1 \leftrightarrow Q_i^2 = \{v_{i_1}^2\}$ ($i = P_r + 1, P_r + 2, \ldots, N$), are firstly matched by the deterministic iterative 1-to-1 node matching algorithm.
- **Implement and vote**. The statistical 1-to-1 node matching algorithm with parameter $p (p < 1)$ is implemented for B ($B \gg M$) rounds and a group of B different 1-to-1 matching results are obtained. All of the correspondences in G_2 of v_i^1 in G_1 are grouped by a node set Z_i^2

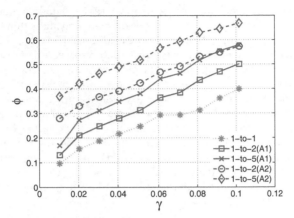

Fig. 4. The matching precision ϕ as the function of the sample ratio γ for $M = 1, 2, 5$ ($M = 1$ means the one-to-one matching result) between the friendship network and the chat network obtained from the database of *Alibaba trademanager* (Du et al., 2010). Here, the chat network is taken as the reference network.

with its size (the number of nodes) satisfying $|Z_i^2| \leq B$. It should be noted that each node in Z_i^2 is attached by a positive integer as its weight representing the times that it is matched to v_i^1 in the total B rounds, and similarly $v_{i_1}^2$ is excluded here.

- **Candidate nodes selection.** The top $M - 1$ nodes with largest weights in Z_i^2 are selected as the $M - 1$ candidate corresponding nodes of v_i^1. Sometimes, there may be only fewer than $M - 1$ nodes in Z_i^2, i.e., $|Z_i^2| \leq M - 1$, in such a situation, other $M - 1 - |Z_i^2|$ nodes can be randomly selected from G_2 to be consistent with Eq. (20).

Similarly, these two one-to-many iterative node matching algorithms are tested on the real-world interactional chat network and friendship network introduced in Section 2.3, and the matching results are shown in Fig. 4. As we can see, both the proposed one-to-many algorithms (especially the random algorithm A2) can significantly improve the matching precision, and thus can be considered to partially overcome the limitation of one-to-one node-matching algorithms.

5. Conclusion

Since an individual may appear in different systems with different identities, many real-world complex systems are considered to be interacted with each other all the time. Revealing these identities of the same individual is a common task in many areas such as sociology, linguistics, biology, etc, by their dedicated methods. When these complex systems are described by networks, this common task can be changed to a node matching problem between different complex networks, and thus can be solved in the framework of graph theory.

In this chapter, we reviewed the overall process to solve such node-matching problems between different networks: We first calculated the similarities between nodes of different networks through their connections to several pairs of preliminarily revealed matched nodes and transferred the node matching problem between two different networks to a maximum weighted bipartite matching problem; then we proposed several node-matching algorithms to solve such problem. By comparison, the iterative node-matching algorithm has approximately linear complexity and behaves much better than the traditional KM algorithm in graph theory. However, it seems that almost all of the network structure-based one-to-one node-matching algorithms lose their efficiencies when the target networks are highly symmetric, e.g., the iterative node-matching results are not that good on real-world chat network and friendship network obtained from the database of *Alibaba trademanager*. Such limitation can be partially overcome by the proposed one-to-many node-matching algorithms, which mainly focus on quickly narrowing down the searching range, rather than revealing exact one-to-one mapping between nodes of different networks. Meanwhile, we also introduced several degree-based revealed matched nodes selecting strategies for optimal and iterative node-matching algorithms, respectively, in order to further improve the matching results. In the future, more information about individuals and connections may be adopted to create more efficient node-matching algorithms.

6. References

Albert, R., Jeong, H., & Barabási, A.-L. (2000). Error and attack tolerance of complex networks, *Nature* 6794(406): 378–382.

Barabási, A.-L. (2009). Scale-free networks: A decade and beyond, *Science* 325(5939): 412–413.

Barabási, A.-L. & Albert, R. (1999). Emergence of scaling in random networks, *Science* 286(5439): 509–512.

Barabási, A.-L. & Oltvai, Z. N. (2004). Network biology: Understanding the cell's functional organization, *Nature* 5(2): 101–113.

Barrat, A., Barthélemy, M., Satorras, R. P. & Vespignani, A. (2004). The architecture of complex weighted networks, *Proceedings of the National Academy of Sciences U.S.A* 101(11): 3747–3752.

Breiman, L. (1996). Bagging predictors, *Machine Learning* 26(2): 123–140.

Cootes, A. P., Muggleton, S. H. & Sternberg, M. J. E. (2007). The identification of similarities between biological networks: Application to the metabolome and interactome, *Journal of Molecular Biology* 369(4): 1126–1139.

Costa, L. D. F., Rodrigues, F. A., Travieso, G. & Boas, P. R. V. (2007). Characterization of complex networks: A survey of measurements, *Advances in Physics* 56(1): 167–242.

Crucitti, P., Latora, V., Marchiori, M. & Rapisarda, A. (2004). Error and attack tolerance of complex networks, *Physica A: Statistical Mechanics and Its Applications* 340(1-3): 388–394.

Dorogovtsev, S. N., Goltsev, A. V. & Mendes, J. F. F. (2008). Critical phenomena in complex networks, *Review of Modern Physics* 80(4): 1275–1335.

Du, F., Xuan, Q. & Wu, T.-J. (2010). One-to-many node matching between complex networks, *Advances in Complex Systems* 13(6): 725–739.

Eguíluz, V. M., Chialvo, D. R., Cecchi, G. A., Baliki, M. & Apkarian, A. V. (2005). Scale-free brain functional networks, *Physical Review Letters* 94(1): 018102.

Freund, Y. & Schapire, R. E. (1997). A decision-theoretic generalization of on-line learning and an application to boosting, *Journal of Computer and System Sciences* 55(1): 119–139.

Giunchiglia, F. & Shvaiko, P. (2004). Semantic matching, *The Knowledge Engineering Review* 18: 265–280.

Jaccard, P. (1901). Étude comparative de la distribution florale dans une portion des alpes et des jura, *Bulletin de la Société Vaudoise des Science Naturelles* 37: 547–579.

Kelley, B. P., Sharan, R., Karp, R. M., Sittler, T., Root, D. E., Stockwell, B. R. & Ideker, T. (2003). Conserved pathways within bacteria and yeast as revealed by global protein network alignment, *Proceedings of the National Academy of Sciences U.S.A* 100(20): 11394–11399.

Krogh, A. & Sollich, P. (1997). Statistical mechanics of ensemble learning, *Physical Review E* 55(1): 811–825.

Kuhn, H. W. (2005). The hungarian method for the assignment problem, *Naval Research Logistics* 52(1): 7–21.

Li, X. & Chen, G. (2003). A local-world evolving network model, *Physica A: Statistical Mechanics and Its Applications* 328(1-2): 274–286.

Lü, L. & Zhou, T. (2011). Link prediction in complex networks: A survey, *Physica A: Statistical Mechanics and Its Applications* 390(6): 1150–1170.

Miyoshi, S., Hara, K. & Okada, M. (2005). Analysis of ensemble learning using simple perceptrons based on online learning theory, *Physical Review E* 71(3): 036116.

Mossa, S., Barthélémy, M., Stanley, H. E. & Amaral, L. A. N. (2002). Truncation of power law behavior in scale-free network models due to information filtering, *Physical Review Letters* 88(13): 138701.

Motter, A. E. & Lai, Y.-C. (2002). Cascade-based attacks on complex networks, *Physical Review E* 66(6): 065102.

Motter, A. E., Nishikawa, T. & Lai, Y.-C. (2003). Large-scale structural organization of social networks, *Physical Review E* 68(3): 036105.

Munkres, J. (1957). Algorithms for the assignment and transportation problems, *Journal of the Society for Industrial and Applied Mathematics* 5(1): 32–38.

Newman, M. E. J. (2001). Clustering and preferential attachment in growing networks, *Physical Review E* 64(2): 025102.

Newman, M. E. J., Forrest, S. & Balthrop, J. (2002). Email networks and the spread of computer viruses, *Physical Review E* 66(3): 035101.

Onnela, J.-P., Saramäki, J., Hyvönen, J., Szabó, G., Lazer, D., Kaski, K., Kertész, J. & Barabási, A.-L. (2007). Structure and tie strengths in mobile communication networks, *Proceedings of the National Academy of Sciences U.S.A* 104(18): 7332–7336.

Ravasz, E., Somera, A. L., Mongru, D. A., Oltvai, Z. N. & Barabási, A.-L. (2002). Hierarchical organization of modularity in metabolic networks, *Science* 297(5586): 1551–1555.

Rozenfeld, H. D., Song, C. & Makse, H. A. (2010). Small-world to fractal transition in complex networks: A renormalization group approach, *Physical Review Letters* 104(2): 025701.

Salton, G. & McGill, M. J. (1983). *Introduction to Modern Information Retrieval*, MuGraw-Hill, Auckland.

Sørensen, T. (1948). A method of establishing groups of equal amplitude in plant sociology based on similarity of species content and its application to analyses of the vegetation on danish commons, *Biologislce Skrifler* 5(4): 1–34.

Watts, D. J. & Strogatz, S. H. (1998). Collective dynamics of 'small-world' networks, *Nature* 393(6684): 440–442.

Xiao, Y., Xiong, M., Wang, W. & Wang, H. (2008). Emergence of symmetry in complex networks, *Physical Review E* 77(6): 066108.

Xuan, Q., Du, F. & Wu, T.-J. (2009). Empirical analysis of internet telephone network: From user id to phone, *Chaos* 19(2): 023101.

Xuan, Q., Du, F. & Wu, T.-J. (2010a). Partially ordered sets in complex networks, *Journal of Physics A: Mathematical and Theoretical* 43(18): 185001.

Xuan, Q., Du, F. & Wu, T.-J. (2010b). Partially ordered sets in complex networks, *Journal of Physics A: Mathematical and Theoretical* 43(39): 395002.

Xuan, Q., Du, F., Wu, T.-J. & Chen, G. (2010). Emergence of heterogeneous structures in chemical reaction-diffusion networks, *Physical Review E* 82(4): 046116.

Xuan, Q., Li, Y. & Wu, T.-J. (2006). Growth model for complex networks with hierarchical and modular structures, *Physical Review E* 73(3): 036105.

Xuan, Q., Li, Y. & Wu, T.-J. (2007). A local-world network model based on inter-node correlation degree, *Physica A: Statistical Mechanics and Its Applications* 378(2): 561–572.

Xuan, Q., Li, Y. & Wu, T.-J. (2008). Does the compelled cooperation determine the structure of a complex network?, *Chinese Physics Letters* 25(2): 363–366.

Xuan, Q. & Wu, T.-J. (2009). Node matching between complex networks, *Physical Review E* 80(2): 026103.

8

Techniques for Analyzing Random Graph Dynamics and Their Applications

Ali Hamlili
ENSIAS, Mohamed V–Souissi University, Rabat,
Morocco

1. Introduction

Graph theory is birth in 1736 with the publication of the work of the Swiss mathematician Leonhard Euler on the problem of finding a round trip path that would cross all the seven bridges of the city of Königsberg exactly once (Euler, 1736). Since then, this theory has known many important developments and has answered to a lot of practical issues. Today, the graph theory is considered as an essential component of discrete mathematics. It aims at analyzing the structure induced by interactions between a set of elements and to study the resulting fundamental properties. Graph theory occurs as a fundamental and theoretical framework for analyzing a wide range of the so-called *real-world networks* in biology, computer sciences, multi-agent systems, chemistry, physics, economy, knowledge management, and sociology. In many works, graph models are employed as constructive descriptions to represent and understand the behavior of different complexe systems (Molloy and Reed, 1998; Mieghem et al., 2000; Newman, 2003; Kawahigashi et al. 2005; Jurdak, 2007). In such models, the graph vertices stand for the components (nodes) of the network that encode information about the values of the state variables of the dynamical system and the edges represent the mutual relationships between the correspondent end-nodes. In practice, random graph theory has become increasingly important for modeling networks whose behaviors exhibit nondeterministic looks. In recent years, many significant results have used random graph models to explain, replicate and simulate the behavior of dynamic real-world networks (Hekmat and Van Mieghem, 2003; Kawahigashi et al., 2005; Durrett, 2006; Onat et al., 2008; Hewer et al., 2009; Hamlili, 2010; Trullols et al., 2010).

To provide a convenient way to represent and analyze dynamic networks by dynamic random graphs, it is very important to clarify how the model of random graphs should explain the behavior of change in the topology of the network. Thus, we introduce some stochastic processes (times of graph change, graph configurations, degree number at a chosen vertex ...) in order to attempt to account for the observed statistical properties in graph dynamics. Therefore, we will try to highlight the basic mathematical operations that transform a graph into other one to make possible describing the dynamic change of graph configurations. In this objective, different concepts and notation are introduced in the preliminary sections and will be used throughout the chapter. A reader familiar with the common topics in general graph theory may skip ahead. However, he may use it as necessary to refer to unknown definitions or unusual notations. Also, a particular attention is agreed to Erdös-Rényi's random graph model (Erdös and Rényi, 1960; Bollobàs, 2001).

2. Preliminary concepts

This section is a short introduction on graph theory. It will review the basic definitions and notation used throughout all this chapter.

2.1 Graphs

A classical graph is a static structure of a set of objects where some pairs of these objects are connected by one or several directed or undirected links. In this chapter, we assume that there is no multiple links between a pair of objects and the orientation of the links doesn't play a decisive role.

Definition 1

An undirected simple graph or simply a graph G can be defined as a pair $G = (V,E)$ of two sets: a nonempty set V of elements called vertices, and a set $E \subset \{(u,v) \in V^2 \ / \ v \neq u\}$ of unordered pairs of vertices. The elements of E are called edges.

Since the graph is undirected, (u,v) and (v,u) designate the same edge which we write simply uv. Furthermore, the assumption "simple" states the fact that between two given vertices, we cannot pass more than a single edge.

Definition 2

If $|S|$ denotes the cardinality of a set S, the number of vertices $N = |V|$ and the number of edges $M = |E|$ of a graph $G = (V,E)$ define respectively the order and the size of this graph.

Furthermore, we assume in this chapter that graphs can be finite or infinite according to their order and such that the sets of vertices and edges can't be jointly or separately empty.

Definition 2

Let $G = (V,E)$ and $G^* = (V^*,E^*)$ be two graphs. We say that G^* is a host for G or equivalently G is a subgraph of G^*, if and only if $V \subset V^*$ and $E \subseteq E^*$.

Definition 3

A graph $G = (V,E)$ is called a weighted graph if and only if a positive function (or weights) can be defined on the set of edges E.

Depending on the underlying area of application, such weights might represent probabilities, costs, lengths, capacities, or other positive quantities having a particular meaning. In the general weighted graph version, both vertices and edges can be weighted.

2.2 Neighbors, neighborhood and connectivity

Routing problems are among the oldest problems in graph theory. They are generally based on the hypothesis of connectivity. Let us note that the concepts of neighborhood and path are the most typical ideas associated to the connectivity assumption.

Definition 4

A neighbor of a vertex u is any vertex v such that $uv \in E$.

We note $N_G(u)$ the set of direct neighbors of u (also called the *neighborhood* of u). e.g.

$$N_G(u) = \{ v \in V \mid uv \in E \} \tag{1}$$

Inversely, two vertices u and v of a graph G are said to be *adjacent*, if $v \in N_G(u)$. The *closed neighborhood* of u is denoted $\overline{N}_G(u) = N_G(u) \cup \{u\}$ and for a set $S \subset V$, the closed neighborhood of S can be defined as $\overline{N}_G(S) = \bigcup_{u \in S} \overline{N}_G(u)$. By analogy, two edges are called neighbors if they have an end-vertex in common. In addition, pairwise non-adjacent vertices or edges are called independent. If all vertices of a subset $S \subseteq V$ are pairwise adjacent, S is called a complete subset or a clique.

Definition 5

On a given graph $G = (V, E)$ we can define a path $\Pi_{u,v}$ between a pair of vertices u and v if and only if there is a sequence of vertices (or walk) $u_i = u, u_{i+1}, \cdots, u_{j-1}, u_j = v$ such that

$$u_k u_{k+1} \in E, \forall k = i..j - 1$$

where the vertices u and v are called end-vertices of the path.

An *elementary path* is a path such that when all the vertices are sequentially visited, a same vertex is never met twice. A path such that the end-vertices coincide is called *cycle*. An *elementary cycle* is a cycle such that all the vertices have exactly two neighbors. The concept of path is behind the notion of connectivity. In the rest of this chapter, we note $\mathcal{P}_{u,v}$ the set of all paths between the vertices u and v.

Definitions 6

Let G be a simple graph,

i. *A pair of vertices (u, v) of G is called connected if G contains a path connecting u to v. Otherwise, they are called disconnected.*
ii. *A graph G is called connected if any pair of vertices of G is connected. Otherwise, it is called non-connected.*

To achieve a *fully connected* graph G, there must exist a path from any vertex to each other vertex in the graph.

The path between the source vertex and the destination vertex may consist of one hop when source and destination are neighbors or several hops if they aren't directly connected by an edge of G. The *hopcount* specifies the number of hops through a path between two vertices. This measure is meaningful only when there is a path between the source and the destination. The *average hopcount* of a graph is the average value of the hopcount between the end-vertices of all the possible paths.

Furthermore, in a non-connected graph there is no path between at least one source-destination pair of vertices. Hence, a non-connected graph consists of several disconnected clusters and/or vertices. Thus, routing is only possible between the different vertices of a same cluster.

Definitions 7

Consider a weighted graph G, (u,v) a pair of vertices of G and let w be the weight function defined on G, the weight assigned to a path $\Pi_{u,v}$ can be computed as the sum of weights assigned to its edges. i.e.

$$\Pi_{u,v} = \left\langle u_i = u, u_{i+1}, \cdots, u_{j-1}, u_j = v \right\rangle \Rightarrow w\left(\Pi_{u,v}\right) = \sum_{k=i}^{j-1} w\left(u_k u_{k+1}\right) \tag{2}$$

2.3 Matrix representation of graphs and degree function of vertices

The topological structure of the graph $G = (V,E)$ can alternatively be described by a $|V| \times |V|$ adjacency matrix $A_G = \left(a_{u,v}\right)_{u,v \in V}$ such that each entry is either 0 or 1

$$a_{u,v} = \begin{cases} 1 & \text{if} \quad v \in N(u) \\ 0 & \text{else} \end{cases} \tag{3}$$

where $a_{u,v} = 1$ signifies that uv is an edge of G. i.e. $uv \in E$.

Definition 8

The degree of a vertex u is the number of its direct neighbors

$$d_G(u) = |N_G(u)| \tag{4}$$

Proposition 1

Consider an undirected graph $G(V,E)$ and $A_G = \left(a_{u,v}\right)_{u,v \in V}$ its adjacency matrix, then

$$d_G(u) = \sum_{v \in V} a_{u,v} \tag{5}$$

The two last equations (4) and (5) are equivalent by definition of the matrix A_G. They induce a function d_G from V to $\mathbf{N}$ (the set of nonnegative integers) called the *degree function*. Particularly, a vertex of degree 0 is called an *isolated* vertex and a vertex of degree 1 is called a *leaf*.

3. Random graphs

Another theory of graphs began in the late 1950s. It was baptized *random graph theory* in several papers by Paul Erdös and Alfréd Rényi. As a real-world network model, the Erdös-

Rényi's random graph model has a number of attractive properties (Bollobàs, 2001). This model is exceptionally quantifiable; it allows an easy calculation of average values of the graph characteristics (Janson et al., 2000; Hamlili, 2010).

In this section, we want introduce in first a generalization of the concepts of the theory of random graphs. This generalization is intended to describe the issues in applicative frameworks where the number of the graph vertices can vary randomly (number of communicating entities in a wireless ad hoc network, number of routers in the Internet, etc). Also, we will show that most classical models, such as those of Erdös-Rényi random graphs and geometric random graphs can be derived as special cases of the model that we put forward as a generalized alternative.

3.1 Generalized random graph model

Intuitively, a generalized random graph representation can be defined in a simple way using the fully weighted graphs.

Definition 9

Consider a non-empty set Ω, *called the set of possible vertices and* $\mathbf{P} = \left(p_{u,v} \right)_{u,v \in \Omega}$ *a symmetric* $|\Omega| \times |\Omega|$ *matrix of probabilities. We call generalized random graph a graph G where each vertex u is generated with the probability* $p_{u,u} \in \left] 0,1 \right[$, *and where for all two existing vertices u and v, the edge uv is built with a probability* $p_{u,v} \in \left] 0,1 \right[$.

Let $\mathscr{G}_P(\Omega)$ be the collection of all possible graphs made on the set of possible vertices Ω such that the graph vertices u are generated independently with the probabilities $p_{u,u}$ and edges uv are built independently in $\Omega \times \Omega$ with the respective probabilities $p_{u,v}$. i.e.

$$\mathscr{G}_P(\Omega) = \left\{ G \mid \forall u \in \Omega : \Pr[u \in V] = p_{u,u} \wedge \forall u,v \in \Omega : v \neq u \Rightarrow \Pr[uv \in E] = p_{u,v} \right\}$$

This definition of random graph models is very general. We should note that, if G is a generalized random graph,

$$G = (V,E) \Rightarrow V \subseteq \Omega \wedge E \subseteq V \times V \tag{6}$$

As will be discussed later, this way of modeling a random graph will represent opportunities for characterizing complex situations where classical models such as Erdös-Rényi model are not satisfactory.

Definition 10

The extended adjacency matrix $A = \left(a_{u,v} \right)_{u,v \in \Omega}$ *associated to a graph* $G = (V,E)$ *of the model* $\mathscr{G}_P(\Omega)$ *is a* $|\Omega| \times |\Omega|$ *matrix such that $a_{u,v}$ is independent equal to 1 if the pair of vertices uv belongs to E knowing that u and v belong to V and 0 otherwise.*

3.2 Practical examples

Different particular cases can be identified and as stated above, in different contexts it may be useful to define the term random graph with different degrees of generality. Hence, the generalized model can describe random geometric graphs (Steele, 1997; Barabasi and Albert, 1999; Penrose, 1999). It suffices to consider G such that $V = \Omega$ and the edge probabilities

$$p_{u,v} = p\big(\|u - v\|\big) = \begin{cases} 1 & \text{if } 0 < \|u - v\| \le R \\ 0 & \text{otherwise} \end{cases} \tag{7}$$

In this model, $p_{u,v}$ depends on the Euclidean distance $\|u - v\|$ between the geometric points locating the vertices u and v.

$$\mathscr{G}_R(V) = \Big\{ G(V,E) \,\Big|\, E \subseteq V \times V \wedge \forall\, u,v \in V : \Pr(vu \in E) = 1 \text{ if } \|u - v\| \le R \text{ and } \Pr(vu \in E) = 0 \text{ if } \|u - v\| > R \Big\}$$

This model is very interesting and as such it can formalize the framework of mobile wireless ad hoc networks where the connectivity of the network depends on the geometric positions of the communicating nodes and a radio coverage range which is generally supposed the same for all the network nodes. In such networks, the random dynamics of the associated graph is induced by the mobility of nodes.

Another example is the random graph model initiated by Erdös and Rényi in the 1950s. This kind of graphs can be represented by a generalized model where the set of vertices is not random (it is constantly the same $V = \Omega$) and all the graph pairs of vertices are connected with the same probability $p_{u,v} = p$. Thus, let V a nonempty set of vertices, we can define the collection $\mathscr{G}_p(V)$ of all possible graphs made on the set of vertices V, such that the graph edges are built independently in $V \times V$ with a probability p. i.e. formally, we can write

$$\mathscr{G}_p(V) = \Big\{ G(V,E) \,\Big|\, E \subseteq V \times V \wedge \forall\, v,w \in V : v \neq w \Rightarrow \Pr[vw \in E] = p \Big\}$$

This model was be used in most areas of science and human activities in biology, chemistry, sociology, computer networks, manufacturing, etc. In a random Erdös-Rényi graph with N vertices, the edges are independently and randomly built with a probability p between the $N(N-1)/2$ possible edges of the full mesh graph. This definition builds the binomial model $\mathscr{G}_p(V)$ of random graphs, also referred as Erdös-Rényi model (Bollobàs, 2001).

Proposition 2

Consider a nonempty set V, a real p in $]0,1[$ *and a random graph G on V. Let* $N = |V|$ *be the order of G and M be the random order of G, then*

$$\Pr\big(M = m \,\big|\, G \in \mathscr{G}_p(V)\big) = p^m (1-p)^{\binom{N}{2} - m} \tag{8}$$

and

$$E[M] = p \frac{N(N-1)}{2} \tag{9}$$

Proof

The set $\mathscr{G}_p(V)$ has $\binom{N}{2}$ random graphs with equal probabilities. In consequence, each one

can be chosen with a probability equal to $\binom{N}{2}^{-1}$. Thus, on one hand,

$$\Pr\left(M = m \mid G \in \mathscr{G}_p(V)\right) = \binom{N}{2}^{-1} \binom{N}{2} p^m (1-p)^{\binom{N}{2}-m}$$

$$= p^m (1-p)^{\binom{N}{2}-m}$$

On the other hand, the number of edges M in a random Erdös-Rényi graph is a random variable with an average value equal to

$$E[M] = \sum_{m=0}^{N(N-1)/2} m\, p^m (1-p)^{\binom{N}{2}-m} = p\,\frac{N(N-1)}{2}$$

3.3 Degree distribution

As defined above, the degree function d_G on the graph vertices returns the number of vertices directly connected to the considered vertex. Thus, the *degree distribution* measures the local connectivity relevance of the graph vertices. In a generalized random graph model, the degree distribution can be set depending on the wished point of view. Its general form is defined by

$$p_k = \Pr\left[d_G(u) = k\right] \tag{10}$$

3.3.1 Binomial model and Poisson approximation

In particular, in the Enrdös–Rényi representation $\mathscr{G}_p(V)$, the theoretical degree distribution of any vertex u is defined by a binomial distribution (Bollobàs, 2001)

$$p_k = \binom{|V|-1}{k} p^k (1-p)^{|V|-1-k} \tag{11}$$

when the number of vertices $|V|$ is small. Otherwise, from the limit central theorem (LCT), the binomial distribution is approximately Gaussian with parameters

$$\begin{cases} \mu = (|V|-1)\,p & \text{(the mean)} \\ \sigma^2 = (|V|-1)\,p\,(1-p) & \text{(the variance)} \end{cases} \tag{12}$$

But the Gaussian distribution is continuous, while the binomial distribution is discrete. Thus, sometimes when $|V|$ is large and under certain special assumptions we prefer the Poisson approximation to the LCT approximation.

Proposition 3

When the graph order is large, the degree distribution is in the order of

$$p_k \approx e^{-\varsigma}\frac{\varsigma^k}{k!} \tag{13}$$

where p is small and $|V|$ is sufficiently large.

Proof

When the graph oder $|V|$ is large, we can also write the degree density as

$$\Pr\big[d(u)=k\big] = \binom{|V|-1}{k}\left[\frac{\varsigma}{|V|-1}\right]^k\left(1-\frac{\varsigma}{|V|-1}\right)^{|V|-1-k} \approx e^{-\varsigma}\frac{\varsigma^k}{k!}$$

where p is small and $|V|$ is sufficiently large. This shows the equation (13).

More generally, this approximation is known in probability theory as the De Moivre's Poisson approximation to the Binomial distribution. Indeed, the Poisson distribution can be applied whenever it is dealing with systems with a large number of possible events such that each of which is rare.

Thus, it is in the order of things to note that the main advantage of the Enrdös–Rényi models comes from the traceability of calculations and the simplicity of parameter estimation. A form or another of this model will be applied as and when it's required.

3.3.2 Power-law models

The traditional model of Erdös and Rényi is not a universal representation for all the random graph behaviors. Sometimes, in many natural scale-free networks (such as World Wide Web pages and their links, Internet, grid computer networking, etc), despite the randomness of the resulting graphs, experimental studies have revealed that the degree distribution can have a pure power-law tail (Albert et al., 1999; Barabàsi and Albert, 1999; Faloutsos et al. 1999)

$$p_k = \alpha\,k^{-\gamma} \tag{14}$$

where $\alpha > 0$ is a scaling constant and $\gamma > 0$ is a constant scaling exponent. Scale-free Barabàsi-Albert random graphs are generally built through a growth process combined to a profile of preferential attachment to existing vertices (Barabàsi and Albert, 1999). Although such graphs have a large number of vertices, the degree distribution deviates significantly from the Poisson law expected for classical random graphs (Barabàsi and Albert, 1999). Other forms of degree distribution with a power-law tail have been studied (Amaral et al., 2000; Newman et al., 2001)

$$p_k = C \ k^{-\gamma} \exp\left(-\frac{k}{\kappa}\right); \ \text{for} \ k > 1 \tag{15}$$

where C is a constant fixed by the requirement of normalization, γ is the constant scaling exponent, and κ is a typical degree size from which the exponential adaptation becomes significant.

In scale-free graphs, the parameters estimation of the degree distribution has to be obviously adapted to each specific case of power-law. In general, the distribution coordinates have to be converted into the logarithmic scales and then apply a method such as least-square method.

4. Random Graph Dynamics (RGD)

In classic graph theory a graph is simply a collection of objects connected to each other in some manner. This description is very restrictive. In fact, the notions of random graph theory have been introduced in the objective to produce better models and more complete tools to represent non deterministic looks of configurations of a dynamic network. Here again, the language of random graphs is used simply to relate the graph structure of the different network situations. This language must be completed by defining a number of indispensable operations in order to introduce a basic mathematical framework for graph dynamics modeling.

Definition 11

A dynamic graph is a graph such that its configuration (or topology) is subject to dynamic changes with time.

Hence, it goes without saying that the topology changes induced by the random graph dynamics are made through a number of fundamental operations that affect only the set of edges E.

4.1 Random change process

Generally, we can define several kinds of graph dynamics. The following contexts are the basic ways of defining this kind of behavior:

- Vertex-dynamic graph model: the set of vertices varies with time. In this context vertices may be added or deleted. However, we must be careful in this case to the edges such that an end-vertex is removed. They should just be deleted too.
- Edge-dynamic graph model: the set of edges E varies with time. Thus, edges may be added or deleted from the graph. In this case, there are no consequences to fear on the set of vertices.

- Vertex-weighted dynamic graph model: the weights on the vertices vary with time.
- Edge-weighted dynamic graph model: the weights on the edges vary with time.
- Fully-weighted dynamic graph model: the weights on both vertices and edges vary with time.

Thus, we consider at first a model of dynamic graphs that combines all these aspects together. A such random dynamic graph G can be defined as a stochastic graph process i.e. a collection of independent random graphs $\mathbf{G} = \{G_t | t \in I\}$ where the parameter t is usually assumed to be time and which take values in a set I which can be continuous or countable (finite or infinite).

In the widest sense, each graph G_t can belong to a different model $\mathscr{G}_{\mathbf{P}_t}(\Omega)$. The set of all possible states is called the state space. If the state space is discrete, we deal with a discrete state stochastic graph process, which is called a chain of graphs. The state space can also be continuous; we then deal with a continuous-state stochastic process. A similar classification can be made regarding whether the index set I is continuous leading to a continuous-time stochastic process or countable leading to a discrete-time stochastic process. Thus, a dynamic graph $\mathbf{G}$ is a representation which assumes that at any time t, there exists an instance $G_t = (V_t, E_t)$ that belongs to a model $\mathscr{G}_{\mathbf{P}_t}(\Omega)$.

To explain the changes in a dynamic random graph, we must often refer to events of presence, absence, addition, deletion, birth, death and structure of objects; where the term "object" refers to both vertices and edges of the graph. The following definitions clarify what do these events really mean and how can they be mathematically defined in the context of graph dynamics.

Definition 12

The weighted graphs $(G_t; \mathbf{P}_t)$ *thus defined are called instantaneous configurations of the random dynamic graph* **G**.

The random graph dynamics can be characterized by the ordered stochastic time process $T_0 = 0 < T_1 < T_2 < \cdots < T_{k-1} < T_k < \cdots$ where the configuration changes of the random dynamic graph G are operated. Thus, a marked random graph process is defined such that, at the kth time of change $T_k = t_k$, a $|\Omega| \times |\Omega|$ symmetric matrix $\mathbf{P}_k = \left(p_{u,v}^{(k)} \right)_{u,v \in \Omega}$ of probabilities is selected and a graph (configuration) $G_k = (V_k, E_k)$ is chosen according to the random graph model $\mathscr{G}_{\mathbf{P}_k}(\Omega)$.

In this model, the diagonal entries of the matrix $\mathbf{P}_k$ represent the probabilities to select, at time t_k, the vertices of V_k individually in Ω and the off-diagonal entries of this symmetric matrix represent the probabilities to activate the edges between two given vertices of V_k. Both the sets of vertices and edges in the graph configuration process are chosen as temporary. Thus, between two existing end-vertices u and v an edge is selected with the temporary probability $p_{u,v}^{(k)}$. Both the vertices and the edges of a dynamic random graph can be seen as subject to dynamic random changes.

This type of model is very interesting for modeling dynamic networks where the parameters are all temporary. As examples, we can mention routers and links of the Internet, friends in web social networking, communicating nodes in mobile wireless mobile ad hoc networks that use essentially the methods of broadcast.

4.2 Change operations

In this section, the studied approaches are classified according to their own definition in the context of graph dynamics. The RGD is said low when only few elements (number of vertices, number of edges, building edges probability, vertices clustering …) change over time. Otherwise, the dynamics is strong. Furthermore, there are several operations that build new graphs from old ones. They might be characterized through a number of descriptor events and basic transforms.

Definition 13 (Presence)

In a dynamic graph **G**, *an object is present at time t if it belongs to the instance* G_t *of* **G**.

Thus, a vertex u is present at time t, if and only if $u \in V_t$. Similarly, an edge uv is present at time t, if and only if $uv \in E_t$ knowing that both u and v belong to V_t.

Definition 14 (absence)

In a dynamic graph **G**, *an object is absent at time t, if it does not belong to the instance* G_t *of* **G**.

Operating dynamic changes on a random graph consists of a series of graph modifications (weight, vertex or edge adaptations). In contrast, the dynamicity analysis is more effective when all combination of additions and deletions of edges and vertices are taken into account.

Definition 15 (Addition)

An object appears in the dynamic graph **G**, *if it transits from the state absent to the state present.*

From an operational perspective, this definition should be clarified. Let u and v be two vertices of G such that uv is not an edge of G, the addition of the edge uv to the graph G is defined by the operation

$$G + uv = \left(V, E \cup \{uv\}\right) \tag{16}$$

Now, there are two ways to add a new vertex to a graph. One way is to add an isolated vertex to the graph G, i.e.

$$G \tilde{\oplus} u = \left(V \cup \{u\}, E\right) \tag{17}$$

and the other is to add a new vertex u which will be connected to an existing vertex v of the graph G, i.e.

$$G \hat{\oplus}_v u = \left(V \cup \{u\}, E \cup \{uv\}\right) \tag{18}$$

From the algorithmic point of view this last operation (18) can be seen as a composition of the two previous ones (16) and (17). Also, note that any instance $G = (V, E)$ of the model $\mathscr{G}_p(\Omega)$ has a host $G^* = (\Omega, E)$ which is defined by adding to the graph G isolated vertices taken in $\Omega - V$

$$G^* = G \underset{u \in \Omega - V}{\breve{\oplus}} u \tag{19}$$

Thus, we conceive that the adjacency matrix of G^* is obtained by completing the adjacency matrix of G by 0. This matrix will be called in the rest of the chapter *extended adjacency matrix of G*. The advantage of working with the host graph G^* can be viewed rather as freeing from the assumption that the set of vertices of a random dynamic graph varies with time. But the downside of this alternative is that the graph model can exhibit needlessly an excessively large order or incomplete information.

Proposition 4

Let $V \subset \Omega$ be two nonempty sets such that $G = (V, E)$ be a graph and $G^* = (\Omega, E)$ be a host for G. Then

$$G \widehat{\oplus}_v u = \left(G \breve{\oplus} u \right) + uv \tag{20}$$

Proof

This property results trivially from the definitions corresponding to the operations $\widehat{\oplus}$ and $\breve{\oplus}$.

Definition 16 (Deletion)

*An object disappears (or is deleted) from the dynamic graph **G**, if it transits from the state present to the state absent.*

In these terms, the edge deletion can be formalized as follow. Let u and v be two vertices of a graph $G = (V, E)$ such that uv is an edge of G, the deletion of the edge uv from the graph G is defined by the operation

$$G - uv = \left(V, E - \{uv\} \right) \tag{21}$$

On another hand, the vertex deletion can be defined

$$G \Delta u = \left(V - \{u\}, E - \bigcup_{v \in N(u)} \{uv\} \right) \tag{22}$$

Indeed, the deletion of the vertex u induces the deletion of all the edges having u as end-vertex.

Definition 17 (Birth)

*The birth of an object is the date of its first appearance in the dynamic graph **G**.*

Formally, the birth of a vertex u is the date τ_u^+ of its first occurrence

$$\tau_u^+ = \min\{t \in \mathbf{R} + | u \in V_t\} \tag{23}$$

and the birth of an edge e is the date τ_u^+ of its first occurrence

$$\tau_e^+ = \min\{t \in \mathbf{R} + | e \in E_t\} \tag{24}$$

Definition 18 (Death)

*Death of an element is the date for the last deletion of this element from the dynamic graph **G**.*

Thus, the death of a vertex is the date τ_u^- of its last occurrence

$$\tau_u^- = \max\{t \in \mathbf{R} + | u \in V_t\} \tag{25}$$

and the birth of an edge e is the date τ_e^- such that

$$\tau_e^- = \max\{t \in \mathbf{R} + | e \in E_t\} \tag{26}$$

Definition 19 (Structure)

A structure S of a graph consists in a dynamical set of elements that satisfies a given property.

A path, a click and a cluster are all examples of structures. Let us note that from the random viewpoint the succession of graph transforms and structure updates that convert a configuration of the dynamic graph in another one are not known in advance.

4.3 Graph topology changes in RGD

Remark that between two consecutive changes of the graph configuration recorded at T_k and T_{k+1}, the extended adjacency matrix $A^{(k)}$ of G_k remains unchanged all through the interval $[T_k, T_{k+1}[$. Thus, the dynamicity of the graph topology can be characterized by the variation of the extended adjacency matrix between T_{k-1} and T_k

$$\Delta A^{(k)} = A^{(k)} - A^{(k-1)} \tag{27}$$

where $A^{(k)}$ is the extended adjacency matrix of the current configuration of the graph at time $T_k = t_k$ and $A^{(k-1)}$ is the extended adjacency matrix of the previous configuration at time $T_{k-1} = t_{k-1}$ (with $T_{k-1} < T_k$).

4.3.1 Characterizing the change of the number of vertices

We can define for successive configurations of the graph, the number of new vertices, the number of lost vertices and the number of maintained vertices respectively at $T_k = t_k$ by

$$\alpha^{(k)} = \sum_{u \in \Omega} \mathbf{1}_{\left\{\Delta a_{u,u}^{(k)} > 0\right\}}, \quad \beta^{(k)} = \sum_{u \in \Omega} \mathbf{1}_{\left\{\Delta a_{u,u}^{(k)} = 0\right\}} \quad \text{and} \quad \gamma^{(k)} = \sum_{v \in \Omega} \mathbf{1}_{\left\{\Delta a_{u,u}^{(k)} < 0\right\}} \tag{28}$$

where $\mathbf{1}_S$ indicates the characteristic function of the set S and $\Delta a_{u,u}^{(k)}$ symbolizes the diagonal term of the matrix $\Delta A^{(k)}$ associated to the eventual vertices of the dynamic graph.

Proposition 5

Let $\mathbf{G}$ be a random dynamic graph following the model $\mathscr{G}_{P_t}(\Omega)$ such that the configuration change is characterized by the stochastic time process $T_0 = 0 < T_1 < T_2 < \cdots < T_{k-1} < T_k < \cdots$. Consider the extended adjacency matrices $A^{(k)} = \left(a_{u,v}^{(k)}\right)_{u,v \in \Omega}$ of $G_k = (V_k, E_k)$ and $\Delta A^{(k)}$ the matrix defined by the equation (27). *Then,*

$$\begin{cases} \Pr\left(\Delta a_{u,u}^{(k)} = 1\right) = p_{u,u}^{(k)}\left(1 - p_{u,u}^{(k-1)}\right) \\ \Pr\left(\Delta a_{u,u}^{(k)} = 0\right) = p_{u,u}^{(k)} p_{u,u}^{(k-1)} + \left(1 - p_{u,u}^{(k)}\right)\left(1 - p_{u,u}^{(k-1)}\right) \\ \Pr\left(\Delta a_{u,u}^{(k)} = -1\right) = \left(1 - p_{u,u}^{(k)}\right) p_{u,u}^{(k-1)} \end{cases} \tag{29}$$

Proof

Since all trials are independent, each of the three probabilities can be decomposed as follows

$$\begin{cases} \mathbf{Pr}\left(\Delta a_{u,u}^{(k)} = 1\right) = \mathbf{Pr}\left(\left\{a_{u,u}^{(k)} = 1\right\} \wedge \left\{a_{u,u}^{(k-1)} = 0\right\}\right) = \mathbf{Pr}\left(a_{u,u}^{(k)} = 1\right)\mathbf{Pr}\left(a_{u,u}^{(k-1)} = 0\right) \\ \mathbf{Pr}\left(\Delta a_{u,v}^{(k)} = 0\right) = \mathbf{Pr}\left[\left(\left\{a_{u,u}^{(k)} = 1\right\} \cap \left\{a_{u,u}^{(k-1)} = 1\right\}\right) \cup \left(\left\{a_{u,u}^{(k)} = 0\right\} \cap \left\{a_{u,u}^{(k-1)} = 0\right\}\right)\right] = \\ \quad = \mathbf{Pr}\left(\left\{a_{u,u}^{(k)} = 1\right\} \cap \left\{a_{u,u}^{(k-1)} = 1\right\}\right) + \mathbf{Pr}\left(\left\{a_{u,u}^{(k)} = 0\right\} \cap \left\{a_{u,u}^{(k-1)} = 0\right\}\right) = \\ \quad = \mathbf{Pr}\left(a_{u,u}^{(k)} = 1\right)\mathbf{Pr}\left(a_{u,u}^{(k-1)} = 1\right) + \mathbf{Pr}\left(a_{u,u}^{(k)} = 0\right)\mathbf{Pr}\left(a_{u,u}^{(k-1)} = 0\right) \\ \mathbf{Pr}\left(\Delta a_{u,u}^{(k)} = -1\right) = \mathbf{Pr}\left(\left\{a_{u,u}^{(k)} = 0\right\} \cap \left\{a_{u,u}^{(k-1)} = 1\right\}\right) = \mathbf{Pr}\left(a_{u,u}^{(k)} = 0\right)\mathbf{Pr}\left(a_{u,u}^{(k-1)} = 1\right) \end{cases}$$

and because $\Pr\left(a_{u,u}^{(k-1)} = 1\right) = p_{u,u}^{(k-1)}$, $\quad \Pr\left(a_{u,u}^{(k-1)} = 0\right) = 1 - p_{u,u}^{(k-1)}$, $\quad \Pr\left(a_{u,u}^{(k)} = 1\right) = p_{u,u}^{(k)}$ and $\Pr\left(a_{u,v}^{(k)} = 0\right) = 1 - p_{u,v}^{(k)}$ we have the result (29).

Corollary 6

Under the assumptions of the above proposition, and by supposing that the configuration process of the dynamic graph is homogeneous and that all the element of Ω can appear with the same probability π in a configuration of the dynamic random graph $\mathbf{G}$, the triplet $\left(\alpha^{(k)}, \beta^{(k)}, \gamma^{(k)}\right)$ follows a trinomial distribution of parameters

$$\left(|\Omega| ; \pi\left(1-\pi\right), \pi\left(1-\pi\right), \pi^2+\left(1-\pi\right)^2\right) \quad i.e.$$

$$\Pr\left(\alpha^{(k)}=n_1, \beta^{(k)}=n_2, \gamma^{(k)}=n_3\right)=\frac{|\Omega|!}{n_1! \, n_2! \, n_3!}\left[\pi\left(1-\pi\right)\right]^{n_1+n_2}\left[\pi^2+\left(1-\pi\right)^2\right]^{n_3}$$

$$(30)$$

Proof

On one hand, there are only three possible outcomes $\left\{\Delta a_{u,u}^{(k)}=1\right\}$, $\left\{\Delta a_{u,u}^{(k)}=0\right\}$ and $\left\{\Delta a_{u,u}^{(k)}=-1\right\}$ such that $\Pr\left(\Delta a_{u,u}^{(k)}=1\right)+\Pr\left(\Delta a_{u,u}^{(k)}=0\right)+\Pr\left(\Delta a_{u,u}^{(k)}=-1\right)=1$ and all trials are independent. On the other hand, we have $|\Omega|$ discrete trials (relating the different situations in the set Ω) and because the occurrence measures of graph vertex situations $\alpha^{(k)}$, $\beta^{(k)}$ and $\gamma^{(k)}$ are interrelated by the equation $\alpha^{(k)}+\beta^{(k)}+\gamma^{(k)}=|\Omega|$, this means that we deal with a multinomial distribution of $|\Omega|$ trials and from the previous proposition (proposition 5) three outcomes with respective probabilities

$$\begin{cases} \Pr\left(\Delta a_{u,u}^{(k)}=1\right)=\pi\left(1-\pi\right) \\ \Pr\left(\Delta a_{u,u}^{(k)}=0\right)=\pi^2+\left(1-\pi\right)^2 \\ \Pr\left(\Delta a_{u,u}^{(k)}=-1\right)=\left(1-\pi\right)\pi \end{cases}$$

This shows the required result in (30).

Following this line of thinking, these results can be applied only if the number of the observed vertices varies with time. That is when the dynamicity affects the vertices of the dynamic graph. Otherwise, while the number observed vertices remain unchanged, we will show similar results in the next subsection under the assumption of dynamicity of edges. These two approaches are complementary.

4.3.2 Characterizing the change of the number of edges

First and foremost let us remain under the assumption of the general model $\mathcal{G}_{P_t}\left(\Omega\right)$ of dynamic graphs. In the context of connectivity, the number of edges connected to the same vertex defines its degree. Thus, from the uncertain connectivity viewpoint of dynamic random graphs, we can define locally for each graph vertex the number of new neighbors, the number of lost neighbors and the number of maintained neighbors respectively at $T_k=t_k$ by

$$\nu_u^{(k)}=\sum_{v\in V_k} 1_{\left\{\Delta a_{u,v}^{(k)}>0\right\}}, \quad \mu_u^{(k)}=\sum_{v\in V_k} 1_{\left\{\Delta a_{u,v}^{(k)}=0\right\}} \quad \text{and} \quad \lambda_u^{(k)}=\sum_{v\in V_k} 1_{\left\{\Delta a_{u,v}^{(k)}<0\right\}} \quad (31)$$

where $\mathbf{1}_S$ indicates the characteristic function of the set S and $\Delta a_{u,v}^{(k)}$ indicates a generic entry of the matrix $\Delta A^{(k)}$. Essentially, the knowledge, step by step, of the evolution of the local metrics $v_u^{(k)}$ and $\lambda_u^{(k)}$ will allow us to determine a propagation equation of degree function at each vertex of the observed dynamic graph.

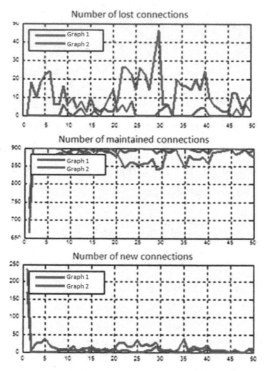

Fig. 1. Connectivity metrics change for two simulated random dynamic graphs

Proposition 7

Let G be a random dynamic graph following the model $\mathscr{G}_{P_t}(\Omega)$ and consider the configurations $(G_k)_{k \in \mathbf{N}}$ of G associated to the stochastic time process $T_0 = 0 < T_1 < T_2 < \cdots < T_{k-1} < T_k < \cdots$. If d_{G_k} denotes the underlying degree function to G_k

$$\forall u \in \Omega: \; d_{G_k}(u) = d_{G_{k-1}}(u) + v_u^{(k)} - \lambda_u^{(k)} \tag{32}$$

Proof

The proof of (32) results directly from the definitions of $v_u^{(k)}$ and $\lambda_u^{(k)}$.

We can also define global graph indices which reflect the global configuration changes of the dynamic random graph (see figure 1)

$$v^{(k)} = \sum_{v \in V_k} v_u^{(k)}, \quad \mu^{(k)} = \sum_{v \in V_k} \mu_u^{(k)} \quad \text{and} \quad \lambda^{(k)} = \sum_{v \in V_k} \lambda_u^{(k)} \tag{33}$$

Under the condition of conservation of the number of vertices from a configuration to a successor one, we can establish a number of results for dynamic graphs under Erdös-Rényi model constrains.

Proposition 8

Let **G** be a random dynamic graph following the model $\mathcal{G}_p(V)$ such that the configuration change is characterized by the stochastic time process $T_0 = 0 < T_1 < T_2 < \cdots < T_{k-1} < T_k < \cdots$. Consider the extended adjacency matrices $A^{(k)} = \left(a_{u,v}^{(k)}\right)_{u,v \in V_k}$ of $G_k = (V, E_k)$ and $\Delta A^{(k)}$ the matrix defined by the equation (27) and such that all the terms of its principal diagonal are equal to zero, then

$$\begin{cases} \Pr\left(\Delta a_{u,v}^{(k)} = 1\right) = p\left(1 - p\right) \\ \Pr\left(\Delta a_{u,v}^{(k)} = 0\right) = p^2 + \left(1 - p\right)^2 \\ \Pr\left(\Delta a_{u,v}^{(k)} = -1\right) = \left(1 - p\right)p \end{cases} \tag{34}$$

Proof

Since all trials are independent, each of the three probabilities can be decomposed such as follows

$$\begin{cases} \Pr\left(\Delta a_{u,v}^{(k)} = 1\right) = \Pr\left(\left\{a_{u,v}^{(k)} = 1\right\} \cap \left\{a_{u,v}^{(k-1)} = 0\right\}\right) = \Pr\left(a_{u,v}^{(k)} = 1\right)\Pr\left(a_{u,v}^{(k-1)} = 0\right) \\ \Pr\left(\Delta a_{u,v}^{(k)} = 0\right) = \Pr\left[\left(\left\{a_{u,v}^{(k)} = 1\right\} \cap \left\{a_{u,v}^{(k-1)} = 1\right\}\right) \cup \left(\left\{a_{u,v}^{(k)} = 0\right\} \cap \left\{a_{u,v}^{(k-1)} = 0\right\}\right)\right] = \\ = \Pr\left(\left\{a_{u,v}^{(k)} = 1\right\} \cap \left\{a_{u,v}^{(k-1)} = 1\right\}\right) + \Pr\left(\left\{a_{u,v}^{(k)} = 0\right\} \cap \left\{a_{u,v}^{(k-1)} = 0\right\}\right) = \\ = \Pr\left(a_{u,v}^{(k)} = 1\right)\Pr\left(a_{u,v}^{(k-1)} = 1\right) + \Pr\left(a_{u,v}^{(k)} = 0\right)\Pr\left(a_{u,v}^{(k-1)} = 0\right) \\ \Pr\left(\Delta a_{u,v}^{(k)} = -1\right) = \Pr\left(\left\{a_{u,v}^{(k)} = 0\right\} \cap \left\{a_{u,v}^{(k-1)} = 1\right\}\right) = \Pr\left(a_{u,v}^{(k)} = 0\right)\Pr\left(a_{u,v}^{(k-1)} = 1\right) \end{cases}$$

and because $\Pr\left(a_{u,v}^{(k)} = 1\right) = \Pr\left(a_{u,v}^{(k-1)} = 1\right) = p$ and $\Pr\left(a_{u,v}^{(k)} = 0\right) = \Pr\left(a_{u,v}^{(k-1)} = 0\right) = 1 - p$, we have the result (34).

Corollary 9

Under the assumptions of the above proposition, and by supposing that the random dynamic graph belongs to the $\mathcal{G}_p(V)$ model, the triplet $\left(v^{(k)}, \mu^{(k)}, \lambda^{(k)}\right)$ follows a trinomial distribution of parameters $\left(|V|(|V| - 1)/2 ; p\left(1 - p\right), p\left(1 - p\right), p^2 + \left(1 - p\right)^2\right)$. i.e.

$$\Pr\left(\nu^{(k)}=n_1,\,\mu^{(k)}=n_2,\,\lambda^{(k)}=n_3\right)=\frac{\binom{|V|}{2}!}{n_1!\,n_2!\,n_3!}\left[p\left(1-p\right)\right]^{n_1+n_2}\left[p^2+\left(1-p\right)^2\right]^{n_3}$$

(35)

Proof

First, note that all pairs of vertices u and v, there are only three possible independent outcomes $\left\{\Delta a_{u,v}^{(k)}=1\right\}$, $\left\{\Delta a_{u,v}^{(k)}=0\right\}$ and $\left\{\Delta a_{u,v}^{(k)}=-1\right\}$. Furthermore, we have

$$\Pr\left(\Delta a_{u,v}^{(k)}=1\right)+\Pr\left(\Delta a_{u,v}^{(k)}=0\right)+\Pr\left(\Delta a_{u,v}^{(k)}=-1\right)=1$$

and all trials are independent. Subsequently, we have $|V|$ discrete trials and since there are no other situations than those described above, the three measures of occurrence of graph edges $\nu^{(k)}$, $\mu^{(k)}$ and $\lambda^{(k)}$ are interrelated and verify

$$\nu^{(k)}+\mu^{(k)}+\lambda^{(k)}=|V|\left(|V|-1\right)/2$$

(36)

Each of the $|V|\left(|V|-1\right)/2$ trials (off-diagonal generic entries of the symmetric matrix $\Delta A^{(k)}$) matches to one of the three possible outcomes with respective probabilities

$$\Pr\left(\Delta a_{u,v}^{(k)}=1\right)=p\left(1-p\right),\;\Pr\left(\Delta a_{u,v}^{(k)}=0\right)=p^2+\left(1-p\right)^2\;\text{and}\;\Pr\left(\Delta a_{u,v}^{(k)}=-1\right)=\left(1-p\right)p\quad(37)$$

Thus, by taking into account the results established by the equation (36) and the system (37), we can conclude that, *the triplet* $\left(\nu^{(k)},\mu^{(k)},\lambda^{(k)}\right)$ *follows a trinomial distribution of parameters* $\left(|V|\left(|V|-1\right)/2;p\left(1-p\right),p\left(1-p\right),p^2+\left(1-p\right)^2\right)$.

Corollary 10

Under the assumptions of the proposition 8, the off-diagonal entries of the correlation matrix associated to $\left(\nu^{(k)},\mu^{(k)},\lambda^{(k)}\right)$ *are such that*

$$\rho\left(\nu^{(k)},\mu^{(k)}\right)=\rho\left(\lambda^{(k)},\mu^{(k)}\right)=-\sqrt{\frac{\left(1-p\right)^2+p^2}{2\left[\left(1-p\right)^2+p\right]}}\;\text{and}\;\rho\left(\lambda^{(k)},\nu^{(k)}\right)=-\frac{p\left(1-p\right)}{\left(1-p\right)^2+p}$$

(38)

Proof

The corollary results from the general properties of the correlation matrix of a trinomial distribution of parameters $\left(|V|\left(|V|-1\right)/2;p\left(1-p\right),p\left(1-p\right),p^2+\left(1-p\right)^2\right)$.

Note that the graph order $|V|\left(|V|-1\right)/2$ drop out of the off-diagonal entries of the correlation matrix associated to $\left(\nu^{(k)},\mu^{(k)},\lambda^{(k)}\right)$.

5. Parameter estimation of the random dynamic graph model

The ultimate objective in random graph dynamics analysis is the estimation of the graph model parameters. In some respects, nearly all types of dynamic changes can be interpreted in this way. In the empirical study, for reasons of traceability of the calculations, we restrict ourselves to an homogeneous model $\mathscr{G}_{P_t}(\Omega)$ such that, there exists a non-negative real

$p \in]0,1[$, for all existing pair of vertices u and v $p_{u,v}^{(k)} = p$.

Furthermore, we assume that by some means a vertex may deduct its neighborhood set directly from information exchanged as part of edge sensing. This section aims to show how the model parameters of the random dynamic graph can be estimated.

5.1 Dynamic graphs with known number of vertices and unknown edge probability

The main goal here is to find the maximum-likelihood estimate (MLE) of the edge probability p when the number of the graph vertices $N = |V|$ is known.

Proposition 11

Let $\mathbf{G}$ be a random dynamic graph following the model $\mathscr{G}_p(V)$ and note $\zeta = p|V|$ when p is small and $|V|$ is large. Consider D the random variable counting the degree in a fixed vertex u of $\mathbf{G}$ and let $D_1, D_2, \cdots, D_s$ be a s-sample formed by the degrees of u for s different configurations of the random dynamic graph at the graph change times $T_1 < T_2 < \cdots < T_s$. Then

$$\begin{cases} \hat{p}_{MLE} = \dfrac{1}{N-1}\bar{D} & \text{if } N \text{ is small} \\ \hat{\zeta}_{MLE} = \bar{D} & \text{if } N \text{ is large} \end{cases} \tag{39}$$

Proof

If the graph order $N = |V|$ is small, from equation (11), the log-likelihood function of the degree distribution in a given vertex u is

$$\ell(p,N;D_1,\cdots,D_s) = \sum_{i=1}^{s}\log\binom{N-1}{D_i} + \log\left(\frac{p}{1-p}\right)\sum_{i=1}^{s}D_i + s\,(N-1)\,\log(1-p)$$

The MLE of p is obtained by solving the partial differential equation $\dfrac{\partial\,\ell(p,N;D_1,\cdots,D_s)}{\partial p} = 0$

i.e.

$$\hat{p}_{MLE} = \frac{\sum_{i=1}^{s}D_i}{s\,(N-1)} = \frac{1}{N-1}\bar{D}$$

Now, if N is large, from equation (13), the log-likelihood function of the degree distribution in a given vertex u is $\ell(\zeta;D_1,\cdots,D_s) = \log(\zeta) \sum_{i=1}^{s} D_i - \sum_{i=1}^{s} \log(D_i!) - s\,\zeta$

The MLE of p is obtained by solving the partial differential equation $\dfrac{\partial\,\ell(\zeta;D_1,\cdots,D_s)}{\partial\,\zeta} = 0$ i.e.

$$\hat{\zeta}_{MLE} = \frac{\sum_{i=1}^{s} D_i}{s} = \overline{D}$$

When s observations of the change of the random dynamic graph topology are accomplished, the variations of $\hat{p}_{MLE}$ and $\hat{\zeta}_{MLE}$ displayed in one vertex or another relates the local knowledge on the values of these parameters taking into account the observed neighborhood behavior at u.

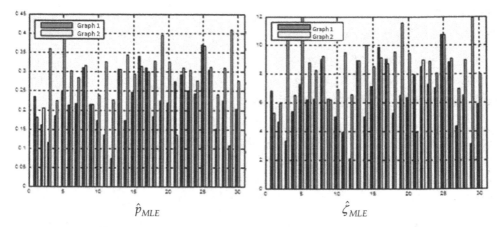

$$\hat{p}_{MLE} \qquad\qquad\qquad \hat{\zeta}_{MLE}$$

Fig. 2. $\hat{p}_{MLE}$ and $\hat{\zeta}_{MLE}$ evaluations for two simulated random graphs of order 30 vertices

Indeed, since $\hat{p}_{MLE}$ depends only on the observed number of neighbors of a selected vertex for different configurations of the random dynamic graph, this quantity is estimated differently in each vertex (see figure 2).

5.2 Dynamic graphs with unknown number of vertices and edge probability

In the statistical common sense, when the graph order is unknown the method proposed in the previous paragraph cannot be applied. Another resourceful method of point estimation called method of moments (Lehmann E. L. and Casella G., 1998) can be used when the number N of the graph vertices and the edge density parameter p are both unknown. Likewise, these parameters should be estimated on the basis of an s-sample $D_1,D_2,\cdots,D_s$ of the degree of a vertex u for s different instances of the random dynamic graph topology.

Proposition 12

Let G be a random dynamic graph following the model $\mathcal{G}_p(V)$ and suppose that $N = |V|$ is small. Consider D the random variable counting the degree in a fixed vertex u of G and let $D_1, D_2, \cdots, D_s$ be a s-sample formed by the degrees of u for s different configurations of the random dynamic graph at the graph change times $T_1 < T_2 < \cdots < T_s$. Then the estimates of moments of the graph parameters are respectively

$$\begin{cases} \hat{p}_{ME} = 1 - \dfrac{S^2}{\overline{D}} \\ \hat{N}_{ME} = 1 + \dfrac{\overline{D}^2}{\overline{D} - S^2} \end{cases} \tag{40}$$

Proof

In this case, the method of moments supposes that the empirical moment $\overline{D}$ is a natural estimate of the theoretical moment of order 1 $E(D)$ and the theoretical centralized moments of order k $\mu_k = E\left[D - (D)\right]^k$ can be estimated by their respective empirical centralized moments $M_k = \dfrac{1}{s}\sum_{i=1}^{s}\left(D_i - \overline{D}\right)^k$. When $N = |V|$ is small, the degree follows a binomial distribution of parameters p an $(N-1)$ and the method of moments in terms of the average and the variance of the observed degrees leads to the system of equations

$$\begin{cases} (N-1)\,p = \overline{D} \\ (N-1)\,p\,(1-p) = S^2 \end{cases} \tag{41}$$

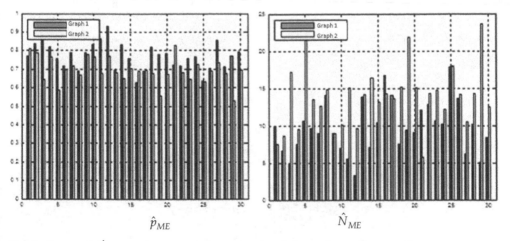

$$\hat{p}_{ME} \qquad\qquad\qquad \hat{N}_{ME}$$

Fig. 3. $\hat{p}_{ME}$ and $\hat{N}_{ME}$ evaluations for two simulated dynamic random graphs

It results from the resolution of the system of equations (41) that the moment estimates of the unknown graph parameters

$$\begin{cases} \hat{p}_{ME} = 1 - \dfrac{S^2}{\bar{D}} \\ \hat{N}_{ME} = 1 + \dfrac{\bar{D}^2}{\bar{D} - S^2} \end{cases}$$

Depending on the neighbors met by each of the random dynamic graph vertices, the evaluation of the estimated parameters will be different.

5.3 Expected degree number of vertices in wide random dynamic graphs

Various random dynamic graph problems (Internet, out vehicle networks, wide ad hoc networks ...) are analyzed and interpreted under the assumption that the order of the resulting graph may be relatively large (with some tens, hundreds or even thousands of vertices). Let ζ be the average degree number

$$\zeta = E(D) \tag{42}$$

As a result of scaling, since we use an approximation of the degree distribution, it is very important to work with good estimates. This shows that the resulting computed values are most likely not due merely to chance.

Proposition 13

Let **G** be a random dynamic graph following the model $\mathscr{G}_p(V)$ and suppose that $N = |V|$ is large. Consider D the random variable counting the degree in a fixed vertex u of **G** and let $D_1, D_2, \cdots, D_s$ be a s-sample formed by the degrees of u for s different configurations of the random dynamic graph at the graph change times $T_1 < T_2 < \cdots < T_s$. Let ζ be the Poisson distribution parameter of the degree. Then, the empirical moment of order 1 $\hat{\zeta} = \bar{D}$ is a good estimate of the average degree. i.e.

i. $\hat{\zeta}$ is unbiased.

ii. $\hat{\zeta}$ realizes the maximum of the likelihood function.

iii. $\hat{\zeta}$ is an efficient estimate of ζ.

Proof

The first property results from

$$E(\bar{D}) = \frac{1}{s} E\left(\sum_{i=1}^{s} D_i \right) = \frac{1}{s} \sum_{i=1}^{s} E(D_i) = E(D) = \zeta$$

because $D_1, D_2, \cdots, D_s$ are independent and identically distributed to D.

Furthermore, the log-likelihood function of a sample $D_1, D_2, \cdots, D_s$ representing the degrees of a given vertex u of G for s different configurations of the random dynamic graph is given by

$$\ell\left(\zeta ; D_1, \cdots, D_s\right) = -\sum_{i=1}^{s} \log D_i! + \sum_{i=1}^{s} D_i \log \zeta - s \zeta$$

The MLE is obtained by solving the partial differential equation $\dfrac{\partial \ell\left(\zeta ; D_1, \cdots, D_s\right)}{\partial \zeta} = 0$ i.e

$$\hat{\zeta} = \frac{\sum_{i=1}^{s} D_i}{s} = \bar{D}$$

which shows the proposition ii. It follows, on one hand,

$$\mathrm{var}\left(\hat{\zeta}\right) = \frac{1}{s^2} \sum_{i=1}^{s} \mathrm{var}\left(D_i\right) = \frac{1}{s} \mathrm{var}\left(D\right) \quad \text{(because are independent and identically distibuted to } D)$$

$$= \frac{\zeta}{s} \qquad\qquad \text{(because } D \text{ is Poisson distributed with parameter } \zeta)$$

and on the other hand,

$$\frac{\partial \ell\left(\zeta ; D_1, \cdots, D_s\right)}{\partial \zeta} = -s + \frac{\sum_{i=1}^{s} D_i}{\zeta} \quad \text{and} \quad \frac{\partial^2 \ell\left(\zeta ; D_1, \cdots, D_s\right)}{\partial \zeta^2} = -\frac{\sum_{i=1}^{s} D_i}{\zeta^2}$$

Thus, the Fisher information quantity of the parameter ζ is such that

$$I\left(\zeta\right) = E\left[-\frac{\partial^2 \ell\left(\zeta ; D_1, \cdots, D_s\right)}{\partial \zeta^2}\right] = \frac{\sum_{i=1}^{s} E\left(D_i\right)}{\zeta^2} = \frac{s}{\zeta} = \frac{1}{\mathrm{var}\left(\hat{\zeta}\right)} \qquad (43)$$

which is the lower bound of the Fréchet-Darmois-Cramer-Rao (FDCR) inequality.

Then, since $\hat{\zeta}$ is unbiased and verifies the lower bound of the FDCR inequality, it is an efficient estimate.

5.4 Routing performance metrics evaluation

5.4.1 Average hopcount estimation

Proposition 14

*Let **G** be a random dynamic graph following the model $\mathscr{G}_p\left(V\right)$ and suppose that $|V|$ is large and known. Consider D the random variable counting the degree in a fixed vertex u of **G** and let $D_1, D_2, \cdots, D_s$ be a s-sample formed by the degrees of u for s different configurations of the random dynamic graph at the graph change times $T_1 < T_2 < \cdots < T_s$. Then, the average hopcount MLE is approximately*

$$\hat{\eta}_{MLE} \approx \frac{\log |V|}{\log \overline{D}}$$

(44)

Proof

Let ζ be the average degree number. As each vertex in the random graph is connected to about ζ other nodes, after η hops, we expect that ζ^{η} vertices must be reached. Thus all the vertices are reached for the value η_V such that $\zeta^{\eta_V} \approx |V|$. Since $|V|$ and ζ are both known

constant parameters, the average hopcount $E[\eta_V] \approx \dfrac{\log |V|}{\log \zeta}$ in large random graphs is

$$E[\eta_V] \approx \frac{\log |V|}{\log \zeta}$$

(45)

From the proposition 11 and the equation (45), and because the MLE of a function of a parameter is equal to the function of the MLE of this parameter, the average hopcount MLE is made by

$$\hat{\eta}_{MLE} \approx \frac{\log |V|}{\log \overline{D}}$$

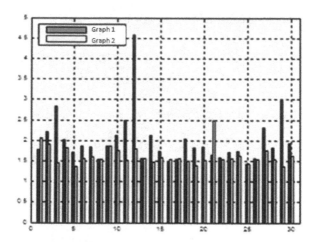

Fig. 4. Hopcount MLE evaluation for two simulated dynamic random graphs

This approximation holds because $\overline{D}$ is a good estimate of ζ in the strict sense of the proposition 13. Under other assumptions than those suggested in this paragraph, different authors have led to further forms of the approximation of the average hopcount number (Mieghem et al., 2000; Bhamidi et al., 2010).

5.4.2 Giant component size estimation

Let us consider **G** a random dynamic graph following the model $\mathscr{G}_p(V)$ and suppose that the graph order $|V|$ is large and known. From the theory of Erdős-Rényi graphs, we know (Molloy and Reed, 1998) that the graph will almost surely have a unique giant component containing a positive fraction of the graph vertices if $p > 1/|V|$ (see figure 5) i.e. the average degree numbet $\zeta > 1$.

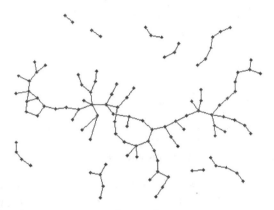

Fig. 5. Existence of the giant component of a graph

Proposition 15

*Let **G** be a random dynamic graph following the model $\mathscr{G}_p(V)$ and suppose that $|V|$ is large and known. Consider D the random variable counting the degree in a fixed vertex u of **G** and let $D_1, D_2, \cdots, D_s$ be a s-sample formed by the degrees of u for s different configurations of the random dynamic graph at the graph change times $T_1 < T_2 < \cdots < T_s$. Then, the average size MLE of the giant component is approximately*

$$\hat{\Theta} = 1 + \frac{1}{\bar{D}} \text{LambertW}\left[-\bar{D}\exp(-\bar{D})\right]$$

(46)

Proof

Let Θ be the giant component size, for large order random graphs, Θ is a solution of the equation

$$\Theta = 1 - \exp(-\zeta\,\Theta)$$

(47)

But, it is well known that there are two possible solutions to this equation

$$\begin{cases} \Theta_1 = 0 \\ \Theta_2 = 1 + \frac{1}{\zeta}\text{LambertW}\left[-\zeta\exp(-\zeta)\right] \end{cases}$$

(48)

where LambertW is the classic "Lambert W " function. Since only the non-zero solution is adequate, the giant component size increases as a function of the average degree ζ of the graph vertices G following the curve $\vartheta : z \mapsto 1 + \dfrac{1}{z}\text{LambertW}\left[-z\exp(-z)\right]$.

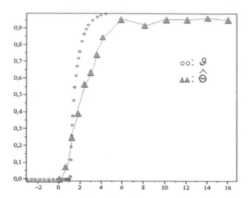

Fig. 6. Comparison of theoretic and empiric solutions ϑ and $\hat{\Theta}$

Thus, from the proposition 11 and because the MLE of a function of an unknown parameter is equal to the function of the MLE of this parameter, when $|V|$ is large the part of the graph occupied by the giant component can be estimated by

$$\hat{\Theta} = 1 + \frac{1}{\bar{D}}\text{LambertW}\left[-\bar{D}\exp(-\bar{D})\right]$$

Elementary dissimilarities exist between random graph models and real-world graphs. Real-world graphs show strong clustering, but Erdös and Rényi's model does not. Many of the graphs, including Internet and World-Wide Web graphs, show a power-law degree distribution (Albert et al., 1999). This means that only a small part of the graph vertices can have a large degree. In fact, Erdös-Rényi assumptions can imply strong consequences on the behavior of the graph (Newman, 2003).

6. Conclusions

In this chapter we have illustrated several basic tools for representing and analyzing dynamic random graph in a general context and a practical approach to estimate the parameters of classical models of such behaviors. The generalized model that we have proposed can describe not only classical real-world networks models but also situations with more complex constraints. In this model, random graph dynamics is outlined by introducing a random time process where the dynamic graph topology changes are recorded. Among the advantages of this model is the possibility to generate successive independent random graphs with potentially different sets of vertices but all belonging to the same basic set of vertices. Otherwise, the fact that the generalized model allows to consider probabilities which are not necessarily all equal gives the possibility to favor the establishment of certain connections over others. This kind of behavior is often observed in wireless mobile networks and social networks. Here, there are ways to implement in the generalized model the ability to prioritize connections to closest vertices. These advantages of modeling are generally not possible through the traditional random graphs of Erdös-Rényi. Also, we have described the global behavior between two consecutive configurations by calculating the probability of change. This is done in the case where the dynamic change concerns only the graph edges, as well as in the case where the dynamic change affects also the graph vertices.

At the end of this chapter we have shown how the parameters of a dynamic random graph can be estimated under the assumptions of the Erdös-Rényi model. Thus, the estimation of these parameters has led to the estimates of the average degree of vertices, the average hopcount and the size of the giant component in large dynamic graphs.

7. References

Albert R., Jeong H., and Barabàsi A.L. (1999). Diameter of the Worldwide Web. Nature, vol. 401, pp.130–131.

Amaral L.A.N., Scala A., Barthélémy M., and Stanley H.E. (2000). Classes of small-world networks. Proc. of Nat. Acad. Sci. USA 97, pp. 11149–11152.
http://polymer.bu.edu/hes/articles/asbs00.pdf

Bollobàs B. (2001). Random Graphs. Cambridge studies in advanced mathematics.

Barabàsi A. L. and Albert R. (1999). Emergence of scaling in random networks, Science, vol. 286, pp. 509–512.

Bhamidi S., van der Hofstad R. and Hooghiemstra G., (2010). First passage percolation on random graphs with finite mean degrees. The Annals of Applied Probability, Vol. 20, No. 5, pp. 1907–1965, DOI: 10.1214/09-AAP666

Durrett R. (2006). Random Graph Dynamics. Cambridge University Press. Cambridge, U.K.

Euler, L. (1736). Solutio problematis ad geometriam situs pertinentis. (The solution of a problem relating to the geometry of position). Commentarii Academiae Scientiarum Imperialis Petropolitanae, vol. 8, pp. 128–140.

Erdös P. and Rényi A. (1960). On the evolution of random graphs. Publications of the Mathematical Institute of the Hungarian Academy of Sciences, vol. 5, pp. 17–61.

Faloutsos M., Faloutsos P., Faloutsos C. (1999). On the power-law relationship of the internet topology. Proceedings of SIGCOMM '99, New York, USA, pp 251–262
http://www.cis.upenn.edu/~mkearns/teaching/NetworkedLife/power-internet.pdf

Hamlili A. (2010). Adaptive Schemes for Estimating Random Graph Parameters in Dynamic Random Networks' Modeling, Proceedings of IEEE/IFP Wireless days, Venice-Italy, 20 - 22 October, 2010.

Hekmat R. and Van Mieghem P. (2003). Degree distribution and hopcount in wireless ad-hoc networks. Proceedings of the 11th IEEE International Conference on Networks (ICON 2003), Sydney, Australia, pp. 603–609, Sept. 28-Oct. 2003.

Hewer T., Nekovee M. and Coveney P. V. (2009). Parameter exploration in parallel for dynamic vehicular network efficiency, International Conference on Advances in Computational Tools for Engineering Applications Proceedings (ACTEA '09), pp. 16 – 20, 15-17 July 2009.

Janson S., Lucak T., and Rucinski A. (2000). Random Graphs, John Wiley and Sons, New York.

Jurdak R. (2007). Wireless Ad Hoc and Sensor Networks: A Cross-Layer Design Perspective, Springer.

Kawahigashi H., Terashima Y., Miyauchi N. and Nakakawaji T. (2005). Modeling ad hoc sensor networks using random graph theory, Proceedings of Consumer Communications and Networking Conference (CCNC'05), pp.104 – 109.

Lehmann E. L. and Casella G. (1998). Theory of Point Estimation. Second Edition, Springer.

Mieghem P.V., Hooghiemstra G., van der Hofstadh R. (2000). A Scaling Law for the Hopcount, Report 2000125. Delft University of Technology, Delft, The Netherlands. http://wwwtvs.et.tudelft.nl/people/piet/telconference.html

Molloy M. and Reed B. (1998). The size of the giant component of a random graph with a given degree sequence, Combinatorics, Probability and Computing, vol. 7, pp. 295–305.

Newman M. E. J., Strogatz S. H., Watts D. J. (2001). Random graphs with arbitrary degree distributions and their applications. Physical Review E 64, 026118. http://arxiv.org/PS_cache/cond-mat/pdf/0007/0007235v2.pdf

Newman M. E. J. (2003). Random graphs as models of networks, in Handbook of Graphs and Networks, Bornholdt S. and Schuster H. G. (eds.), Wiley-VCH, Berlin.

Onat F. A., Stojmenovic I. and Yanikomeroglu H. (2008). Generating random graphs for the simulation of wireless ad hoc, actuator, sensor, and internet networks, Pervasive and Mobile Computing, Vol. 4, Issue 5, pp. 597-615.

Penrose M. D. (1999). On k-connectivity for a geometric random graph, Random Structures and Algorithms, vol. 15, pp. 145-164.

Steele M. (1997). Probability theory and combinatorial optimization. Vol. 69, SIAM.

Trullols O., Fiore M., Casetti C., Chiasserini C. F., and Barcelo Ordinas J. M. (2010). Planning Roadside Infrastructure for Information Dissemination in Intelligent Transportation Systems. Comput. Commun., 33(4):432-442.

9

The Properties of Graphs of Matroids[*]

Ping Li[1] and Guizhen Liu[2,†]

¹Department of Mathematics, West Virginia University, Morgantown, WV
²School of Mathematics and System Science, Shandong University, Jinan
¹USA
²China

1. Introduction

Let E be a finite set of elements. For $S_1, S_2 \subseteq E$, set $S_1 - S_2 = \{x | x \in S_1 \text{ and } x \notin S_2\}$. Let $\mathcal{C}$ be a collection of non-null subsets of E which satisfies the following two axioms.

(**C1**) A proper subset of a member of $\mathcal{C}$ is not a member of $\mathcal{C}$.

(**C2**) If $a \in C_1 \cap C_2$ and $b \in C_1 - C_2$ where $C_1, C_2 \in \mathcal{C}$ and $a, b \in E$, then there exists a $C_3 \in \mathcal{C}$ such that $b \in C_3 \subseteq (C_1 \cup C_2) - \{a\}$.

Then $M = (E, \mathcal{C})$ is called a matroid on E. We refer to the members of $\mathcal{C}$ as circuits of matroid M. The set of bases of a matroid M is a nonempty collection $\mathcal{B}$ of subsets of E such that the following condition is satisfied. For any $B, B' \in \mathcal{B}$, $|B| = |B'|$ and for any $e \in B \setminus B'$, there exists $e' \in B' \setminus B$ such that $(B \setminus \{e\}) \cup \{e'\} \in \mathcal{B}$, We also write as $M = (E, \mathcal{B})$. Each member of $\mathcal{B}$ is called a base of M. The rank r of a matroid is the number of elements in a base and the co-rank r^* is the number of its basic circuits. If $e \in E \setminus B$, then $B \cup \{e\}$ contains a unique basic circuit and denoted by $C(e, B)$. For a given base B of M, the set of basic circuits with respect to B is denoted by $\mathcal{C}_B$. We use $\mathcal{B}_e$ and $\overline{\mathcal{B}_e}$ to denote the set of bases containing e and avoiding e, respectively. Let $M = (E, \mathcal{C})$ be a matroid. If $X \subseteq E$, then the matroid on $E - X$ whose circuits are those of M which are contained in $E - X$ is called the restriction of M to $E - X$ (or the matroid obtained by deleting X from M) and is denoted by $M \setminus X$ or $M | (E - X)$. There is another derived matroid of importance. If $X \subseteq E$, then the family of minimal non-empty intersections of $E - X$ with circuits of M is the family of circuits of a matroid on $E - X$ called the contraction of M to $E - X$. If $X = \{e\}$, we use $M \setminus e$ and M / e to denote the matroid obtained from M by deleting and contracting e, respectively. A matroid obtained from M by limited times of contractions and limited times of deletions is called a minor of M. A subset S of E is called a separator of M if every circuit of M is either contained in S or $E - S$. Union and intersection of two separators of M is also a separator of M. If $\varnothing$ and E are the only separators of M, then M is said to be connected. The minimal non-empty separators of M are called the

*This research was supported by NNSF(61070230) of China.
†Corresponding Author

components of M. The base graph of matroid M is a graph $G_B(M)$ with vertex set $V(G_B)$ and edge set $E(G_B)$ such that $V(G_B) = B$ and $E(G_B) = \{BB' \mid B, B' \in \mathcal{B}, \mid |B - B'| = 1\}$.

Let G be a graph. The vertex set and edge set of a graph G are denoted by $V(G)$ and $E(G)$, respectively. If $A \subseteq V(G)$, then $G[A]$ denotes the induced subgraph of G by A. A k-path is a path of k-edges and denoted by P_k. A k-circuit is a circuit of k-edges and denoted by C_k. K_n denotes the complete graph of order n. A graph is *Hamiltonian connected* if for any two vertices there is a Hamilton path connects them. A graph is *Hamiltonian* if it contains a Hamilton circuit. A graph G is positively Hamiltonian, written $G \in H^+$, if for every edge of G, there is a Hamilton circuit containing it. G is negatively Hamiltonian, written $G \in H^-$, if for every edge of G, there is a Hamilton circuit avoiding it. When $G \in H^+$ and $G \in H^-$, we say that G is *uniformly Hamiltonian*. If for every edge e of G, there is a k-circuit containing it for any k, $3 \le k \le |V(G)|$, then G is called edge-pancyclic. A graph G is called E_2-Hamiltonian if every two edges of G are contained in a Hamilton cycle of G. Let G be a simple graph of order at least 3 vertices. Then graph G is called p_3-Hamilton, if for any path P with 3 vertices, there exists a Hamilton cycle of G which contains P. If for any two vertices v_1 and v_2 and any edge v_2v_3 where $v_1 \ne v_3$, graph G has a Hamilton path from v_1 to v_2 and such that edge v_2v_3 in this path, then we say that graph G is 1-Hamilton connected. Terminology and notations not defined here can be found in [1] and [2].

Maurer defined the base graph of a matroid , and discussed the graphical properties of the base graph of a matroid [3-4]. Cummins showed that every matroid base graph with at least three vertices has a Hamilton circuit [5]. Holzmann and Harary showed that for every edge in a base graph there is a Hamilton circuit containing it and another Hamilton circuit avoiding it [6]. Alspach and Liu studied the properties of paths and circuits in base graphs of matroids [7]. The connectivity of the base graph of matroids is investigated by Liu [8]. The other graphical properties of the base graphs of matroid have also been investigated by Liu [9-16].

Now we give a new concept as follows. The circuit graph of a matroid M is a graph $G_C = G(M)$ with vertex set $V(G_C)$ and edge set $E(G_C)$ such that $V(G_C) = \mathcal{C}$ and $E(G_C) = \{CC' \mid C, C' \in \mathcal{C}, \mid C \cap C' \mid \ne 0\}$, where the same notation is used for the vertices of G and the circuits of M. We give another new graph related to the bases of matroids as follows. The intersection graph of bases of matroid $M = (E, \mathcal{B})$ is a graph $G_I(M)$ with vertex set $V(G_I)$ and edge set $E(G_I)$ such that $V(G_I) = \mathcal{B}$ and $E(G_I) = \{BB' : |B \cap B'| \ne 0, B, B' \in \mathcal{B}(M)\}$, where the same notation is used for the vertex of G_I and the base of M. The properties of paths , cycles and the connectivity of circuit graphs of matroids are discussed in this chapter. In particular, some new results obtained by us are given.

2. Preliminary results

To prove the main theorem we need the following preliminary results.

Lemma 2.1. [17] A matroid M is connected if and only if for every pair e_1, e_2 of distinct elements of E, there is a circuit containing both e_1 and e_2.

Lemma 2.2. [17] If M is a connected matroid, then for every $e \in E$, either M/e or $M \backslash e$ is also connected.

Lemma 2.3. [17] Let C and C^* be any circuit and co-circuit of a matroid M. Then $|C \cap C^*| \neq 1$.

Lemma 2.4. [1] If $a \in C_1 \cap C_2$ and $b \in C_1 - C_2$ where $C_1, C_2 \in C$, then there exists a $C_3 \in C$ such that $b \in C_3 \subseteq (C_1 \cup C_2) - \{a\}$.

Let $M = (E, C)$ be a connected matroid. An element e of E is called an essential element if $M \backslash e$ is disconnected. Otherwise it is called an inessential element. A connected matroid each of whose elements is essential is called a critically connected matroid or simply a critical matroid.

Lemma 2.5. [17] A critical matroid of rank 2 contains a co-circuit of cardinality two.

A matroid M is trivial if it has no circuits. In the following matroids will be nontrivial. Next we will discuss the properties of the matroid circuit graph. To prove the main results we firstly present the following result which is clearly true.

Lemma 2.6. [17] Let M be any nontrivial matroid on E and $e \in E$. If G_C and G_{Ce} are circuit graphs of M and $M \backslash e$, respectively, then G_{Ce} is a subgraph of G induced by V_1 where $V_1 = \{C \mid C \in C, e \notin C\}$.

Obviously the subgraph G_{C2} of G induced by $V_2 = V - V_1$ is a complete graph. By Lemma 2.6 G_{C1} and G_{C2} are induced subgraphs of G and $V(G_{C1})$ and $V(G_{C2})$ partition $V(G)$.

Lemma 2.7. [17] For any matroid $M = (E, C)$ which has a 2-cocircuit $\{a, b\}$, the circuit graph of M is isomorphic to that of M/a.

Proof. Since $|C \cap \{a, b\}| \neq 1$ for any circuit C, by Lemma 2.3, the circuits of M can be partitioned into two classes, those circuits containing both a and b and those circuits containing neither a nor b. Likewise, the circuits of M/a can be partitioned into two classes: those containing b and those not containing b, clearly there is a bijection between $\mathscr{C}(M)$ and $\mathscr{C}(M/a)$. Hence $G(M) \cong G(M/a)$, the lemma is proved.

Lemma 2.8. [17] Suppose that $M = (E, C)$ is a connected matroid with an element e such that the matroid $M \backslash e$ is connected and $G = G_C(M)$ is the circuit graph of matroid M. Let $G_1 = G(M \backslash e)$ be the circuit graph of $M \backslash e$ and G_2 be the subgraph of G induced by V_2 where $V_2 = \{C \mid C \in C, e \in C\}$. If the matroid $M \backslash e$ has more than one circuit, then for any edge $C_1 C_2 \in E(G)$, there exists a 4-cycle $C_1 C_2 C_3 C_4$ in graph G such that one edge of the 4-cycle belongs to $E(G_1)$ and one belongs to $E(G_2)$ and C_1, C_2 are both adjacent to C_3.

Proof. By Lemma 2.6, $V(G_1)$ and $V(G_2)$ partition $V(G)$. There are three cases to distinguish.

Case 1. $e \in E - (C_1 \cup C_2)$. Thus $C_1 C_2$ is an edge of $M \backslash e$. By Lemma 2.1, there are at least three vertices in $G(M \backslash e)$. There is an element e_1 such that $e_1 \in C_1 \cap C_2$. Let G_1 and G_2 be the graphs defined as above. Note that G_2 is a complete graph. By Lemma 2.1, there is a vertex C_3 in G_2 containing both e_1 and e. Thus in G, C_3 is adjacent to both C_1 and C_2. Since $C_1 \not\subseteq C_3$, there exists e_2 such that $e_2 \in C_1$, but $e_2 \notin C_3$. By Lemma 2.1, there is a circuit C_4 in G_2 containing e_2 and e and $C_3 \neq C_4$. Thus C_4 is adjacent to C_1.

Case 2. $e \in C_1 - C_2$ or $e \in C_2 - C_1$. Suppose that $e \in C_2 - C_1$, $e_1 \in C_1 \cap C_2$. By Lemma 2.4, there is a circuit $C_3 \subseteq (C_1 \cup C_2) - \{e_1\}$ containing e. We assume that $e_2 \in C_1 \cap C_3$, $e_3 \in E - (C_1 \cup \{e\})$. Note that e_3 exists because, by hypothesis, $M \backslash e$ has more than one circuit. By Lemma 2.1, in $G_1 = G(M \backslash e)$ there is a circuit C_4 containing e_2 and e_3. $C_1 C_2 C_3 C_4$ is the 4-cycle we wanted. C_1 and C_2 are both adjacent to C_3.

Case 3. $e \in C_1 \cap C_2$. C_1 and C_2 are both in G_2. If there are only two circuits containing e, it is easy to see that $C_1 \cup C_2 = E(M)$ by Lemma 2.6. We prove that $C_1 \cap C_2 = \{e\}$. If $C_1 \cap C_2 = \{e, e'\}$, then $\{e, e'\}$ is a co-circuit of M because by Lemma 2.4, if there is a circuit containing e' does not contain e, then there is a circuit containing e does not contain e'; which is a contradiction to the hypothesis. Thus $C_1 \cap C_2 = \{e\}$. Then there is only one circuit $C_3' = (C_1 \cup C_2) - \{e\}$ in $M \backslash e$. For if there is a circuit $C_4' \neq C_3'$ in $M \backslash e$, then there exists $e_1 \in E(M) - (\{e\} \cup C_4')$ and $e_1 \in C_1 - C_2$ or $e_1 \in C_2 - C_1$. We assume that $e_1 \in C_1 - C_2$. By Lemma 2.4, there is a circuit C_5 such that $e \in C_5 \subseteq (C_2 \cup C_4') - \{e_2\}$ where $e_2 \in C_2 \cap C_4'$. $e_1 \notin C_5$, $C_5 \neq C_1$, thus there are at least three circuits containing e, a contradiction. So there are more than two circuits containing e, we assume that $e_3 \in C_1 - C_2$. By Lemma 2.4, there is C_3 in G_1 such that $e_3 \in C_3 \subseteq (C_1 \cup C_2) - \{e\}$. By Lemma 2.1, there is a vertex C_4 in G_1 containing e_3 and e_4 where $e_4 \in E - (C_3 \cup \{e\})$. We get the 4-cycle $C_1 C_2 C_3 C_4$. The proof is completed.

Lemma 2.9. [17] Suppose that $M = (E, \mathcal{C})$ is a connected matroid with an element e such that the matroid $M \backslash e$ is connected and $G = G(M)$ is the circuit graph of matroid M. Let $G_1 = G(M \backslash e)$ be the circuit graph of $M \backslash e$ and G_2 be the subgraph of G induced by V_2 where $V_2 = \{C \mid C \in \mathcal{C}, e \in C\}$. If $C_1 \in V(G_1)$, $C_2 \in V(G_2)$ and $d(C_1, C_2) = 2$, there exists a 3-path $C_1 C_3 C_4 C_2$ in graph G such that $C_3 \in V(G_1)$, $C_4 \in V(G_2)$ and C_1, C_2 are both adjacent to C_3 and C_4.

Proof. By Lemma 2.6, $V(G_1)$ and $V(G_2)$ partition $V(G)$. By assumption, $|C_1 \cap C_2| = 0$. We assume that $e_1 \in C_1$, $e_2 \in C_2$. By Lemma 2.1, there is C_4 in G_2 containing both e_1 and e. Thus in G, C_4 is adjacent to C_1 and C_2. By Lemma 2.1, there is C_3 in G_1 containing both e_1 and e_2. Thus in G, C_3 is adjacent to C_1, C_2 and C_4

Let P_4 be the Hamilton path connecting C_1 and C_2 in $G(M \backslash e)$ which traverses $C_1' C_2'$ and P_2 be the m-path connecting C_3 and C_4 in G_2 ($1 \leq m \leq n_2 - 1$), respectively. $P_3 + C_1' C_3 + C_3 C_2' + P_4$ is a n_1-path of $G(M)$ that joins C_1 and C_2. $P_3 + C_1' C_4 + P_2 + C_3 C_2' + P_4$ is a $n_1 + m$-path of $G(M)$ that joins C_1 and C_2.

Subcase 1.2. $e \in C_1 - C_2$ or $e \in C_2 - C_1$. Suppose that $e \in C_2 - C_1$. Thus $C_1 \in V(G_1)$, $C_2 \in V(G_2)$. If $d(C_1, C_2) = 1$, by Lemma 2.8, there is a 4-cycle $C_1 C_2 C_3 C_4$ in G such that $C_3 \in V(G_2)$ and $C_4 \in V(G_1)$ and $C_1 C_2 C_3$ is a 3-cycle of G. By induction, for any k_1, $1 \leq k_1 \leq n_1 - 1$, there is a k_1-path in G_1 connecting C_4 and C_1. Note that G_2 is a complete graph. For any k_2, $1 \leq k_2 \leq n_2 - 1$, there is a k_2-path in G_2 connecting C_2 and C_3. Let P_1 be the k_1-path in G_1 connecting C_1 and C_4 and P_2 be the k_2-path in G_2 connecting C_3 and C_2, respectively. $P_1 + C_4 C_3 + P_2$ is a $k_1 + k_2 + 1$-path of $G(M)$ that connects C_1 and C_2. If $d(C_1, C_2) = 2$ and $e \in C_2 - C_1$, by Lemma 2.9, there is a 3-path $C_1 C_3 C_4 C_2$ in G such that $C_3 \in V(G_1)$ and $C_4 \in V(G_2)$ and $C_1 C_3 C_2$ is a 2-path of G. Then the proof is similar to the case when $d(C_1, C_2) = 1$.

Subcase 1.3. $e \in C_1 \cap C_2$. Thus $d(C_1, C_2) = 1$. If there are only two circuits containing e, then there is only one circuit in $M \backslash e$. The result holds obviously because $G = K_3$. Assume that there are more than two circuits containing e. Note that G_2 is a complete graph. For any m, $1 \leq m \leq n_2 - 1$, there is a path of length m connecting C_1 and C_2. Choose $C_3 \in V(G_2)$

such that $C_3 \neq C_1$ and $C_3 \neq C_2$. By Lemma 2.8, there is a 4-cycle $C_1C_3C_4C_5$ in G such that $C_4C_5 \in E(G_1)$ and $C_1C_3C_4$ is a 3-cycle of G. By induction, for any k, $1 \leq k \leq n_1 - 1$, there is a k-path in G_1 connecting C_4 and C_5. Note that G_2 is a complete graph. Let P_1 be the k-path in G_1 connecting C_4 and C_5 and P_2 be the Hamilton path in G_2 connecting C_1 and C_2 which traverses the edge C_1C_3. Let $P_2 = C_1C_3 + P_3$. $C_1C_4 + C_4C_3 + P_3$ is a n_2-path that connects C_1 and C_2. $C_1C_5 + P_1 + C_4C_3 + P_3$ is the $n_2 + k$-path we wanted.

Case 2. The matroid M is critically connected. By Lemma 2.2, for any element e in M, M/e is connected. By Lemma 2.5, M has a 2-cocircuit $\{a, b\}$. By Lemma 2.7, the circuit graph of M/a is isomorphic to that of M. By induction hypothesis, the result holds.

Thus the theorem follows by induction.

Lemma 2.10. [2] A graph G is k-edge-connected if and only if any two distinct vertices of G are connected by at least k edge- disjoint paths.

3. The properties of circuit graphs of matroids

We have known that a matroid M is connected if and only if for every pair e_1, e_2 of distinct elements of E, there is a circuit containing both e_1 and e_2. The following results are the new results obtained by Li and Liu.

Theorem 3.1. [17] For any connected matroid $M = (E, \mathcal{C})$ which has at least three circuits, the circuit graph $G = G(M)$ is positively Hamiltonian, that is, for every edge of G, there is a Hamilton circuit containing it.

We shall prove the theorem by induction on $|E|$. When $|E| = 3$, each element in M is parallel to another. It is easy to see that $G = K_3$. The theorem is clearly true. Suppose that the result is true for $|E| = n - 1$. We prove that the result is also true for $|E| = n > 3$. Let C_1C_2 be any edge in G.

There are two cases to distinguish.

Case 1. There is an element e in M such that $M\backslash e$ is connected. Let G_1 and G_2 be the graphs defined as above. We assume that $|V(G_1)| = n_1$ and $|V(G_2)| = n_2$.

There are three subcases to distinguish.

Subcase 1.1. $e \in E - (C_1 \cup C_2)$. Thus C_1C_2 is an edge of $M\backslash e$. By Lemma 2.4, there are at least three vertices in $G(M\backslash e)$. By induction, $G(M\backslash e)$ is edge-pancyclic. For any m, $3 \leq m \leq n_1$, there is a cycle of length m containing C_1C_2. By Lemma 2.8, for any edge $C_1'C_2'$ in the Hamilton cycle of $G(M\backslash e)$ containing C_1C_2 where $C_1'C_2' \neq C_1C_2$, there is a 4-cycle $C_1'C_2'C_3C_4$ in G such that $C_3C_4 \in E(G_2)$ and $C_1'C_2'C_3$ is a 3-cycle in G. If there are only three vertices in $G(M\backslash e)$, let $C_1 = C_1'$. Note that G_2 is a complete graph. Let P_1 be the Hamilton path connecting C_1' and C_2' in $G(M\backslash e)$ which traverses C_1C_2 and P_2 be the k-path connecting C_3 and C_4 in G_2 ($1 \leq k \leq n_2 - 1$), respectively. $P_1 + C_2'C_3 + C_3C_1'$ is a $n_1 + 1$-cycle of $G(M)$ that contains C_1C_2. $P_1 + C_2'C_3 + P_2 + C_4C_1'$ is a $n_1 + k + 1$-cycle of $G(M)$ that contains C_1C_2.

Subcase 1.2. $e \in C_1 - C_2$ or $e \in C_2 - C_1$. Suppose that $e \in C_2 - C_1$. Thus $C_1 \in V(G_1)$, $C_2 \in V(G_2)$. By Lemma 2.8, there is a 4-cycle $C_1C_2C_3C_4$ in G such that $C_3 \in V(G_2)$ and $C_4 \in V(G_1)$ and $C_1C_2C_3$ is a 3-cycle of G. By induction, for any k_1, $1 \leq k_1 \leq n_1 - 1$, there is a k_1-path in G_1 connecting C_4 and C_1. Note that G_2 is a complete graph. For any k_2, $1 \leq k_2 \leq n_2 - 1$, there is a k_2-path in G_2 connecting C_2 and C_3. Let P_1 be the k_1-path in G_1 connecting C_1 and C_4 and P_2 be the k_2-path in G_2 connecting C_3 and C_2, respectively. $P_1 + C_4 C_3 + P_2 + C_2 C_1$ is a $k_1 + k_2 + 2$-cycle of $G(M)$ that contains $C_1 C_2$.

Subcase 1.3. $e \in C_1 \cap C_2$. If there are only two circuits containing e, then there is only one circuit in $M \backslash e$. The result holds obviously because $G = K_3$. Assume that there are more than two circuits containing e. Note that G_2 is a complete graph. For any m, $3 \leq m \leq n_2$, there is a cycle of length m containing $C_1 C_2$. Choose $C_3 \in V(G_2)$ such that $C_3 \neq C_1$ and $C_3 \neq C_2$. By Lemma 2.8, there is a 4-cycle $C_1 C_3 C_4 C_5$ in G such that $C_4 C_5 \in E(G_1)$ and $C_1 C_3 C_4$ is a 3-cycle of G. By induction, for any k, $1 \leq k \leq n_1 - 1$, there is a k-path in G_1 connecting C_4 and C_5. Note that G_2 is a complete graph. Let P_1 be the k-path in G_1 connecting C_4 and C_5 and P_2 be the Hamilton path in G_2 connecting C_3 and C_1 which traverses the edge $C_2 C_1$. $C_4 C_3 + P_2 + C_1 C_4$ is a $n_2 + 1$-cycle that contains $C_1 C_2$. $P_1 + C_4 C_3 + P_2 + C_1 C_5$ is the $n_2 + k + 1$-cycle we wanted.

Case 2. The matroid M is critically connected. By Lemma 2.2, for any element e in M, M/e is connected. By Lemma 2.5, M has a 2-cocircuit $\{a, b\}$. By Lemma 2.7, the circuit graph of M/a is isomorphic to that of M. By induction hypothesis, the result holds.

Thus the theorem follows by induction.

Theorem 3.2. [21] Let M be any connected matroid which has at least four circuits and let $G = G(M)$ be the circuit graph of M. Then for each edge of $G = G(M)$ there is a Hamilton cycle avoiding it. that is, the circuit graph $G = G(M)$ is negatively Hamiltonian.

Proof. We prove the theorem by induction on $|E|$. It is easy to see that $|E| \geqslant 4$. When $|E| = 4$ and $r(M) = 1$, $M = U_{1,4}$ [1]. It is easy to see that $G(M) \in H^-$. When $|E| = 4$ and $r(M) = 2$, M has at most three circuits except when $M = U_{2,4}$. Obviously $G(U_{2,4}) = K_4$ and $K_4 \in H^-$. Then suppose that the theorem is true for $|E| = n - 1$. We shall prove the theorem holds for $|E| = n > 4$. Let $C_1 C_2$ be any edge of G. There are two cases to consider.

Case 1. There exists an element e such that $M \backslash e$ is connected. Let G_1 and G_2 be the graphs defined as above.

There are three subcases to consider.

Subcase 1.1. If $e \in E - (C_1 \cup C_2)$, then $C_1 C_2$ is an edge of $G(M \backslash e)$. By Theorem 3.1, there is a Hamilton cycle in $G(M \backslash e)$ containing $C_1 C_2$. By the proof of Lemma 2.8, there is a 4-cycle $C_1 C_2 C_3 C_4$ in G such that $C_3 C_4 \in E(G_2)$. As in the proof of Theorem 3.1, let P_1 be a Hamilton path in G_1 connecting C_1 and C_2 and P_2 be a Hamilton path in G_2 connecting C_3 and C_4, respectively. Then the cycle $P_1 + C_2 C_3 + P_2 + C_4 C_1$ is a Hamilton cycle of G avoiding $C_1 C_2$.

Subcase 1.2. Let $e \in C_1 - C_2$ or $e \in C_2 - C_1$. It is easy to see that there are more than two vertices in $G(M \backslash e)$, and there are at least three vertices in G_2. We assume that $e \in C_2 - C_1$ and $e_1 \in C_1 - C_2$. By Lemma 2.1, there is C_3 in G_2 containing e and e_1. By the proof of Lemma 2.8, in G there is a 4-cycle $C_1 C_3 C_4 C_5$ such that $C_4 \in V(G_2)$, $C_5 \in V(G_1)$. Let P_1 be a Hamilton path in $G(M \backslash e)$ connecting C_1 and C_5 and P_2 be a Hamilton path in G_2 connecting C_4 and C_3, respectively. $P_1 + C_5 C_4 + P_2 + C_3 C_1$ is a Hamilton cycle avoiding $C_1 C_2$.

Subcase 1.3. If $e \in C_1 \cap C_2$, as in the proof of subcase 1.3 in Theorem 3.1, let $C_2 = C_3$ and P_2 be a Hamilton path connecting C_1 and C_2 in G_2, then $P_1 + C_4C_2 + P_2 + C_1C_5$ is a Hamilton cycle we wanted.

Case 2. For every $e \in E(M)$, $M \backslash e$ is disconnected. Then the matroid M is critically connected. By Lemma 2.2, for any element e in M, M/e is connected. By Lemma 2.5, M has a cocircuit $\{a, b\}$. By Lemma 2.7, $G(M) \cong G(M/a)$. By induction hypothesis, the result holds.

The proof of the theorem is completed.

Corollary 3.3. For any connected matroid M, the circuit graph $G_C(M)$ is uniformly Hamilton whenever $G_C(M)$ contains at least four vertices. That is for any edge e of $G_C(M)$, there is a Hamilton cycle containing e and there is another Hamilton cycle excluding e.

Theorem 3.4. [20] Let $G = G(M)$ be the circuit graph of a connected matroid $M = (E, C)$. If $|V(G)| = n$ and $C_1, C_2 \in V(G)$ with $d(C_1, C_2) = r$, then there is a path of length k joining C_1 and C_2 for any k satisfying $r \le k \le n - 1$.

Proof. We shall prove the theorem by induction on $|E|$. When $|E| = 3$, each element in M is parallel to another. It is easy to see that $G = K_3$. The theorem is clearly true. Suppose that the result is true for $|E| = n - 1$. We prove that the result is also true for $|E| = n > 3$. It is easy to see that for any vertices C_1, C_2 in $V(G)$, we have $d(C_1, C_2) = r$ where $r = 1$ or $r = 2$ by the definition of the circuit graph of a matroid.

There are two cases to distinguish.

Case 1. There is an element e in M such that $M \backslash e$ is connected. Let G_1 and G_2 be the graphs defined as above. We assume that $|V(G_1)| - n_1$ and $|V(G_2)| = n_2$.

There are three subcases to distinguish.

Subcase 1.1. $e \in E - (C_1 \cup C_2)$. Thus $C_1 \in V(G_1)$, $C_2 \in V(G_1)$. Clearly, there are at least three vertices in $G(M \backslash e)$. By induction, for any k, $r \le k \le n_1 - 1$, there is a path of length k joining C_1 and C_2 in $G(M \backslash e)$. By Lemma 2.8, for any edge $C_1' C_2'$ in the Hamilton path connecting C_1 and C_2 in $G(M \backslash e)$, there is a 4-cycle $C_1' C_2' C_3 C_4$ in G such that $C_3 C_4 \in E(G_2)$ and $C_1' C_2' C_3$ is a 3-cycle in G. If there are only three vertices in $G(M \backslash e)$, let $C_1 = C_1'$. Note that G_2 is a complete graph. Let $P_1 = P_3 + C_1' C_2' + P_4$ be the Hamilton path connecting C_1 and C_2 in $G(M \backslash e)$ which traverses $C_1' C_2'$ and P_2 be the m-path connecting C_3 and C_4 in G_2 ($1 \le m \le n_2 - 1$), respectively. $P_3 + C_1' C_3 + C_3 C_2' + P_4$ is a n_1-path of $G(M)$ that joins C_1 and C_2. $P_3 + C_1' C_4 + P_2 + C_3 C_2' + P_4$ is a $n_1 + m$-path of $G(M)$ that joins C_1 and C_2.

Subcase 1.2. $e \in C_1 - C_2$ or $e \in C_2 - C_1$. Suppose that $e \in C_2 - C_1$. Thus $C_1 \in V(G_1)$, $C_2 \in V(G_2)$. If $d(C_1, C_2) = 1$, by Lemma 2.8, there is a 4-cycle $C_1 C_2 C_3 C_4$ in G such that $C_3 \in V(G_2)$ and $C_4 \in V(G_1)$ and $C_1 C_2 C_3$ is a 3-cycle of G. By induction, for any k_1, $1 \le k_1 \le n_1 - 1$, there is a k_1-path in G_1 connecting C_4 and C_1. Note that G_2 is a complete graph. For any k_2, $1 \le k_2 \le n_2 - 1$, there is a k_2-path in G_2 connecting C_2 and C_3. Let P_1 be the k_1-path in G_1 connecting C_1 and C_4 and P_2 be the k_2-path in G_2 connecting C_3 and C_2, respectively. $P_1 + C_4 C_3 + P_2$ is a $k_1 + k_2 + 1$-path of $G(M)$ that connects C_1 and C_2.

If $d(C_1, C_2) = 2$ and $e \in C_2 - C_1$, by Lemma 2.9, there is a 3-path $C_1 C_3 C_4 C_2$ in G such that $C_3 \in V(G_1)$ and $C_4 \in V(G_2)$ and $C_1 C_3 C_2$ is a 2-path of G. Then the proof is similar to the case when $d(C_1, C_2) = 1$.

Subcase 1.3. $e \in C_1 \cap C_2$. Thus $d(C_1, C_2) = 1$. If there are only two circuits containing e, then there is only one circuit in $M \backslash e$. The result holds obviously because $G = K_3$. Assume that there are more than two circuits containing e. Note that G_2 is a complete graph. For any m, $1 \leq m \leq n_2 - 1$, there is a path of length m connecting C_1 and C_2. Choose $C_3 \in V(G_2)$ such that $C_3 \neq C_1$ and $C_3 \neq C_2$. By Lemma 2.8, there is a 4-cycle $C_1 C_3 C_4 C_5$ in G such that $C_4 C_5 \in E(G_1)$ and $C_1 C_3 C_4$ is a 3-cycle of G. By induction, for any k, $1 \leq k \leq n_1 - 1$, there is a k-path in G_1 connecting C_4 and C_5. Note that G_2 is a complete graph. Let P_1 be the k-path in G_1 connecting C_4 and C_5 and P_2 be the Hamilton path in G_2 connecting C_1 and C_2 which traverses the edge $C_1 C_3$. Let $P_2 = C_1 C_3 + P_3$. $C_1 C_4 + C_4 C_3 + P_3$ is a n_2-path that connects C_1 and C_2. $C_1 C_5 + P_1 + C_4 C_3 + P_3$ is the $n_2 + k$-path we wanted.

Case 2. The matroid M is critically connected. By Lemma 2.2, for any element e in M, M/e is connected. By Lemma 2.5, M has a 2-cocircuit $\{a, b\}$. By Lemma 2.7, the circuit graph of M/a is isomorphic to that of M. By induction hypothesis, the result holds.

Thus the theorem follows by induction.

Theorem 3.5. [20] Suppose that $G = G_C(M)$ is the circuit graph of a connected matroid M and C_1 and C_2 are distinct vertices of G. Then C_1 and C_2 are connected by $d = min\{d(C_1), d(C_2)\}$ edge-disjoint paths where $d(C_1)$ and $d(C_2)$ denote the degree of vertices C_1 and C_2 in G, respectively.

Proof. We shall prove the theorem by induction on $|E|$. When $|E| = 3$, each element in M is parallel to another. It is easy to see that $G = K_3$. The theorem is clearly true. Suppose that the result is true for $|E| = n - 1$. We prove that the result is also true for $|E| = n > 3$. Let C_1 and C_2 be any two vertices in G.

There are two cases to distinguish.

Case 1. $(C_1 \cup C_2) = E$. It is easy to see that C_1 and C_2 are both adjacent to any circuit in $\mathscr{C} - \{C_1 \cup C_2\}$ and the conclusion is obviously true.

Case 2. $(C_1 \cup C_2) \neq E$.

There are two subcases to distinguish.

Subcase 2.1. There is an element $e \in E - (C_1 \cup C_2)$ such that $M \backslash e$ is connected. Let $G_1 = G(M \backslash e)$ be the circuit graph of $M \backslash e$ and G_2 be the subgraph of G induced by V_1 where $V_1 = \{C \mid C \in \mathscr{C}, e \in C\}$. Thus C_1 and C_2 are in G_1. By induction, in G_1, C_1, C_2 are connected by $d_1 = min\{d_1(C_1), d_1(C_2)\}$ edge-disjoint paths where $d_1(C_1)$ and $d_1(C_2)$ denote the degree of vertices C_1 and C_2 in G_1, respectively. Let $\mathscr{P}_1 = \{P_1, P_2, \cdots, P_{d_1}\}$ be the family of shortest edge-disjoint paths connecting C_1 and C_2 in G_1. Without loss of generality, we may assume that $d_1(C_1) \geq d_1(C_2)$. There are two subcases to distinguish.

Subcase 2.1 a. $d_1(C_1) = d_1(C_2)$. Thus $d_1 = min\{d_1(C_1), d_1(C_2)\} = d_1(C_1) = d_1(C_2)$. We assume that there are m vertices $A_1, A_2, \cdots, A_m$ in G_2 that are adjacent to C_1 and n vertices $D_1, D_2, \cdots, D_n$ in G_2 that are adjacent to C_2 where m, n are integers. G_2 is a complete graph, so A_i is adjacent to $D_j(i = 1, 2, \cdots, m; j = 1, 2, \cdots, n)$. Here maybe $A_i = D_j$ for some $1 \leq i \leq m; 1 \leq j \leq n$. Let $q = min\{m, n\}$. $C_1 A_i D_i C_2 (i = 1, 2, \cdots, q)$ are q edge-disjoint paths in G. It is easy to see that $d(C_1) = d_1(C_1) + m, d(C_2) = d_1(C_2) + n$ and $d = min\{d(C_1), d(C_2)\} = min\{d_1(C_1) + m, d_1(C_2) + n\} = d_1(C_1) + min\{m, n\} = d_1(C_1) + q$. $\mathscr{P} = \mathscr{P}_1 \cup \{C_1 A_1 D_1 C_2, C_1 A_2 D_2 C_2, \cdots, C_1 A_q D_q C_2\}$ are d edge-disjoint paths connecting C_1 and C_2 in G.

Subcase 2.1 b. $d_1(C_1) > d_1(C_2)$. By induction, in G_1 there are $d_1 = min\{d_1(C_1), d_1(C_2)\} = d_1(C_2)$ edge -disjoint paths connecting C_1 and C_2. Let $\mathscr{P}_1 = \{P_1, P_2, \cdots, P_{d_1(C_2)}\}$ be the family of shortest edge-disjoint paths connecting C_1 and C_2 in G_1. It is obvious that each $P_i(i = 1, 2, \cdots, d_1(C_2))$ contains exactly one vertex adjacent to C_1 and one vertex adjacent to C_2. Let $A_1, A_2, \cdots, A_{d_1(C_1)-d_1(C_2)}$ be the vertices in G_1 that are adjacent to C_1 but not contained in d_1 edge-disjoint paths. By Lemma 2.1, for any element e' in A_i $(i = 1, 2, \cdots, d_1(C_1) - d_1(C_2))$ there is a circuit A'_i in G_2 containing e and e', thus $A_i A'_i$ is an edge in $G(M)$. Let $D_1, D_2, \cdots, D_m$ denote the vertices in G_2 that is adjacent to C_2. G_2 is a complete graph, so A'_i is adjacent to $D_j(i = 1, 2, \cdots, d_1(C_1) - d_1(C_2); j = 1, 2, \cdots, m)$. If $m \leq d_1(C_1) - d_1(C_2)$, $C_1 A_i A'_i D_i C_2$ are m edge-disjoint paths connecting C_1 and C_2 where A'_i can be D_i $(i = 1, 2, \cdots, m)$. Here it is possible that $A'_i = A'_j (i \neq j; i, j = 1, 2, \cdots, d_1(C_1) - d_1(C_2))$. But it is forbidden that $D_i = D_j (i \neq j; i, j = 1, 2, \cdots, m)$. $d(C_2) = d_1(C_2) + m \leq d_1(C_1) < d(C_1)$, thus $d = min\{d(C_1), d(C_2)\} = d(C_2)$. $\mathscr{P} = \mathscr{P}_1 \cup \{C_1 A_1 A'_1 D_1 C_2, C_1 A_2 A'_2 D_2 C_2, \cdots, C_1 A_m A'_m D_m C_2\}$ are d edge-disjoint paths connecting C_1 and C_2 in G. If $m > d_1(C_1) - d_1(C_2)$, let $\mathscr{P}_2 = \{C_1 A_1 A'_1 D_1 C_2, C_1 A_2 A'_2 D_2 C_2, \cdots, C_1 A_{d_1(C_1)-d_1(C_2)} A'_{d_1(C_1)-d_1(C_2)} D_{d_1(C_1)-d_1(C_2)} C_2\}$ be $d_1(C_1) - d_1(C_2)$ edge-disjoint paths connecting C_1 and C_2 where A'_i can be D_i $(i = 1, 2, \cdots, d_1(C_1) - d_1(C_2))$. Let $L_1, L_2, \cdots, L_n$ denote the vertices in G_2 that is adjacent to C_1. G_2 is a complete graph, so L_i is adjacent to $D_j(i = 1, 2, \cdots, n; j = d_1(C_1) - d_1(C_2) + 1, d_1(C_1) - d_1(C_2) + 2, \cdots, m)$. If $m > n + d_1(C_1) - d_1(C_2)$, $d(C_1) = d_1(C_1) + n \leq d_1(C_2) + m < d(C_2)$, thus $d = min\{d(C_1), d(C_2)\} = d(C_1)$. $\mathscr{P}_3 = C_1 L_i D_{d_1(C_1)-d_1(C_2)+i} C_2$ are n edge-disjoint paths connecting C_1 and C_2 where L_i can be $D_{d_1(C_1)-d_1(C_2)+i}$ $(i = 1, 2, \cdots, n)$. Thus $\mathscr{P} = \mathscr{P}_1 \cup \mathscr{P}_2 \cup \mathscr{P}_3$ are $d = d(C_1)$ edge-disjoint paths in G. If $d_1(C_1) - d_1(C_2) < m \leq n + d_1(C_1) - d_1(C_2)$, $\mathscr{P}'_3 = C_1 L_i D_{d_1(C_1)-d_1(C_2)+i} C_2$ are $m - (d_1(C_1) - d_1(C_2))$ edge-disjoint paths connecting C_1 and C_2 where L_i can be $D_{d_1(C_1)-d_1(C_2)+i}$ $(i = 1, 2, \cdots, m - (d_1(C_1) - d_1(C_2)))$ but $D_{d_1(C_1)-d_1(C_2)+i} \neq D_{d_1(C_1)-d_1(C_2)+j}$ $(i \neq j; i, j = 1, 2, \cdots, m - (d_1(C_1) - d_1(C_2)))$. $d(C_2) = d_1(C_2) + m \leq d_1(C_1) + n = d(C_1)$, thus $d = min\{d(C_1), d(C_2)\} = d(C_2)$. $\mathscr{P} = \mathscr{P}_1 \cup \mathscr{P}_2 \cup \mathscr{P}'_3$ are $d = d(C_2)$ edge-disjoint paths connecting C_1 and C_2 in G. The conclusion holds.

Subcase 2.2. There is no element $e \in E - (C_1 \cup C_2)$ such that $M \backslash e$ is connected. If $E - (C_1 \cup C_2) = \{e\}$ and $M e$ is disconnected, it is easy to see that $C_1 \cap C_2 = \emptyset$ and C_1, C_2 are the two components of $M \backslash e$. Thus any circuit of M intersecting both C_1 and C_2 contains e. Then C_1 and C_2 are both adjacent to any circuit in $\mathscr{C} - \{C_1 \cup C_2\}$ and the conclusion is obviously true. Suppose that $|E - (C_1 \cup C_2)| \geq 2$ and for any $e \in E - (C_1 \cup C_2)$, $M \backslash e$ is disconnected. By Lemma 2.5, M has a 2-cocircuit $\{a, b\}$. By Lemma 2.7, the circuit graph of M/a is isomorphic to that of M. By induction hypothesis, the result holds. Thus the theorem follows by induction.

By Theorem 3.5, we can get the following corollary.

Corollary 3.6. Suppose that $G = G(M)$ is the circuit graph of a connected matroid M with minimum degree $\delta(G)$. Then the edge connectivity $\kappa'(G) = \delta(G)$.

Proof. By Theorem 3.5, we know that $\kappa'(G) \geq \delta(G)$. Since for any graph G, we have $\kappa'(G) \leq \delta(G)$, then $\kappa'(G) = \delta(G)$.

Theorem 3.7. [20] Let G be the circuit graph of a connected matroid $M = (E, \mathscr{C})$. If $|V(G)| = n$ and $k_1 + k_2 + \cdots + k_p = n$ where k_i is an integer, $i = 1, 2, \ldots, p$, then there is a partition of $V(G)$ into p parts $V_1, V_2, \ldots, V_p$ such that $|V_i| = k_i$ and the subgraph H_i induced by V_i contains a k_i -cycle when $k_i \geq 3$, H_i is isomorphic to K_2 when $k_i = 2$, H_i is a single vertex when $k_i = 1$.

Proof. We shall prove the theorem by induction on $|E|$. When $|E| = 3$ and $|V(G)| = 1$, the result holds clearly. When $|E| = 3$ and $|V(G)| = 3$, $M = U_{1,3}[1]$. It is easy to see that $G = K_3$. The theorem is clearly true. Suppose that the result is true for $|E| = m - 1$. We prove that the result is also true for $|E| = m > 3$.

Case 1. There is an element e in M such that $M \backslash e$ is connected. Let G_1 and G_2 be the graphs defined as above. We assume that $|V(G_1)| = n_1$ and $|V(G_2)| = n_2$. There exists an q such that $k_1 + k_2 + \cdots + k_{q-1} < n_1$ and $k_1 + k_2 + \cdots + k_q \geq n_1$. By induction, the vertices of $G(M \backslash e)$ can be partitioned into q parts $V_1, V_2, \ldots, V_q'$ such that $|V_1| = k_1, |V_2| = k_2, \ldots, |V_{q-1}| = k_{q-1}, |V_q'| = n_1 - (k_1 + k_2 + \cdots + k_{q-1}) = k_q'$ and the subgraph H_i (H_q') induced by V_i $(i = 1, 2, \ldots, q-1)$ (V_q') contains a k_i (k_q') -cycle when $k_i \geq 3$ $(k_q' \geq 3)$, H_i (H_q') is isomorphic to K_2 when $k_i = 2$ $(k_q' = 2)$, H_i (H_q') is a single point when $k_i = 1$ $(k_q' = 1)$. When $k_q = k_q'$, the result holds clearly because G_2 is a complete graph. When $k_q > k_q'$, there are three subcases to consider.

Subcase 1.1. $k_q' = 1$. Suppose that C_1 is the subgraph in $G(M \backslash e)$ induced by V_q'. When $k_q = 2$, obviously there is a vertex in G_2 that is adjacent to C_1. When $k_q \geq 3$, we prove that there is a 3-cycle $C_1 C_2 C_3$ in G such that $C_2 C_3 \in E(G_2)$. For any $e_1 \in C_1$, by Lemma 2.1, there is a circuit $C_2 \in G_2$ containing e_1 and e. Let $e_2 \in C_1 - C_2$. There is a circuit $C_3 \in G_2$ containing e_2 and e. We get the 3-cycle $C_1 C_2 C_3$. Note that G_2 is a complete graph. In G_2 there is a $k_q - 2$ path P connecting C_2 and C_3. $C_1 C_2 + P + C_3 C_1$ is a k_q -cycle in G. Because G_2 is a complete graph, the subgraph induced by $V(G_2) - V(P)$ is also a complete graph. Thus the vertices of the subgraph induced by $V(G_2) - V(P)$ can be partitioned into $p - q$ parts $V_{q+1}, V_{q+2}, \ldots, V_p$ such that $|V_i| = k_i$, $i = q + 1, q + 2, \ldots, p$, and the subgraph H_i induced by V_i contains a k_i -cycle when $k_i \geq 3$, H_i is isomorphic to K_2 when $k_i = 2$, H_i is a single point when $k_i = 1$. The result holds.

Subcase 1.2. $k_q' = 2$. Suppose that $C_1 C_2$ is the subgraph in $G(M \backslash e)$ induced by V_q'. When $k_q = 3$, by Lemma 2.8, the result holds. When $k_q \geq 4$, by Lemma 2.8, there is a 4-cycle $C_1 C_2 C_3 C_4$ in G such that $C_3 \in V(G_2)$ and $C_4 \in V(G_2)$ and $C_1 C_2 C_3$ is a 3-cycle of G. In G_2 there is a $k_q - 3$ -path P connecting C_3 and C_4. $C_1 C_2 + C_2 C_3 + P + C_4 C_1$ is a k_q -cycle in G. Because G_2 is a complete graph, the subgraph induced by $V(G_2) - V(P)$ is also a complete graph. Thus the result holds.

Subcase 1.3. $k_q' > 2$. The subgraph H_q' in $G(M \backslash e)$ induced by V_q' contains a k_q' -cycle. Let $C_1 C_2$ be any edge in this cycle and P_1 be the Hamilton path in H_q' connecting C_1 and C_2. By Lemma 2.8, there is a 4-cycle $C_1 C_2 C_3 C_4$ in G such that $C_3 C_4 \in E(G_2)$ and $C_1 C_2 C_3$ is a 3-cycle of G. When $k_q - k_q' = 1$, $P_1 + C_2 C_3 + C_3 C_1$ is a k_q -cycle in G. When $k_q - k_q' \geq 2$, in G_2 there is a $k_q - k_q' - 1$ -path P_2 connecting C_3 and C_4. $P_1 + C_2 C_3 + P_2 + C_4 C_1$ is a k_q -cycle in G. Note that

G_2 is a complete graph. The subgraph induced by $V(G_2) - V(P_2)$ is also a complete graph. Thus the result holds.

Case 2. The matroid M is critically connected. By Lemma 2.2, for any element e in M, M/e is connected. By Lemma 2.5, M has a 2-cocircuit $\{a, b\}$. By Lemma 2.7, the circuit graph of M/a is isomorphic to that of M. By induction hypothesis, the result holds.

Thus the theorem follows by induction.

From Theorem 3.7 we have the following theorem holds.

Theorem 3.8. Let $G = G(M)$ be the circuit graph of a connected matroid $M = (E, C)$. If $|V(G)| = n$ and $k_1 + k_2 + \cdots + k_p = n$ where k_i is an integer and $k_i \geq 3, i = 1, 2, \ldots, p$, then G has a 2-factor F containing p vertex-disjoint cycles $D_1, D_2, \ldots, D_p$ such that the length of D_i is k_i $(i = 1, 2, \ldots, p)$.

By the similar methods we can prove that the above theorems also holds for the intersection graph of bases of matroids.

Finally, We present the following open problems to be considered.

Problem 1. Let $G = G(M)$ be the circuit graph of a connected matroid $M = (E, C)$. If $|V(G)| = n$ and $C_1, C_2 \in V(G)$ with $d(C_1, C_2) = r$, how many Hamilton paths connect C_1 and C_2 in G? Furthermore, how many k-paths connect C_1 and C_2 in G $(r \leq k \leq n-1)$?

Problem 2. Let $G = G(M)$ be the intersection graph of bases of matroid $M = (E, B)$. If $|V(G)| = n$ and $B_1, B_2 \in V(G)$ with $d(B_1, B_2) = r$, how many Hamilton paths connect B_1 and B_2 in G? Furthermore, how many k-paths connect B_1 and B_2 in G $(r \leq k \leq n-1)$?

Other related results of graphs on matroids can be found in [22-60].

4. References

[1] J. G. Oxley, Matroid theroy, Oxford University Press, New York, 1992.
[2] J. A. Bondy, U. S. R. Murty, Graph Theory With Applications, American Elsevier, New York, 1976.
[3] S. B. Maurer, Matroid basis graphs I, J.Comb. Theory B, 14 (1973), 216-240.
[4] S. B. Maurer, Matroid basis graphs II, J. Comb. Theory B, 15 (1973) 121-145.
[5] R. L. Cummins, Hamilton circuits in tree graphs. IEEE Trans Circuit theory, 1 (1966), 82-90.
[6] C. A. Holzmann, P. G. Norton and M. D. Tobey. A graphical representation of matroids. SIAM J Appl Math., 29 (1973), 618-672.
[7] B. Alspach and G. Liu. Paths and cycles in matroid base graphs, Graphs and combinatorics, 1989, 5(3), 207-211.
[8] G. Liu. The connectivities of matroid base graphs, J. Operations Research, 3(1) (1984), 67-68.
[9] G. Liu, Matroids complexes–geometrical representations on matroids. Acta Math. Scientia, 5 (1985), 35-42.

[10] G. Liu, The Connectivities of Adjacent Tree Graphs. *Acta Mathematicae Applicatae Sinica,* 3(4) (1987), 313-317.

[11] G. Liu. A lower bound on connectivities of matroid base graphs, *Discrete Math.,* 69(1) (1988), 55-60.

[12] G. Liu. On connectivities of base graph of some matroids, *J. Sys. Sci. and Math. Scis.,* 1(1) (1988), 18-21.

[13] G. Liu, On connectivities of tree graphs, *J. Graph Theory,* 12(3) (1988), 453-459.

[14] G. Liu, The proof of a conjecture on matroid basis graphs. *Scince Sinica,* 1990: 593-599.

[15] Liu Guizhen and Chen Qinghua, Matroids, Publishing House of National University Defense Technology, Changsha, (in chinese) 1994.

[16] Guizhen Liu and L. Zhang, Forest graphs of graphs, Chinese Journal of Engineering Mathmatics, 22 (6) (2005), 1100-1109.

[17] P. Li, Guizhen Liu, Cycles in circuit graphs of matroids, *Graphs and Combinatorics* 23 (2007), 425-431.

[18] Yinghao Zhang and Guizhen Liu, On Properties of the intersection graphs of matroids, to appear in *Frontiers of Mathematics.*

[19] H. Deng and F. Xia. The P_3-Hamiltonian property of matroid base graphs. *Jour Nat Sci Hunan Norm Uni.,* 1 (2000), 5-8.

[20] D. Deng and R. Li. The 1-Hamiltonian property of matroid base graphs. *Acta. Scinat Univ. Norm. Hunan.,* 22 (3) (1999), 1-5..

[21] L. Li, Matroids and Graphs, Ph.D. Dissertation, Shandong University, (2005).

[22] Ping Li, Guizhen Liu , Hamilton cycle in circuit graphs of matroids, Computer and mathematics with Applications 55 (2008), 654-659

[23] Ping Li, Some Properties of Circuit Graphs of Matroids, Doctoral Dissertation, Shandong University, (2010)

[24] B. Bollobas, A lower bound for the number of nonisomorphic Matroids, *J. Comb. Theory,* 1969, 7, 366-368.

[25] J. A. Bondy, Transversal matroids, base orderable matroids and graphs, *Quart. J. Math.,* 1972, 23, 81-89.

[26] J. A. Bondy, A.W. Ingleton, Pancyclic graph II, *J.Comb. Theory,* B20,1976, 41-46

[27] R. A. Brualdi, On foundamental transversal matroids, *Proc. Amer. Math. Soc.,* 45 (1974), 151-156.

[28] R. A. Brualdi, G. W. Dinolt, Characterization of transversal matroids and their presentations, *J. Comb. Theory,* 12 (1972), 268-286.

[29] M. Cai, A solution of Chartand's problem on spanning trees, *Acta Mathematicae Applicatae Sinica* (English Ser.), 1:2 (1984), 97-98.

[30] P. A. Catlin, J. W. Grossman, A. M. Hobbs, and H. J. Lai. Fractional arboricity, strength, and principal partitions in graphs and matroids. *Disc. Appl. Math.,* 40(1992), 285-302.

[31] H. H. Crapo, Single element extensions of Matroids, *J. Res. Nat. Bur. Stand.,* 69B (1965), 57-65.

[32] J. Donald, C. Holzmann and M. Tobey. A characterization of complete matroid base graphs, *J. Comb. Theory Ser.,* 22B (1977), 139-193.

[33] J. R. Edmonds, Minimum partition of a matroid into indedent subsets. *J.Res. Natl. Bur. Stand.,* 69B(1965), 67-72.

[34] J. R. Edmonds, Matroids and the greedy algorithm, *Math Programming*, 1, (1971), 127-136.

[35] V. Estivill-Castro, M. Noy and J. Urrutia, On the chromatic number of tree graphs. *Discrete Math.*, 223 (2000), 363-366.

[36] J. Gao, Quasi-1-Hamilton connectedness of tree graphs. *Math. Res. and Exposition*, 7(3) (1987), 498.

[37] J. Gao, 1-Hamilton connectedness of tree graphs. *Mathematica Applicata.*, 6(2) (1993), 136-144.

[38] H. Harary, R. J. Mokken and M. J. Plantholt, Interpolation theorem for diameters of spanning trees. *IEEE Trans Circuits and System*, 30 (1983), 429-432.

[39] F. Harary and M. J. Plantholt, Classification of interpolation theorems for spanning trees and other families of spanning subgraphs. *J. Graph Theory*, 13(6)(1989), 703-712.

[40] R. Rado, A theorem on independence relations. *Quart. J. Math. Oxford Ser.*, 13 (1942), 83-89.

[41] C. A. Holzmann, F. Harary, On the tree graph of a matroid. *SIAM J. Appl. Math.*, 22 (1972), 187-193.

[42] A. W. Ingleton, Gammoids and transversal matroids, *J. Comb. Theory*, 15 (1973), 51-68.

[43] T. Kamae. The existence of Hamilton circuit in a tree graph. *IEEE Trans Circuit theory*, 14 (1967), 279-283.

[44] G. Kishi and Y. Kajitani. On Hamilton circuits in a tree graphs. *IEEE Trans Circuit theory*, 15 (1968), 42-50.

[45] E. Lawler. Combinatorial Optimization. *John Wiley, New York*, 1972.

[46] L. Li, The Hamilton properties of tree graphs. *J. Shandong Unv. of Technology*, 27 (3) (1997), 261-263.

[47] L. Li and G. Liu, The connectivities of the adjacency leaf exchange forest graphs, *J. Shandong University*, 39 (6) (2004), 49-51.(in Chinese)

[48] L. Li, Q. Bian and G. Liu, The matroid incidence graphs, *J. Shandong University*, 40 (2) (2005), 24-40.(in Chinese)

[49] L. Li, Matroids and Graphs, Ph.D. Dissertation, Shandong University, (2005).

[50] X. Li. The connectivities of the *SEE*-graph and the *AEE*-graph for the connected spanning k-edge subgraphs of a graph *Discete Math.*, 183 (1998), 237-245.

[51] X. Li, V. Neumann-Lara and E. Rivera-Campo, Two Approaches for the Generalization of Leaf Edge Exchange Graphs on Spanning Trees to Connected Spanning k-Edge Subgraphs of a Graph. *Ars. combin.*, 75 (2005), 257-265.

[52] L. Lovasz, A. Recski, On the sum of Matroids, *Acta Math. Acad. Sci. Hung*, 24 (1973), 329-333.

[53] S. B. Maurer, Intervals in matroid basis graphs. *Discrete Math.*, 11 (1975), 147-159.

[54] U. S. R. Murty, On the number of bases of matroid, *Proc. Second Louisiana Conference on Combinatorics*, (1971), 387-410.

[55] U.S.R. Murty, Extremal critically connected matroids, *Discrete Math.*, 8 (1974), 49-58.

[56] D. Naddef and W. R. Pulleyblank. Hamiltonicity and combinatorial polyhedra. *J. Combin. theory B*, 31 (1981), 279-312.

[57] C. St. J. A. Nash-Williams, Edge-disjiont spanning trees of finite graphs. *J. London Math. Soc.*, 36 (1961), 445-450.

[58] C. St. J. A. Nash-Williams, Decompositions of finite graphs. *J. London Math. Soc.*, 39 (1964), 12.

[59] J. G. Oxley, Matroid Theory, Oxford University Press, New York, 1992.

[60] M. J. Piff, An upper bound for the number of matroids. *J. Comb. Theory*, 13 (1973), 241-245.

10

Path-Finding Algorithm Application for Route-Searching in Different Areas of Computer Graphics

Csaba Szabó and Branislav Sobota
Technical University of Košice
Slovak Republic

1. Introduction

Common graphical notations for graphs use a set of points interconnected by lines. Points are called vertices. The lines called edges can also have in some cases a direction (orientation) specified.

Paths are sequences of vertices and edges between them. Path-finding between two locations in a graph is used in many areas like GPS navigation, CAD systems (auto-routing in design of circuit boards), and applications of artificial intelligence, computer games, virtual reality, military forces or robotics. The main aim of path-finding is to avoid collisions with obstacles and safely pass through the virtual world (or the real one) along the path found by the selected algorithm. Examples of this need are autonomous robots on distant planets without the possibility to be controlled in real time because of latencies in signal sending, or automatic vehicles control presented in (Simon et al., 2006) and (Kuljić et al., 2009). Another examples of path-finding are strategic computer games, mostly with computer opponent. Path-finding is also widely used in finding best routers interconnection for data transmission in many kinds of computer networks.

To be successful in path-finding, the need of reliable algorithm and reliable implementation of that algorithm is enormous. Another important thing is the need of some user-friendly visualization of results of path-finding, which is realized by application of computer graphics techniques (Vokorokos et al., 2010).

A geographical information system (GIS) as defined in (Tuček, 1998) is an information system (IS) designed to work with data that represent geographical or spatial coordinates. These data can be further analyzed and statistically evaluated.

Another way to specify a GIS is that it is a specialized IS defined as a collection of computer hardware (as noted by (Vokorokos et al., 2004)) and software and geographical data designed for effective gathering, retaining, editing, processing, analysis and visualization of all kinds of geographical information. Generally, it is an IS designed as specialized for spatial data, or an IS developed to such specialization during its life cycle by extensions and/or refactoring steps presented by (Szabó & Sobota, 2009).

Being an IS, GIS has to provide information in graphical (e.g. location of the hotel on the map) and non-graphical (e.g. fees, room equipment) way that supports fast searching and actualization of data. Nowadays, graphical information are stored as vector graphics (two-dimensional), but the trends are clearly forecasting the second generation GIS as an IS working with 3D objects and surfaces as in (Szabó et al., 2010).

It is useful to extend the implementation of the above-mentioned searching of objects by route-searching between two selected objects. As the map is the core of any GIS, a possible way of implement path-finding on graphs within such a system is the mapping of these maps onto search-able graphs that will be shown in next sections of this chapter.

Section 2 presents the theoretical background needed in algorithm descriptions in the other following sections. In Sections 3–4, terms such as map and route are defined. Blind-search and A* algorithms are introduced in Section 5. Next section (Section 6) presents the test cases and results: results on static two-dimensional maps, results from tests where three-dimensional (city) maps were used, results of testing with dynamics in the maps. Tests also demonstrate the tested 3D city information system, especially its user interface, and hardware configurations of host computers used during testing. Section 7 ends the chapter by concluding it and pointing out future work.

2. Graph theory basics

Leonhard Euler is considered the founder of graph theory, since he solved the problem known as the Seven Bridges of Königsberg in 1736. Graph theory as a science discipline started with first graph theory monographs written by mathematicians (Kőnig, 1936) and (Berge, 1958). The next part of this section relies on definitions in (Bučko & Klešč, 1999).

Let V be a final non-empty set, i.e. the set of vertices, and let E be the set of edges defined as:

$$E = \{e | e = \{v_1, v_2\} ; v_{1,2} \in V\}. \tag{1}$$

Obviously, $E \subseteq \binom{V}{2}$, where $\binom{V}{2}$ is a set of all two-element subsets of V:

$$\binom{V}{2} = \{A | A \subset V \wedge |A| = 2\}. \tag{2}$$

Considering these two sets presented in Equation 2 and Equation 1, the couple $G = (V, E)$ defines graph G with the corresponding vertices and edges.

Path is a sequence of vertices where an edge exists for any two consecutive vertices of this sequence. Degree of a vertex is the count of edges outgoing from this vertex, e.g. the degree of any internal vertex in a path is not less than two, the end vertices have a degree at least one.

The graph G is a coherent graph if all vertices are reachable, i.e. for all vertices exists at least one path to any other vertex.

Let $G = (V, E)$ be a coherent graph. Distance $d(u, v)$ between vertices u and v of graph G is the length of shortest route linking vertices u, v, i.e. the shortest path. Vertices distance $d(u, v)$ of graph G has the following properties:

$$d(u, v) \geq 0; d(u, v) = 0 \Leftrightarrow u = v$$
$$d(u, v) = d(v, u) \tag{3}$$
$$\forall z \in V : d(u, v) \leq d(u, z) + d(z, v)$$

These properties are metrics axioms; so ordered pair (V, d) is a metrical space. Nevertheless V is a vertical set of graph $G = (V, E)$, d is a metric – vertices distance in graph G. The advantage of these properties is the fact that they consist even after isomorphism is done. One result of this is the fact, that also notions defined by metric have the same advantage.

Let $G = (V, E)$ be a coherent graph, then:

1. For every vertex $u \in V$ the number $e(u, G) = \max d(u, v)$, $v \in V$ is called eccentricity of vertex u.
2. Number $P(G) = \max e(v, G)$, $v \in V$ is average of graph G.
3. Number $r(G) = \min e(v, G)$, $v \in V$ is radius of graph G.
4. The center of graph G is every vertex u of graph G with eccentricity equal to radius $e(u, G) = r(G)$.

Graph diameter can also be defined as $P(G) = \max d(u, v)$; $u, v \in V$.

Graph $G = (V, E)$ is an Eulerian graph, if this graph is coherent and every one of its vertices has even degree. Graph can be covered by a single enclosed stroke (i.e. Eulerian circuit) only if it is an Eulerian graph. Graph $G = (V, E)$ can be covered by a single unenclosed stroke only if it is a coherent graph with two vertices of degree 2 (opened stroke starts in one of these vertices and ends in another).

Let $G = (V, E)$ be a graph, $|V| = n$, $n \geq 3$. Let the degree of every vertex of graph G be at least $n/2$. Then G is a Hamilton graph. If graph $G = (V, E)$ is a Hamilton graph, this graph has to be terminal, non-empty, coherent and may not consist of bridges. Nevertheless fulfillment of these conditions does not guarantee existence of Hamilton graph.

Oriented graph G is defined as ordered pair (V, E), where V is the set of vertices and E is set of ordered element pairs of set V. Elements of V are called graph vertices and elements of E are called oriented graph edges.

3. Maps

The core element of any GIS is the map. All objects have spatial coordinates defining their placement on this map. On other hand, maps are also basic elements if it comes to path-finding or animations. Architecture of such a system for animations based on drawing objects on a 2D map is shown on Fig. 1, its output is presented on Fig. 2.

The specifics and role of maps in the selected domain areas are presented in the following subsections.

3.1 Maps in virtual reality

Simulations use 2D and 3D maps that are dynamically changing due simulation state changes. This type of maps is also used in any interactive graphics such virtual reality city walkthroughs or computer games.

Dynamic of the maps' change is expressed by structural changes in the maps, i.e. at least one object's spatial coordinates have been changed. Changes near found paths might indicate the need of re-running the path-finding algorithm on the new map segments.

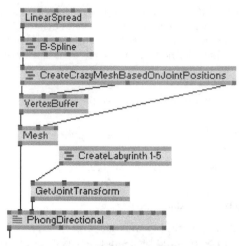

Fig. 1. Architecture of example system for 3D maze drawing using a 2D map

Fig. 2. Example system output with animated actor

3.2 Maps in GIS

Geographical information systems mainly use static maps. These maps could be both 2D or 3D or combined. Mostly, these maps are segments of cadastral maps, where the most important factor is the proper positioning of the segment root (see Fig. 3) that could be achieved by manual operation or using image filters (Póth, 2007).

3.3 Circuit boards as maps

Printed Circuit Boards (PCB) are well tested and proven technology. A manufacturer of electronics cannot imagine manufacturing of electronic equipment without this technology. Reliable operation of the current integrated circuits with high operating frequencies requires a minimum length of printed circuit conductors. Therefore, it is necessary to pay close attention to the PCB design mainly in the case of the optimal length of the printed circuit conductors. The conductive pattern may be on one side or both sides of the board and might be linked

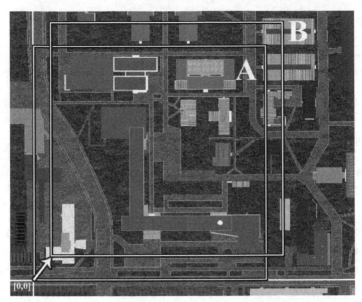

Fig. 3. Correct (A) and incorrect (B) position of a segment of the city map designated for GIS internal map creation

in various ways. In the case of multilayer PCB, it might be between a few plates of copper conductive patterns. The creation and routing of conductors on PCB is the key of PCB design. In principle, the addressed issues are the issues discussed in this chapter.

The algorithm used in PCB design applications is the principal factor of their success. The most used algorithms in this area are heuristic algorithms, channel algorithm, rip-up algorithm (rip and reroute) and maze algorithms. These algorithms include, for example Lee router or also algorithms such as Blind search or A* described and compared later in this chapter.

In short, Lee router was first published more than 40 years ago and it was successfully implemented 20 years ago as a maze algorithm (Brown & Zwolinsky, 1990). In its simplest form, this router represents the algorithm to find a path between two points in 2D area with various obstacles. Its main features are ability to found always a path between the points, if any, within the limits of the work area size (PCB size) and this path is always the shortest route between two points.

However, it is very slow and, moreover, cannot organize its data structure to minimize gaps, and so it is being replaced by above mentioned algorithms now. Lee router organizes PCB space to the network of points. All unfilled points in the basic network are unmarked during algorithm initialization. The algorithm consists of two phases: an exploratory phase and a processing phase. Exploratory phase is similar with seed-fill algorithm used in computer graphics. The target point is on last place in the list and we travel back over the list of unfilled points to the original seeds in the subsequent processing phase. This original seed represents a starting point. Very simple and efficient improvement is the possibility to use 45 degrees angles too.

3.4 From map to graph representation

The basic working structure of algorithms is the map, which is represented as a set $M = N \times N$ (two-dimensional quadratic grid). Square is a member of set M. Every square represents position in map and has its own coordinates $[x, y]$ according to beginning P_m. Squares $[x1, y1]$ and $[x2, y2]$ are neighbors, if ($|x1 - x2| = 1$ and $|y1 - y2| <= 1$) or ($|x1 - x2| <= 1$ a $|y1 - y2| = 1$) (they are neighbors geometrically). The beginning in map P_m is a defined square. This representation is called map representation.

Next, a function is defined:

$$w_m : M \to \mathbb{R}^+ \cup \{+\infty\}. \tag{4}$$

This function assigns regress (e.g. weight, terrain cost) to every square. This function determines the difficulty of input for each square (or square traversing). If the regress is $+\infty$ then the square is impassable and contains an obstacle (black color nodes in Fig. 4, white color indicates passable vertices). This function is simple only if it contains a two-element set 1 and $+\infty$. In the map, the start square of the route is marked A, the target square is marked B.

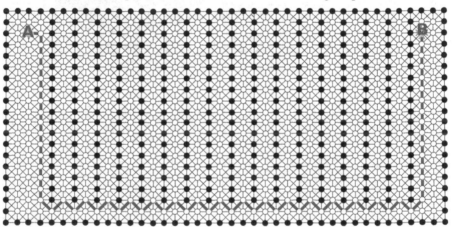

Fig. 4. Example graph of a maze with walls (black color), lanes (white color) and a route (red color)

4. Routes

Route in map is a sequence of map squares $(m_1, m_2, \ldots, m_n)$ in which every next square is neighbor to previous square on the route. There are no squares on the route that are included more than once. The route length is the sum of regresses of all route squares:

$$d = \sum_i w_m(m_i). \tag{5}$$

5. Algorithms

The algorithm task is to find the shortest route in map from the start to the target, i.e. from A to B.

The map representation is a special graph example. Vertices correspond to squares. The bijection transforming the graph to squares will be called *map*. Transition between neighbor squares corresponds to graph edge and initial vertex in graph P_g corresponds to initial vertex in map P_m. Next function to define is w, which will perform weight calculation on the graph. Function w_g is defined as representation from set of edges to set of non-negative real numbers with $+\infty$:

$$w_g : E \to \mathbb{R}^+ \cup \{+\infty\}. \tag{6}$$

For edge $e = (p, q)$, function w_g is defined as:

$$w_g(e) = w_m(map(q)). \tag{7}$$

Heuristics is defined as function $h : V \to \mathbb{R}^+ \cup \{+\infty\}$, which assigns the expected distance to target to vertex v. It estimates the length of shortest route from vertex v to the target.

The most common heuristics used for path-finding in 2D static maps are:

- Euclidean distance or Euclidean metric,
- Manhattan method that uses $dx + dy$, and
- the method of the maximum that uses $\max(dx, dy)$ as heuristics value.

There are several algorithms for route-searching in static maps:

- blind-search
- divide & conquer
- breadth-first search
- bidirectional breadth-first search
- depth-first search
- iterative depth-first search
- Dijkstra's algorithm
- best-first search
- Dijkstra's algorithm with heuristics (A*)

In our tests, the blind-search algorithm and Dijkstra's algorithm with Manhattan method of heuristics (A*) were used.

5.1 Blind-search algorithm

The blind-search algorithm (example results are shown in Fig. 5(a)–5(c)) is based on progress to the target, until it collides with obstacle, then the direction of movement is changed and the algorithm tries to move along nearest to get around the obstacle. This algorithm works with aim on one square and usually does not take into consideration the whole evaluation function w, but only its two states represented by values $+\infty$ or 1 (i.e. other than $+\infty$). So it works only with simple evaluating function and does not count with regress. This algorithm does not guarantee the finding of shortest route and can be used on special maps because of its speed, low memory requirements and simple implementation.

The evaluation process is as follows:

$$abs\left[(e_x - t_x) + (e_y - t_y)\right], \tag{8}$$

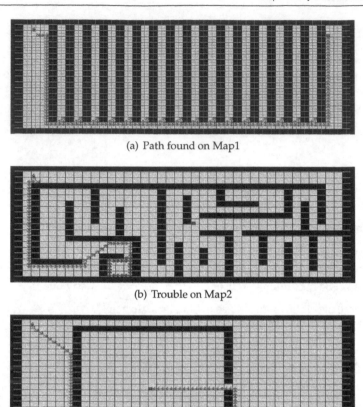

(a) Path found on Map1

(b) Trouble on Map2

(c) Path found on Map3

Fig. 5. Example paths found by the blind-search algorithm

where $e_{x,y}$ are the $[x, y]$ coordinates of the actually evaluated square and $t_{x,y}$ are coordinates of target square on the map.

The square with the lowest evaluation is chosen and set as new starting square for the next iteration. Continuing using this mechanism, the algorithm moves to target with the fastest possible approach. Because of the use of a simple evaluation, this algorithm is one of the fastest algorithms for path-finding.

Problem occurs when the evaluated square is impassable. If this situation occurs, the algorithm stops to evaluate squares and remembers the obstacle direction and then tries to avoid this obstacle. The avoid process is done by choosing the nearest available square, which is gained by turning the movement direction into right. This square is the new starting square, which is not evaluated but only checked if the direction with obstacle is not released. If yes, then the square in this direction is set as new starting square and the evaluation process starts again. If not, then the algorithm continues with new direction and tests first route. By this movement the algorithm can collide with new obstacles. It remembers again only the direction with obstacle and sets new course. This new course is tested and if it is released, then the original direction is tested. If the algorithm collides with obstacle it stops to evaluate squares and tests transitivity of only one square. This can fasten the algorithm but can also become the algorithm into endless loop (see Fig. 5(b)).

5.2 A* algorithm

The A* algorithm (test results on Fig. 6(a)–6(c)) is a natural generalization of Dijkstra's algorithm and the scanning is based only on heuristics. In A* the element classification process is done by a binary heap. All square near the evaluated square are also evaluated in eight directions. Function g(x) is evaluated from the starting square. To work with integer values, a modification is made to the representation of Euclidean distance as follows:

- If the evaluated square is in the diagonal, then the number 14 is assigned to this square, and
- if not, then number 10 is assigned to this square.

As heuristics, the Manhattan method is used:

$$10 * \left[abs(e_x - t_x) + abs(e_y - t_y) \right] , \qquad (9)$$

i.e. the distance between $dx + dy$ is computed. Sequence function is defined as $f(x) = g(x) + h(x)$. The algorithm description follows.

Assume that graph $G = (V, E)$ has already evaluated edges, i.e. the mapping and weight computation is already done. Two values for every already processed vertex will be stored: $d[v]$ will store the length of the shortest path to the vertex, $p[v]$ will be the vertex before v on the path. The algorithm also uses two sets: $OPEN$ and $CLOSED$. $OPEN$ is the set of vertices to process, $CLOSED$ includes vertices already processed as presented in (Sobota, Szabó & Perháč, 2008):

1. At the beginning, the starting vertex is inserted into $OPEN$.
2. The iteration cycle starts with the choose of the vertex n with best value of $f(n)$. This vertex will be inserted into set $CLOSED$ and all from n reachable vertices v are selected, and if they are not in set $OPEN$, they will be inserted with value $p[v] = n$ and the corresponding value of $d[v] = d[n] + l(n, v)$, where $l(n, v) = w_g(v)$ is the cost function for the selected edge from n. If v was already a member of $OPEN$, a test is needed if the new path is shorter than the old one by comparing $d[v] < d[n] + l(n, v)$. If the comparison fails, update of the values $p[v]$ and $d[v]$ is performed.
3. The presented cycle continues until the target is found or the set $OPEN$ is empty.
4. After the algorithm had finished, the path is reconstructed using values of $p[v]$.

Fig. 6(a)–6(c) show examples of usage of the A* algorithm on 2D static maps to compare with the results of the blind-search algorithm from the previous section.

There are possible modifications to this algorithm that can decrease memory usage and fasten the A* algorithm. The most common techniques optimize the number of vertices within the sets $OPEN$ and $CLOSED$ by discarding the less useful ones. This technique limits the count of the elements of each set separately to get best performance but do not miss some possible paths. The common problem is the grid resolution selection: high resolution rapidly slows down the algorithm execution; low resolution may cause loss on important information as smaller buildings and/or gates. The maximum sizes of sets $OPEN$ and $CLOSED$ were

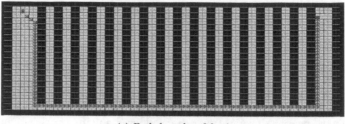

(a) Path found on Map1

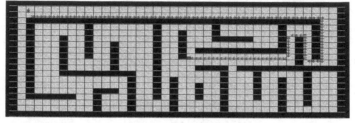

(b) Path found on Map2

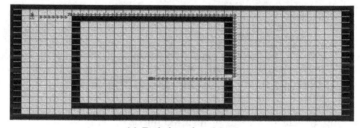

(c) Path found on Map3

Fig. 6. Example paths found by the A* algorithm

examined and experimentally defined in (Sobota, Szabó & Perháč, 2008) and (Vlasova, 2008) as 1000 and 3000.

6. Algorithm use cases and evaluation results

6.1 Test results for static 2D maps - the maze example continued

A couple of tests have been run on two-dimensional static maps to compare the blind-search and A* algorithms, which implementations were based on (Adams, 2003; Barron, 2003). Ten (10) maps were tested.

Routes found differ in length (compare results on Fig. 5(c) and Fig. 6(c)). The route always starts with red arrow. If the route is returning by the same route backward, then the arrow is changed to yellow and the route continues with yellow arrow. If the route crosses itself (blind-search algorithm), then it is in cycle and the arrow will be marked by boxed blue color (see 5(b)). Yellow square is start square, blue square is target square.

The efficiency of blind-search algorithm is only 40% because it found a route only in four of 10 maps. A* algorithm searched in all maps 14 895 squares together. Blind-search needed to search 83 928 squares, because it used to cycle itself (see Fig. 5(b)) and the condition of cycle detection was at 10000 cycles. The blind-search algorithm searched 5.64 times more squares than A* algorithm. Considering only those maps, where both algorithms succeeded, blind-search algorithm would be the better one in speed and number of squares searched, respectively.

We present the test results of both algorithms on Fig. 7. It is obvious that blind-search was not so effective like the A* algorithm when compared the count of squares searched on all maps. The speeds on all maps are not comparable due to the low efficiency of the blind-search algorithm.

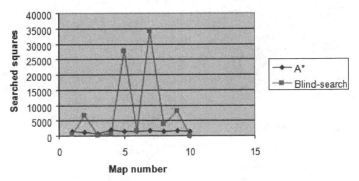

Fig. 7. Statistical results comparison for both algorithms gathered on 10 maps

6.2 Testing with static 3D city maps

Two-dimensional maps, as presented above, are a good testing tool, but in real systems three-dimensional ones are more frequently asked. Representing higher fidelity to real world, virtual reality techniques help to build more user-friendly and more precise information systems.

Dealing with geographical information, a three-dimensional map (or more precisely, the visualization of it) introduces a few new problems into the path-finding by extending the amount of information to be searched.

Therefore, tests on more complicated surfaces and maps with building objects, hills etc. were also run. The result is shown on Fig. 8 in the form of a route found from point A to point B between the buildings of a randomly generated city.

Tests on optimizing of sizes for sets $OPEN$ and $CLOSED$ were also performed. As already mentioned, in (Vlasova, 2008) were made some experiments under the supervision of the authors. The testing conditions were as follows:

- each test case was run separately on the same hardware (no sequences etc.),
- visualization (rendering) speed was set to minimum, and
- each test case was run ten times and results were averaged.

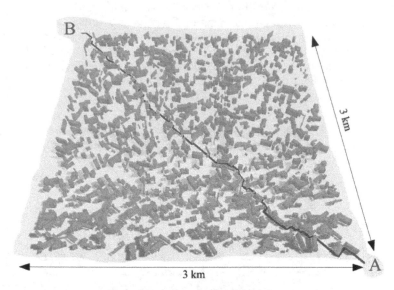

Fig. 8. Path found by A* algorithm on a random 9km² city map

Optimization technique	Average execution time [s]
none	489
CLOSED	458
OPEN	368
OPEN & CLOSED	346

Table 1. Speed test results for A* algorithm optimization

The test results are shown on Table 1 and Table 2. The best performance was achieved by dual optimizing (i.e. using both limits), but the optimization of size of set *OPEN* had stronger effect on execution speed than the size optimization of set *CLOSED*.

Optimization technique	Path length [m]	$\|OPEN\|$	$\|CLOSED\|$
none	5210	6022	5210
CLOSED	5210	6022	3000
OPEN	5210	1000	5263
OPEN & CLOSED	5210	1000	3000

Table 2. Set sizes and path lengths during A* algorithm optimization

6.2.1 The 3D City IS test case

The tested A* algorithm was implemented within a GIS called the 3D city IS, details of the system can be found in (Sobota et al., 2010; Sobota, Korečko & Perháč, 2009; Sobota, Perháč & Petz, 2009; Sobota, Szabó, Perháč & Ádám, 2008; Sobota, Szabó & Perháč, 2008).

Fig. 9(a) shows the main screen of the tested GIS. There are a few building objects shown on the base map. In the upper right corner are navigation buttons (ref. no. 3 for minimize, ref.

no. 4 – close). Upper left corner contains also two buttons: ref. no. 1 is for opening a map; ref. no. 2 point on the options menu. 'Options' are e.g. map resolution. The ref. number 5 indicates the search button, which functionality was tested in the test cases.

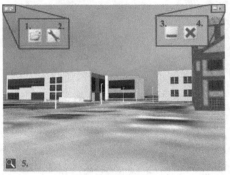 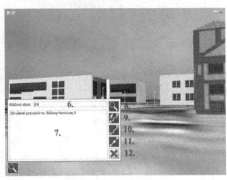

(a) The main screen   (b) Path searching dialog

Fig. 9. 3D city IS test case setup

Selecting this last option, a search dialog opens that is shown on Fig. 9(b). As any classical search dialogs, it has a field for the search query (reference number ref. no. 6), results displaying area (ref. no. 7), search execution button (ref. no. 8), animation button (ref. no. 9), path display button (ref. no. 10), object display button (ref. no. 11), and close dialog button (ref. no. 12).

The testing procedure is simple:

1. Open a map, e.g. in this case each of the following maps was used:
   • a map with some known labeled objects randomly generated by other tool,
   • the real (but yet only partial) city map, and
   • the map of the university campus.
2. Type a search phrase and evaluate what has been found and displayed.
3. Now follows the test of path-finding abilities.
   • Function go-to-object only moves the camera to the selected target.
   • Function show-path opens after a few minutes the resulting route from the actual position on the map to the selected target as shown on Fig. 8.

The speed is faster with less resolution, but some narrow streets have been missed. With the campus map are nearly real-time results due to the small count of objects.

The virtual reality oriented part of the tested geographical information system is the animation function. The path-finding module was shared with the other already mentioned parts. The test case had to ensure the quality of the obstacle detection. This part offers path visualization by circles and the animated walkthrough of the scene rendered from the map and additional object information stored in the GIS database. Fig. 10 shows the animation screen. Buttons labeled 13–16 are classical navigation buttons for movement within the animation. This test case is evaluated by expressing the quality and feel by human testers; and were used in the debugging phase of development of the path animation part of the system.

Fig. 10. Walk-through animation of 3D city information system

6.3 Path-finding results in dynamic 2D maps

In virtual reality applications (also in computer simulators and games), the effect of reality is significantly achieved by dynamics of the scene.

For problem simplification, the next test case uses pseudo 3D solution, meaning that the scene is a composition of 2D (map) and 3D objects. Note that in this case path-finding algorithms have to deal with less data, i.e. weight function and final calculations could be performed faster. The same simplification could be made in the case of any type of systems listed above.

There were 10 maps created with different amount and positions of static obstacles. Tests were provided for 1, 8, 16, 24, 32, 40, 48, 56, 64, 72, 80, 88, 96, 192 and 384 active moving objects called units. Test cases were oriented on calculation time (e.g. speed of path-finding in relation to CPU) and visualization time (e.g. the role and impact of GPU) while comparing the two selected algorithms.

Tables 3 and 4 show the contrast between the algorithms during the testing process. While considering all maps as in Table 3 A* algorithm seems to be better in every case, the fact is that there are averaged values for all maps, where blind-search algorithm failure to find a path incorrectly implies being this algorithm the slower one. Considering only calculation time on maps, where both algorithms succeed shown in Table 4 reflect that blind-search algorithm is faster but less stable.

Next group of tests was designed to compare speeds of both path-finding algorithm implementations on different hardware CPUs. Table 5 shows the test configurations used.

Although the implementations should not be affected by the hardware and results are not used as CPU metrics, Tables 6–7 present the values. The only conclusion to make is that the slower A* algorithm could be evaluated faster on a fast CPU than the faster blind-search algorithm using a slower CPU.

Units	1	8	16	24	32	40	48	56
A* [s]	0.022	0.146	0.281	0.381	0.514	0.658	0.849	0.966
BS [s]	0.043	0.42	0.853	1.262	1.707	2.303	2.756	3.233
Units	64	72	80	88	96	192	384	
A* [s]	1.077	1.126	1.324	1.453	1.665	3.169	6.343	
BS [s]	3.717	4.247	4.705	5.173	5.671	11.376	23.799	

Table 3. Average time for 10 maps for A* and blind-search (BS) algorithm in relation to No. of units

Units	1	8	16	24	32	40	48	56
A* [s]	0.009	0.058	0.112	0.152	0.206	0.263	0.340	0.386
BS [s]	0.001	0.008	0.016	0.025	0.032	0.044	0.052	0.058
Units	64	72	80	88	96	192	384	
A* [s]	0.431	0.450	0.530	0.581	0.666	1.268	2.537	
BS [s]	0.065	0.075	0.081	0.087	0.097	0.189	0.390	

Table 4. Average time for A* and blind-search (BS) algorithm in relation to No. of units (for maps, where BS succeed)

Name	CPU	GPU	RAM
TConf1	AMD Duron 1200 MHz	Geforce MX400/64MB	256 MB
TConf2	AMD Athlon 1700+	Radeon 9200SE	512 MB
TConf3	AMD Athlon 2000+	Geforce MX460/64MB	512 MB
TConf4	AMD Barton 2500+	Geforce 5900	2 GB
TConf5	Intel P4 Centrino 1.7 GHz	Intel GMA 900	512 MB

Table 5. Hardware test configurations

Units	1	8	16	24	32	40	48	56
TConf1 [s]	0.000	0.016	0.031	0.047	0.047	0.062	0.094	0.094
TConf2 [s]	0.002	0.016	0.020	0.032	0.042	0.053	0.066	0.074
TConf3 [s]	0.002	0.010	0.018	0.029	0.039	0.046	0.056	0.065
TConf4 [s]	0.002	0.007	0.017	0.023	0.031	0.039	0.047	0.055
TConf5 [s]	0.000	0.016	0.015	0.016	0.031	0.031	0.047	0.047
Units	64	72	80	88	96	192	384	
TConf1 [s]	0.109	0.125	0.125	0.140	0.156	0.328	0.640	
TConf2 [s]	0.086	0.090	0.106	0.117	0.125	0.252	0.504	
TConf3 [s]	0.074	0.085	0.095	0.104	0.113	0.226	0.452	
TConf4 [s]	0.065	0.073	0.078	0.085	0.092	0.189	0.377	
TConf5 [s]	0.047	0.062	0.062	0.078	0.078	0.157	0.297	

Table 6. Average times for A* algorithm using different hardware configurations

Last group of tests was aimed to measure average time for animations of unit movement on path from A to B. In test realization, for places A and B constant locations were used. Time was measured including path-finding, but the additional condition was to achieve the same formation of the units in both locations, e.g. for 96 units a 8×12 rectangle. Test results are shown in Table 8 and Table 9.

Units	1	8	16	24	32	40	48	56
TConf1 [s]	0.000	0.000	0.000	0.016	0.031	0.031	0.031	0.031
TConf2 [s]	0.000	0.007	0.014	0.012	0.017	0.021	0.027	0.035
TConf3 [s]	0.000	0.004	0.008	0.012	0.017	0.022	0.027	0.032
TConf4 [s]	0.000	0.003	0.006	0.009	0.013	0.016	0.020	0.023
TConf5 [s]	0.000	0.000	0.000	0.000	0.000	0.000	0.000	0.015

Units	64	72	80	88	96	192	384
TConf1 [s]	0.047	0.047	0.063	0.063	0.063	0.109	0.250
TConf2 [s]	0.040	0.044	0.045	0.052	0.055	0.117	0.229
TConf3 [s]	0.035	0.040	0.042	0.046	0.051	0.104	0.209
TConf4 [s]	0.029	0.032	0.035	0.039	0.040	0.084	0.168
TConf5 [s]	0.015	0.031	0.031	0.031	0.031	0.078	0.156

Table 7. Average times for blind-search algorithm using different hardware configurations

Units	1	96	192	384
TConf1 [s]	39	63	84	124
TConf2 [s]	31	31	35	50
TConf3 [s]	15	19	30	35
TConf4 [s]	13	13	25	28
TConf5 [s]	22	28	43	50

Table 8. Average animation times for A* algorithm using different hardware configurations

Units	1	96	192	384
TConf1 [s]	55	92	117	167
TConf2 [s]	43	43	49	70
TConf3 [s]	21	27	42	49
TConf4 [s]	18	19	34	39
TConf5 [s]	31	39	60	68

Table 9. Average animation times for blind-search algorithm using different hardware configurations

7. Conclusion

This chapter presents a view on graph theory from point of view of virtual reality. Applications and testing of blind-search algorithm and the modified Dijkstra's A* algorithm of path-finding in graphs within a geographical information system were presented.

From the test results some conclusions can be made. The blind-search algorithm is fast but is not accurate enough. It can be used in maps with low number of obstacles only when the results have to be gained as fast as possible and the route does not have to be the shortest one.

A* algorithm is the most used algorithm and showed its qualities in our tests as well. Despite time results (the speed of this algorithm is not the fastest), this algorithm is reliable and has provided better results. The optimization techniques of set size restrictions were useful and brought higher speed into execution of path-finding.

The implementation of both algorithms showed up as useful, but the significant difference of applicability between them excludes the blind-search algorithm from the set of candidates of route-searching algorithms for a GIS.

Algorithms were implemented into a 3D city information system as a prototype of a GIS. The included functionalities except those presented include the object browser and map editor. The future plan is to interconnect the system with other systems for geographical data processing such points clouds analyzers, video or ortophoto processors etc.

The similarity to today's GPS in the case of route- searching is intentional. According to the trends in the area, future navigation systems might offer three-dimensional interfaces as well. The only difference is in the type of the map, because our system uses a static map while a good navigation system uses a dynamic one that might future GIS use too.

As mentioned in the introducing section, the use of graph theory becomes common in modern computer graphics and virtual reality systems. Collision detection is the main area of it, but, as this article implies, the information system area should follow these trends as well by implementing its algorithms into the visualization and searching modules.

8. Acknowledgements

Authors thank R. Šváby, M. Šváby (M. Vlasova) and V. Kríž for their constructive work in the implementation and debugging phase of development of the presented computer graphics software applications.

This work was supported by VEGA grant project No. 1/0646/09: "Tasks solution for large graphical data processing in the environment of parallel, distributed and network computer systems."

9. References

Adams, J. (2003). *Advanced Animation with DirectX*, Premier Press, a division of Course Technology.

Barron, T. (2003). *Strategy Game Programming with DirectX 9.0*, Wordware Publishing Inc.

Berge, C. (1958). *Théorie des graphes et ses applications*, Vol. II of *Collection Universitaire de Mathématiques*, Dunod, Paris 1958 (English edition, Wiley 1961; Methuen & Co, New York 1962; Russian, Moscow 1961; Spanish, Mexico 1962; Roumanian, Bucharest 1969; Chinese, Shanghai 1963; Second printing of the 1962 first English edition. Dover, New York 2001).

Brown, A. & Zwolinsky, M. (1990). Lee router modified for global routing, *CAD* (5): 296–300.

Bučko, M. & Klešč, M. (1999). *Discrete Mathematics, Diskrétna matematika (orig. Slovak title)*, Academic Press elfa, Košice.

Kőnig, D. (1936). *Theorie der Endlichen und Unendlichen Graphen: Kombinatorische Topologie der Streckenkomplexe*, Akad. Verlag, Leipzig.

Kuljić, B., Simon, J. & Szakáll, T. (2009). Pathfinding based on edge detection and infrared distance measuring sensor, *Acta Polytechnica Hungarica* 6(1): 103–116.

Póth, M. (2007). Comparison of convolutional based interpolation techniques in digital image processing, *Proc. of the 5th Int. Symposium on Intelligent Systems and Informatics, SISY 2007*, Subotica, Serbia, pp. 87–90.

Simon, J., Szakáll, T. & Čović, Z. (2006). Programming mobile robots in ANSI C language for PIC MCU's, *Proc. of the 4th Serbian-Hungarian Joint Symposium on Intelligent Systems, SISY 2006*, Subotica, Serbia, pp. 131–137.

Sobota, B., Hrozek, F. & Szabó, Cs. (2010). 3D visualization of urban areas, *Proc. of the 8th Int. Conf. on Emerging eLearning Technologies and Applications, ICETA 2010*, Stará Lesná, Slovakia, pp. 277–280.

Sobota, B., Korečko, Š.. & Perháč, J. (2009). 3D modeling and visualization of historic buildings as cultural heritage preservation, *Proc. of the Tenth Int. Conf. on Informatics, INFORMATICS 2009*, Herlany, Slovakia, pp. 94–98.

Sobota, B., Perháč, J. & Petz, I. (2009). Surface modelling in 3D city information system, *J. of Comp. Sci. and Control Systems* 2(2): 53–56.

Sobota, B., Szabó, Cs., Perháč, J. & Ádám, N. (2008). 3D visualisation for city information system, *Proc. of Int. Conf. on Applied Electrical Engineering and Informatics, AEI 2008*, Athens, Greece, pp. 9–13.

Sobota, B., Szabó, Cs. & Perháč, J. (2008). Using path-finding algorithms of graph theory for route-searching in geographical information systems, *SISY 2008 – Proc. of the 6th International Symposium on Intelligent Systems and Informatics*, Subotica, Serbia, pp. 1–6.

Szabó, Cs. & Sobota, B. (2009). Some aspects of database refactoring for GIS, *Proc. of the Fourth Int. Conf. on Intelligent Computing and Information Systems, ICICIS 2009*, Cairo, Egypt, pp. 352–356.

Szabó, Cs., Sobota, B. & Kiš, R. (2010). Terrain LoD in real-time 3D map visualization, *8th Joint Conference on Mathematics and Computer Science, Selected Papers*, J. Selye University, Komárno, Slovakia, pp. 395–402.

Tuček, J. (1998). *Geographical Information Systems, Geografické informační systémy (orig. Czech title)*, Computer Press, Praha.

Vlasova, M. (2008). *3D city information system*, Master's thesis, DCI FEEaI TU, Košice, Slovakia.

Vokorokos, L., Blišťan, P., Petrík, S. & Ádám, N. (2004). Utilization of parallel computer system for modeling of geological phenomena in GIS, *Metalurgy J.* 43(4): 287–291.

Vokorokos, L., Danková, E. & Ádám, N. (2010). Parallel scene splitting and assigning for fast ray tracing, *Acta Electrotechnica et Informatica* 10(2): 33–37.

11

Application of the Graph Theory in Managing Power Flows in Future Electric Networks

P. H. Nguyen[1], W. L. Kling[1], G. Georgiadis[2],
M. Papatriantafilou[2], L. A. Tuan[2] and L. Bertling[2]
[1]Eindhoven University of Technology,
[2]Chalmers University of Technology,
[1]The Netherlands
[2]Sweden

1. Introduction

Electrical power system is one of the largest and most crucial engineering systems which spreads everywhere in countries to supply electricity for hundreds of millions of consumers from hundreds of thousands of producers. In its more than one century history, the development of the power system proceeded through evolutional stages from local isolated networks with small-scale load and generation to large interconnected networks with myriads of consumers vertically fed by centralized generation, such as coal, hydro, or nuclear power plants. System stability and reliability have been continuously improved along this development of the power systems.

Nowadays, the society not only demands a high level of reliability of electricity supply but also concerns with the environmental impacts from the electrical power system. To achieve a reliable and sustainable electricity supply, there is an increasing need to use energy from renewable sources such as wind, solar, or biomass. The development of many intermittent and inverter-connected Renewable Energy Sources (RES) will require having new ways of planning, operating, and managing the entire process. In other words, the power system is moving into a new development era.

Due to the expected large-scale deployment of distributed generation (DG), the electrical power system is changing gradually from a vertically controlled and operated structure to a horizontally one. Fig. 1 illustrates one of the expected changes in the power system with the large penetration of DG. The integration of RES related to both large-scale production (e.g., wind and solar farms) and massively distributed production (e.g., micro combined heat and power plants and photovoltaic systems at residential and tertiary buildings) causes a number of challenges in fluctuating/unpredictable and bidirectional power flows in distribution networks. Future electrical power systems must be able to manage these bidirectional power flows, and has to deal with the uncertainties of renewable power generation. Power flow management will be needed in order to cope with these challenges.

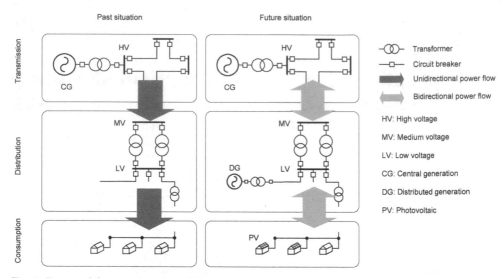

Fig. 1. Past and future situation of the power system

In this chapter, applications of the graph theory to handle the function of power flow management will be introduced. Detail descriptions of these methods can be found in (Nguyen et al., 2010(a); Nguyen et al., 2010(b)).

2. Power flow management

The use of the Optimal Power Flow (OPF) framework is a common practice for managing power flow within the electric transmission network, where the algorithm is centralized and deployed at the economic dispatch stage (Huneault & Galiana, 1991). The mathematical model of the OPF problem can be presented as follows:

$$\text{Minimize } f(x, u) \tag{1}$$

subject to:

$$g(x, u) = 0 \tag{2}$$

$$h(x, u) \leq 0 \tag{3}$$

where $f(x, u)$ is the objective function that can be formulated to represent different operational goals, e.g., minimization of total power production cost or total power loss. The vector of independent variables u represents the state of the system, i.e., the phase angles and voltage magnitudes. The vector of dependent variables x represents the control variables, for example, power generations or tap ratios of On-Load Tap Changer (OLTC) of transformers. The equality constraint represents the power balance between supply and demand while the inequality constraint shows the operational limits of the network components.

The OPF requires a large-scale control overview of the whole network that is difficult to implement in the future situation with high penetrations of DG units. Although some distributed OPF techniques have been proposed, they need complex input information and take relatively long processing times (Kim & Baldick, 2000). Along with OPF, stability constraints and congestion problems can implicitly be investigated (Gan et al., 2000; Bompard et al., 2003). However, those procedures are most suitable for the transmission networks which have a limited number of network components, e.g., generators, transmission lines, and substations.

Since the electric networks of the future are expected to include numerous DGs dispersed over wide areas, other solutions for power flow management must be founded. A price-based control method, which can also be considered as a distributed OPF solution, has been proposed for systems with high level of DG penetration (Jokic, 2007). By converting the power system parameters into desired market signals, the solution yields nodal prices for generators that help to mitigate the network congestion problem and also contribute to other so called ancillary services. This can be presented in a mathematical model as follows:

$$\text{Minimize } \sum_{i=1}^{n} f(P_i, P_i^{ex}, A_i, A_i^{ex}) \tag{4}$$

subject to:

$$P_i - P_i^{ex} - P_i^{load} = 0 \tag{5}$$

$$A_i - A_i^{ex} - A_i^{req} \geq 0 \tag{6}$$

$$g(P_i, A_i) \leq 0 \tag{7}$$

where $f(P_i, P_i^{ex}, A_i, A_i^{ex})$ is the aggregated cost function of an autonoumous network part i; P_i and A_i are the generated power and the provided ancillary service respectively in the network i, and P_i^{ex} and A_i^{ex} present the generation and services coming from outside of the network i; the equality constraint represents for power balance; the upper bound condition denotes requirements of ancillary services while the lower bound condition shows the operational limits of network components.

The method, however, concerns only the supply and demand of the system in which actors can be influenced by price signals. Other controllable devices of the system, e.g., electronic-based power flow controllers, are not considered. In (Dolan et al., 2009), the power flow management is represented as constraint satisfaction. Algorithms in that research determine the level of DG curtailment because of loading constraints in a small test network. Based on a local information network, a distributed two-level control scheme was proposed to adjust power output from clusters of photovoltaic (PV) generators when disturbances occur (Xin et al., 2011).

3. Power routing – A new way for power flow management

The function of distributed power routing is to fully exploit the potential of the local resources in managing the power flow. The function deals with transport optimization related to the actual load and generation schedules of the market parties. Price signals yielded by the market clearing conditions can be used as an input for the routing algorithms to achieve certain operational objectives, e.g., relief of network congestion, maximization of the reliability of the network, minimization of losses and if desired minimization of the production cost and maximization of serving high-priority customers.

In this section, the function of distributed power routing is proposed for the future electric network as a new way to manage power flows. Basically, the function of power routing is the same as the optimization of the power flow which can be formulated in a mathematical model as follows:

$$\text{Minimize } \mathcal{F} =$$
$$\sum_{i \in \mathcal{G}}(\omega_{Gi}^{+}\Delta P_{Gi}^{+} + \omega_{Gi}^{-}\Delta P_{Gi}^{-}) + \sum_{(i,j) \in \mathcal{T}}(\omega_{Tij}^{+}\Delta P_{Tij}^{+} + \omega_{Tij}^{-}\Delta P_{Tij}^{-}) + \sum_{k \in \mathcal{D}}(\omega_{Dk}^{+}\Delta P_{Dk}^{+} + \omega_{Dk}^{-}\Delta P_{Dk}^{-}) \quad (8)$$

subject to:

$$\sum_{i \in \mathcal{G}} P_{Gi} = \sum_{(i,j) \in \mathcal{T}} P_{Tij} + \sum_{k \in \mathcal{D}} P_{Dk} \tag{9}$$

$$P_{Gi}^{min} \leq P_{Gi} \leq P_{Gi}^{max}, \qquad \forall i \in \mathcal{G} \tag{10}$$

$$\left| P_{Tij} \right| \leq P_{Tij}^{max}, \qquad \forall (i,j) \in \mathcal{T} \tag{11}$$

$$\left| P_{Tij} \right| \pounds P_{Tij}^{max}, \qquad "(i,j) \hat{I} T \tag{12}$$

where

$$P_{Gi} = P_{Gi}^{0} + \Delta P_{Gi}^{+} - \Delta P_{Gi}^{-}, \qquad \forall i \in \mathcal{G} \tag{13}$$

$$P_{Tij} = P_{Tij}^{0} + \Delta P_{Tij}^{+} - \Delta P_{Tij}^{-}, \quad \forall (i,j) \in \mathcal{T} \tag{14}$$

$$P_{Dk} = P_{Dk}^{0} + \Delta P_{Dk}^{+} - \Delta P_{Dk}^{-}, \qquad \forall k \in \mathcal{D} \tag{15}$$

The objective function (8) is the total cost for re-routing power when a disturbance occurs in the system. As the power routing function might change power generation at each bus i (ΔP_{Gi}^{+}, ΔP_{Gi}^{-}), power flow through each network device i-j (ΔP_{Tij}^{+}, ΔP_{Tij}^{-}), and demand at each bus k (ΔP_{Dk}^{+}, ΔP_{Dk}^{-}) different from the original market clearing conditions (P_{Gi}^{0}, P_{Tij}^{0}, P_{Dk}^{0}), the objective function takes these representative costs into account. Due to the fact that most of the renewable generation can participate only in down regulation (with cost ω_{Gi}^{-}), integration of storage devices becomes important to give the power routing function flexibility in up regulation (with cost ω_{Gi}^{+}) of generation. Power flow change on network devices influences the total power losses and reliability of the system. However, the associated transmission costs ($\omega_{Tij}^{+}, \omega_{Tij}^{-}$) as considered to be low compared to the other costs. Since the demand side becomes more active with mechanisms of Demand Side Management (DSM) and Demand Response (DR), its potential in regulating demand up (with cost ω_{Dk}^{+}) and down (with cost ω_{Dk}^{-}) are considered in the objective function.

The power balance condition is represented in the equality constraint (9). The inequality constraints in (10) and (11) show the generation and consumption boundaries. The transmitted power needs to be within the device's thermal limits in the inequality constraint (12). This optimization model assumes that voltage is autonomously controlled and reactive power is not considered in the formulation.

4. Graph-based algorithms

Graph theory has been utilized in some power system applications, such as wholesale cross-border trading by using a shortest path algorithm (Wei et al., 2001), and achieving maximum power transmission with FACTS devices (Armbruster et al., 2005). This section will focus on solving minimum cost flow problem by using successive shortest path and scaling cost-relabel algorithm.

The power system, firstly, is converted to a graph $G(V,E)$, where V represents the set of vertices (buses in the network) and E represents arcs (interconnection lines among buses in the network). The arc length (arc cost) c_{ij} and residual (available) capacity r_{ij} associated with each arc (i, j) are derived from the transmission costs $(\omega^+_{Tij}, \omega^-_{Tij})$ and the transmission line capacity P^{max}_{Tij}. Two additional vertices are then added to trigger algorithms: a virtual source node s and a sink node t. For each bus i with generation, a pair of arcs is connected to the source node s with residual capacities equivalent to generation capacities $(P^{max}_{Gi}, P^{min}_{Gi})$ and arc costs equivalent to generation costs $(\omega^+_{Gi}, \omega^-_{Gi})$. For each bus k with load demand, a pair of arcs is connected to the sink node t with residual capacities equivalent to the bounds of power demand $(P^{max}_{Dk}, P^{min}_{Dk})$ and arc costs equivalent to demand costs $(\omega^+_{Dk}, \omega^-_{Dk})$.

Fig. 2 shows an example of a 3-bus system and its representative directed graph. It is assumed in this example that there is no difference between costs for up and down power regulation and flow. Values of parameters of the system are given in table 1.

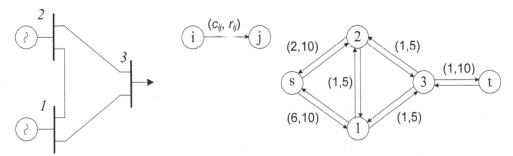

Fig. 2. Single-line diagram and representative directed graph of the 3-bus k network

Components	Cost [p.u.]	Capacity [MW]
Generator at bus 1	$\omega^+_{G1} = \omega^-_{G1} = 6$	$P^{max}_{G1} = 10;\ P^{min}_{G1} = 0$
Generator at bus 2	$\omega^+_{G1} = \omega^-_{G1} = 2$	$P^{max}_{G2} = 10;\ P^{min}_{G2} = 0$
Load at bus 3	$\omega^+_{D3} = \omega^-_{D3} = 1$	$P^{max}_{D3} = P^{min}_{D3} = 10$
Transmission lines	$\omega^+_{Tij} = \omega^-_{Tji} = 1$	$P^{max}_{Tij} = 5$

Table 1. Data for the 3-bus test network

By representing the electric network as a directed graph, the function of power flow management can be considered as a minimum cost flow problem. In this chapter, the Successive Shortest Path (SSP) and the Scaling Push-Relabel (SPR) algorithm are used to solve the minimum cost flow.

4.1 Successive Shortest Path algorithm

Successive Shortest Path (SSP) is one of the basic algorithms to deal with the minimum cost flow problem in graph theory. The algorithm starts by searching for the shortest path to augment the flow from the source node s to the sink node t. A node potential, $\left(\pi_j \geq \pi_i - c_{ij}; \forall(i,j) \in E\right)$, associated with each node and a reduced cost, $\left(c_{ij}^\pi = c_{ij} - \pi_i + \pi_j\right)$, associated with each arc (i, j) are used to find the shortest path. After augmenting the flow and updating the information, the process is repeated until there is no possible path anymore from s to t.

Regarding the above example of the 3-bus network, SSP initially finds the shortest path to be s-2-3-t, as shown in Fig. 3. The flow along this shortest path is augmented until the rated capacity of arc 2-3 (5 MW) is reached. When the information is updated, the algorithm searches again for a new shortest path s-2-1-t.

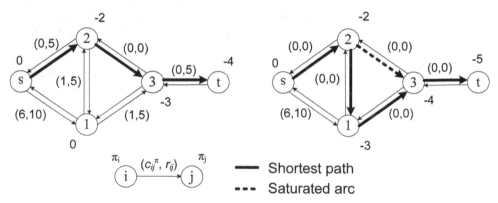

Fig. 3. Illustration of the successive shortest path algorithm for the 3-bus network

4.2 Scaling Push-Relabel algorithm

Scaling Push-Relabel (SPR) belongs to the set of the polynomial-time algorithms to solve the minimum cost flow problem in complex networks. It is different from capacity scaling which is a scaled version of the SSP algorithm.

The algorithm includes two main processes: cost scaling and push-relabel. The cost scaling step determines a boundary for the ε-optimal condition in which the scaling factor ε is initially set at the maximum cost of the graph. The boundary will gradually approach the optimal solution by scaling $\varepsilon \leftarrow \varepsilon/2$ after each iterative loop. Within the procedure of cost scaling, the push-relabel step is used. This step aims to push as much excess flow e as possible from a higher node to a lower node. The height of each node is computed from the node potential π_i. Since there is an excess in node i ($e_i > 0$) while it is lower than its neighbors, the node will be relabeled ($\pi_i = \pi_i + \varepsilon/2$) to be higher than at least one neighbor node to push the excess.

In Fig. 4, the push-relabel procedure is illustrated by displaying the first steps of the SPR algorithm in the above example of the 3-bus network. As the source node s is the first active node, it will be relabelled before it pushes flows down to node 1 and 2. When there is no excess at node s anymore, node 2 as the next active node initiates the push-relabel procedure. However, it needs to be relabeled to be higher than at least one neighbor. After relabeling, node 2 can push flow to nodes 1 and 3. A similar step occurs at node 1.

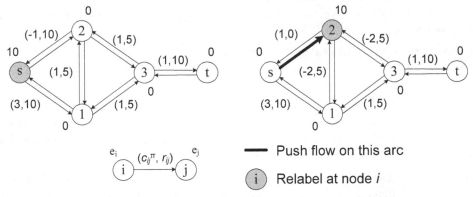

Fig. 4. Illustration of the push-relabel procedure for the 3-bus network

4.2.1 Decentralized and self-healing extensions of SPR

This section presents an extension of the SPR algorithm. In particular, we show that it is possible to implement the SPR method in a distributed way by using only local information and coordination (in contrast to the widely used centralized methods). This property of the algorithm has the additional benefit of allowing the grid operation to recover/react autonomously and faster from changes in demand/supply, cost or capacity. Note that the implementation used in the remaining of this chapter has certain centralized characteristics, since it was used for the comparison between SPR and SSP.

The SPR algorithm described above consists of two main phases: an initial phase to calculate feasible flows and a subsequent phase to convert the initial flow into a minimum cost flow (refinements). It is identified in (Goldberg & Tarjan, 1988) a variety of methods (centralized, parallel or distributed) which can be used in these two phases. Specifically for the refinement phase, a completely distributed Goldberg Tarjan (DGT) scheme that uses the concept of blocking flows and travelling atoms has been developed in (Goldberg & Tarjan, 1989). A flow is called blocking when every path from source to destination contains at least one saturated arc, while an atom at some point in time is defined as a maximal quantity of excess that has travelled as a whole up to that point. The DGT refinement scheme starts with an initialization phase, where the arcs leaving the source are saturated with atoms, and pushes atoms forward until no more pushes are possible.

Lemma 1: The DGT refinement scheme converges under the presence of changes in demand/supply of nodes or capacity of arcs, within $O(n)$ steps after the last change occurs.

Proof: Changes in the demand/supply of nodes affects the initial saturation of the outgoing arcs of the source and load content of nodes, and as a result more atoms are being created at the source or fewer atoms are being accepted at the destination. In the first case, the new atoms are being propagated forward and the algorithm proceeds normally. In the second case, the nodes at the destination become blocked and some atoms are being returned backwards, again resuming the normal operation of the algorithm. In addition, changes in the capacity of arcs affect their residual capacities. As a result, increased (decreased) residual capacity in an arc may unblock (block) the end-node closer to the source. In both cases, the algorithm resumes its normal course, sending atoms forward or backward accordingly.

Since the normal operation of the DGT algorithm is resumed after the last change, in the worst case it will terminate in the same number of steps, which is $O(n)$ for the distributed case.

Using the above refinement scheme along with a distributed maximum flow (DMF) algorithm that can tolerate changes in node demand/supply and arc capacities as proposed in (Ghosh et al., 1995) for the initial phase, it is easy to see that the resulting algorithm has certain self-healing properties since it can tolerate these changes.

Theorem 1: Consider a minimum cost flow algorithm that consists of an initial phase of a feasible flow calculation using the DMF algorithm and subsequent refinement phases using the DGT algorithm. This algorithm can be implemented in a completely distributed way and can converge under the presence of changes in demand/supply of nodes or capacity of arcs, within a certain number of steps after the last change occurs.

Note that the initial and refinement phases must be separated in order to tolerate changes, thus some form of synchronization between nodes is required. This is discussed in (Goldberg & Tarjan, 1988) since it is also a challenge for the original algorithm, and the proposed termination conditions there can also be applied here. Note also that the amount of steps needed for the algorithm to converge under the presence of changes depends on the implementation of this synchronization.

4.3 Complexity of the algorithms

Though the SSP algorithm is straightforward to be implemented, its computational complexity is $O(n^2 mB)$ (Ahuja et al., 1993), where n is the number of the network nodes, m is the number of network arcs and B is the upper bound on the largest supply (demand) of any node. In the worst case, each augmentation phase carries a very small amount of flow, resulting in a fairly large number of iterations. A modification of the scaling algorithm can reduce the number of iterations to $O(m \log B)$.

The SPR algorithm can be implemented in a variety of methods (centralized, parallel or distributed) for both the initial and subsequent refinement phases. The scheme followed in this chapter for both phases is distributed with centralized characteristics, i.e., all active nodes are kept in set S, which is passed from node to node and controls the activation sequence. This corresponds to a sequential variant of the algorithm, which yields $O(n^2 m \log nC)$ convergence time, where C is the maximum cost of an arc (Ahuja et al., 1993). However, other distributed variants are also possible, and the previously mentioned refinement scheme based on blocking flows and travelling atoms achieves $O(n^2 \log n \min\{\log n\, C, m \log n\})$ total convergence time (Goldberg & Tarjan, 1988).

5. Agent-based implementation

For the power routing function it is important to implement the graph-based algorithms in a distributed environment. This section introduces Multi-Agent System (MAS) as a suitable platform for that because it can facilitate distributed control and perform monitoring functions in the power system.

5.1 Multi-agent system technology

Agent-oriented programming is a relatively new technique to implement artificial intelligence into distributed system operation (Ahuja et al., 1993). An agent can be created by a short program (software entity) to operate autonomously with its environment. Moreover, the agent can interact with each other to form a Multi-Agent System. With characteristics of reactivity, proactiveness, and social ability, the MAS technology can offer numerous benefits in distributed power networks.

Actually, a part of the agent's features has been revealed in some applications of the power system before. As an example, an Intelligent Electronic Device (IED) performs various control and protection functions according to changes in their environment, e.g., voltage drop and current increase. Recently, applications of MAS in power systems have been explored in many aspects such as disturbance diagnosis, restoration, protection, and power flow and voltage control. Several research projects have begun to investigate MAS as an approach to manage distributed generation, virtual power plants and micro grids.

5.2 Algorithm implementation

To utilize the discussed graph-based algorithms in the power routing function, it is assumed that agents are available representing nodes (buses) of an active distribution network (ADN). Each agent A_i can obtain current state variables of bus i from the power network such as power flow in incoming (outgoing) feeders P_{Tij}^0 with $\forall (i,j) \in \mathcal{T}$, power generation P_{Gi}^0, and load demand P_{Dk}^0. The limits of power generation, transport, and load demands can be pre-defined or updated during a communication period of A_i. In addition, A_i is provided with information about the costs for adjusting power production $(\omega_{Gi}^+, \omega_{Gi}^-)$, consumption $(\omega_{Dk}^+, \omega_{Dk}^-)$ as well as transmission costs $(\omega_{Tij}^+, \omega_{Tij}^-)$. Besides managing autonomous control actions, A_i can route messages to communicate with same-level agents via the MAS platform. Two additional agents, A_s and A_t, are created to represent the source node s and the sink node t of the graph $G(V, E)$ respectively. By using the MAS platform, the SSP and SPR algorithms can be implemented in a distributed environment.

5.2.1 Implementation of the SSP algorithm

Fig. 5 illustrates a simplified agent sequence diagram for the above example of the 3-bus network using the SSP algorithm. There are three main types of actions, which are defined for each agent as follows:

- Update information: Node agents (A_1, A_2, and A_3) update information from the power network and market conditions. By using *Information()* messages, they share the information with agents A_s and A_t.

- Update node potential: After updating information, A_s initiates to an update of the node potential. A message of *Get_label()* is sent to its neighbors (A_1, A_2). When receiving the *Get_label()* command, a node agent compares the proposal with its node potential to get the cheapest (smallest) one and its predecessor. The shortest path is determined when A_t updates its node potential.
- Augmentation: A_t starts augmenting flow along the new shortest path. An *Augment_flow()* message is sent from node agent to its predecessor which is to track back the shortest path. When A_s receives the message, it begins looking for a new shortest path by repeating the above action of updating node potentials.

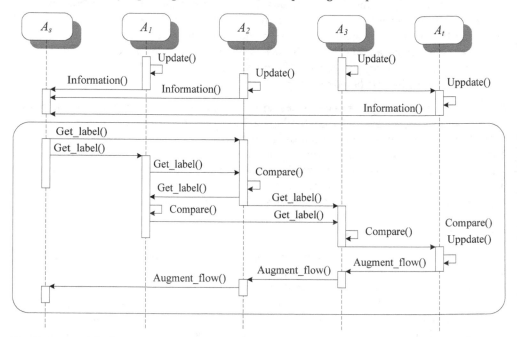

Fig. 5. A simplified agent sequence diagram for deploying the SSP algorithm

5.2.2 Implementation of the SPR algorithm

A simplified agent sequence diagram for deploying the SPR algorithm is shown in Fig. 6. Main types of actions for each agent are described as follows:

- Update information: The procedure is similar to the case of the SSP algorithm. A slight difference is the information that A_s uses for initiating the SPR algorithm. It is the total generated power instead of total load demand, as in the SSP algorithm.
- Cost scaling: The scaling factor ε is initial set by the maximum value of the costs updated from the power network. When the function *Cost_Scaling()* is called, it transformers ε-optimal flow into a $\frac{1}{2}$ε-optimal flow.
- Relabel node potential: The relabel phase is called by each agent when it has flow excess and its height is lower than its neighbors. The function adds a value of $\frac{1}{2}$ε to the height.

- Push flow: When the node agent finds a feasible neighbor in which to push flow, it increases as much as possible the power with respect to its flow excess and the connecting arc's residual capacity.

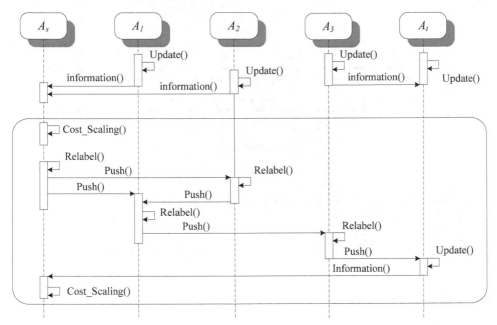

Fig. 6. A simplified agent sequence diagram for deploying the SPR algorithm

6. Simulation results and discussions

This section investigates the performance of the power routing function of the two algorithms in the simulation cases. The power network is simulated in MATLAB/Simulink environment while Java Agent Development Framework (JADE) (Telecom Italia S.p.A., 2010) is used for creating a Multi-Agent System (MAS) platform. The protocol for communication between two environments is based on client/server socket communication.

6.1 Typical radial distribution network

A simulation example is implemented to investigate the performances of the algorithms on a typical radial medium voltage (MV) network. This test network includes two feeders with 10 buses on each feeder, connected to the same substation. The two ends of the feeders can be connected through a normally open point (NOP). Parameters of the test network are given as follows:

- Line section: π-equivalent circuit, line section parameters: $Z = 0.25 + j0.178\ \Omega$; $B = 1\mu S$; $P_{Tij}^{max} = 10$ MW.
- Base load: Each bus has a base load of 1 MW + $j0.48$ MVAr.
- Distributed generation: DG units are available at bus 1, 3, and 5 of feeder 1 and 16, 18, and 20 of feeder 2.

Fig. 7 shows a single-line diagram and the representative directed graph of the radial test network. In the representative graph, the square symbols show nodes having DG units which are connected to the virtual source node s. The remaining nodes have only load demand.

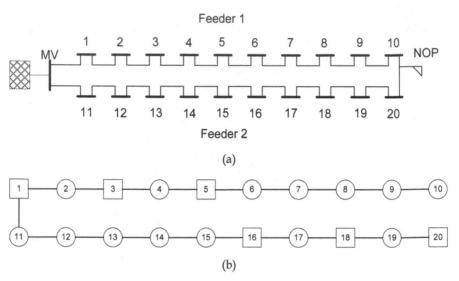

(a)

(b)

Fig. 7. Single-line diagram and representative directed graph of the radial test network

6.1.1 Case 1 - Update information and start up algorithms

The simulation starts with an initial state as shown in table 2. The three DG units of feeder 1 provide more power than the ones connected to feeder 2, but with higher marginal costs.

Generator	$\omega_{Gi}^+ = \omega_{Gi}^-$ [p.u.]	P_{Gi}^0 [MW]	P_{Gi}^{min} [MW]	P_{Gi}^{max} [MW]
1	15	11	0	15
3	11	3	0	10
5	9	3	0	10
16	5	2	0	3
18	5	2	0	3
20	5	2	0	3

Table 2. Initial state of the radial test network

At $t = 5\ s$, each agent starts collecting and sharing information across the MAS platform. At $t = 10\ s$, new reference values are set for the DG units. The goal in this case is to minimize generation cost. Fig. 8 shows the behaviour of the DG units before and after receiving new set points from MAS using the SPR algorithm. In this case, P_{gen2} and P_{gen3} increases from 3 MW to 4 MW and 10 MW respectively, while the other generators of feeder 2 increases from 2 MW to 3 MW. These controls are implemented to avoid the most expensive generation from bus 1 ($P_{gen1} = 0$ MW).

The total saving cost in money-based unit before and after utilizing the power routing function is also shown in Fig. 8. A major part of the total cost (76.97 p.u.) is saved from changes of power generation mentioned above while the contribution from the power transmission on the total cost is 4.02 p.u.

The SSP algorithm gives almost the same results as the SPR algorithm. Minor differences are the set points for P_{gen2} set to 4.3 MW and P_{gen3} set to 9.7 MW. These come from the fact that SSP is based on the total load demand which is smaller than the total generation capacity used in the SPR algorithm.

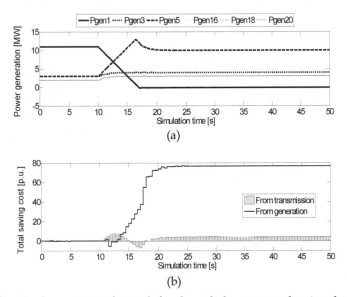

(a)

(b)

Fig. 8. Case of optimal generation dispatch for the radial test network using the SPR algorithm

6.1.2 Case 2 – Load demand increases

After settling at this new optimal operation mode, the simulation is continued by increasing the demand by $2\,MW + j0.2\,MVAr$ at bus 2, 4, 17, and 19. Fig. 9 shows the dynamic operations of the test network. At $t = 1s$, the load demand increases 20%. It is assumed that the generator at bus 1 will be responsible for the primary control to compensate initially for the amount of load increase. After receiving information from the power network at $t = 15\,s$, MAS gives back the new set points at $t = 30\,s$. Because of the increase in demand, all generators at bus 3, 5, 16, 18, and 20 operate at their maximum capacities while the generator at bus 1 is set to 2 MW.

In this case, the total saving cost is referred to the cost value before applying the power routing function, in period from 25 s to 30 s. Therefore, the prior state of the system with lower load demand has a lower operating cost (with high total saving cost, 120 p.u.). This saving cost starts decreasing when the primary control of the generator at bus 1 imitates.

After $t = 30\ s$, the change of power generation contributes to decreasing the total cost 23.85 p.u. Meanwhile, the charge for the power transmission increases 4.42 p.u.

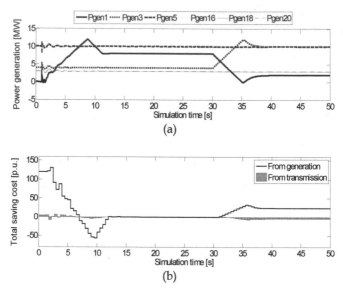

(a)

(b)

Fig. 9. Case of load demand increases for the radial test network using the SPR algorithm

6.1.3 Case 3 – Network configuration change

In this case, a contingency is considered where line section 5-6 is taken out of service. To supply power to the rest of feeder 1, the NOP will be closed. The new configuration of the network is shown in Fig. 10. Stand-by storages are enabled at bus 16, 18 and 20 which increases their capacities up to 5 MW, 7 MW and 9 MW respectively.

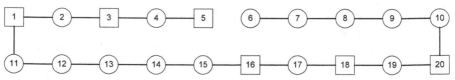

Fig. 10. Representative directed graph for the new configuration of the radial test network

Fig. 11 shows the changes of power generation and power flow, as well as the total operating cost in this case. In feeder 2, the power generation at bus 16, 18, and 20 are increased to their new maximum capacity (5 MW, 7 MW, and 9 MW). While P_{gen3} is constant, P_{gen1} and P_{gen5} are decreased to limit the exchanged flow below 10 MW. The total saving cost in this case has relatively equal contributions from the power generation and power transmission, 48.62 p.u. and 48.59 p.u. respectively.

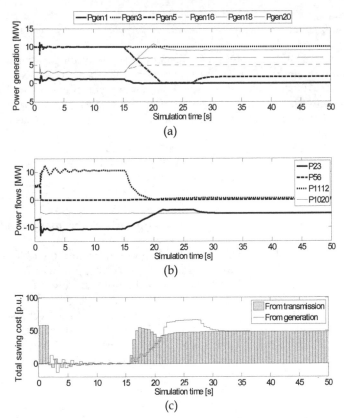

Fig. 11. Case of network configuration change for the radial test network using the SPR algorithm

7. Conclusion

This chapter has presented an application of the graph theory in power systems for the function of power routing. The performance of the power routing function using Successive Shortest Path (SSP) and Scaling Push Relabel (SPR) algorithms are investigated on a simulation of a radial network. Simulation results show that the algorithm has self-stabilizing and self-healing properties in response to changes in the cost and capacity for power demand/supply. Simulations reveal also the ability of the power routing function to deal with issues of network variations and constraints. Moreover, the function can be implemented in an online, real-time environment which is emerging in the development of the future power systems.

8. References

Ahuja, R.K., Magnanti, T.L., & Orlin, J.B. (1993) *Network flows: theory, algorithms, and applications*, Prentice Hall.

Armbruster, A. et al., 2005. Power transmission control using distributed max-flow. In *29th Annual International Computer Software and Applications Conference, COMPSAC*, Vol. 2, pp. 256-263

Bompard, E. et al. (2003). Congestion-management schemes: a comparative analysis under a unified framework. *IEEE Transactions on Power Systems*, Vol. 18, No. 1, pp. 346-352.

Dolan, M. et al. (2009). Techniques for managing power flows in active distribution networks within thermal constraints. *Proceeding of the 20th International Conference and Exhibition on Electricity Distribution - CIRED 2009*, Prague, Czech Republic, June 18-11, 2009.

Gan, D., Thomas, R.J. & Zimmerman, R.D. (2000). Stability-constrained optimal power flow. *IEEE Transactions on Power Systems*, Vol. 15, No. 2, pp. 535-540.

Ghosh, S., Gupta, A., & Pemmaraju, S.V. (1995). A self-stabilizing algorithm for the maximum flow problem. *Proceeding of the IEEE Fourteenth Annual International Phoenix Conference on Computers and Communications*, pp. 8-14.

Goldberg, A.V. & Tarjan, R.E. (1988). A new approach to the maximum-flow problem. *Journal of the ACM*, Vol. 35, pp. 921-940.

Goldberg, A.V. & Tarjan, R.E. (1989). A parallel algorithm for finding a blocking flow in an acyclic network. *Information Processing Letters*, Vol. 31, pp. 265-271.

Huneault, M. & Galiana, F.D. (1991). A survey of the optimal power flow literature. *IEEE Transactions on Power Systems*, Vol. 6, No. 2, pp. 762-770.

Jokic, A. (2007). *Price-based optimal control of electrical power systems*. Technische Universiteit Eindhoven, ISBN 978-90-386-1574-5, Eindhoven, The Netherlands.

Kim, B.H. & Baldick, R. (2000). A comparison of distributed optimal power flow algorithms. *Power Systems, IEEE Transactions on*, Vol. 15, No. 2, pp. 599-604.

Nguyen, P.H., Kling, W.L., & Myrzik, J.M.A. (2010). An application of the successive shortest path algorithm to manage power in multi-agent system based active networks. *European Transactions on Electrical Power*, Vol. 20, No. 8, pp. 1138-1152.

Nguyen, P. H., Kling, W.L., Georgiadis, G., Papatriantafilou, M., Anh-Tuan, L. & Bertling, L. (2010). Distributed routing algorithms to manage power flow in agent-based active distribution network. *Proceedings of the 2010 IEEE PES Conference on Innovative Smart Grid Technologies Conference Europe (ISGT Europe)*, 11-13 October 2010, Gothenburg, Sweden. (pp. 1-7).

Telecom Italia S.p.A., 2010. Java Agent Development Framework. Available from http://jade.tilab.com/index.html [Accessed May 2010].

Wei, P. et al., 2001. A decentralized approach for optimal wholesale cross-border trade planning using multi-agent technology. *Power Systems, IEEE Transactions on*, Vol. 16, No. 4, pp. 833-838.

Xin, H. et al. (2011). A Self-Organizing Strategy for Power Flow Control of Photovoltaic Generators in a Distribution Network. *IEEE Transactions on Power Systems*, Vol. 26, No. 3, pp. 1462-1473.

12

Symbolic Determination of Jacobian and Hessian Matrices and Sensitivities of Active Linear Networks by Using Chan-Mai Signal-Flow Graphs

Georgi A. Nenov

Higher School of Transport "T. Kableshkov", Sofia,
Bulgaria

1. Introduction

Every network synthesis procedure normally includes a first-order or (more rarely) second-order network sensitivity analysis. The main problem here is the evaluation of the corresponding first- or second-order derivatives of network functions with respect to the circuit element values. These derivatives form the network Jacobian (J) and Hessian (H) matrices, respectively. A variety of methods exist for such an evaluation but most of them are intended for the sensitivity of one network transfer function only. Besides this in many cases it is desirable to find the symbolic expressions of the sensitivities because such a presentation facilitates the element value influence determination. An other useful and important application of the matrices J and H is in the tasks for optimization of synthesized networks with respect to their sensitivities or other parameters (Korn & Korn, 1968; Wilde,1978).

As it is well known all linear active networks can be modeled by using passive elements and nullator-norator pairs (nullors). The presented paper deals with the application of Chan-Mai signal-flow graphs (CMG) to the determination of the matrices J and H elements, having in mind the peculiarities of nullors and their influence on the passive element network admittance matrix and on the corresponding CMG. The method developed here is an improved and enlarged version of the approach in (Nenov, 2004). One demonstrates that the method reduces to the obtaining of two (for the elements of J) or four (for the elements of H) isomorphic Chan-Mai signal-flow graphs.

2. Chan-Mai signal flow graph

It was introduced in graph theory in 1967 (Chan & Mai, 1967). Compared with other kinds of oriented graphs (especially Mason and Coates graphs) the Chan-Mai graph (CMG) holds out a simplest way to the representation the relationships between the dependent and independent quantities in an algebraic equation set. In order to make easier the understanding of the following sections of the paper further we give the procedure for drawing of CMG and the basic formulae related.

Assume the algebraic set

$$A.X = Y \tag{1}$$

is given, where

$$A = \begin{bmatrix} a_{11} & a_{12} & \cdots & a_{1n} \\ a_{21} & a_{22} & \cdots & a_{2n} \\ \cdot & \cdot & \cdot & \cdot \\ a_{n1} & a_{n2} & \cdot & a_{nn} \end{bmatrix} \tag{2}$$

is a square matrix with real or complex entries and

$$\left. \begin{aligned} X &= \begin{bmatrix} x_1 & x_2 & \cdots & x_n \end{bmatrix}_t; \\ Y &= \begin{bmatrix} y_1 & y_2 & \cdots & y_n \end{bmatrix}_t \end{aligned} \right\} \tag{3}$$

are the vectors of the dependent and of the independent variables, respectively. The CMG consists of n vertices with sink signals $y_1, y_2, \ldots, y_n$, n vertices with source signals $x_1, x_2, \ldots, x_n$ and maximum n^2 edges with transmission coefficients a_{ji} directed from the vertex x_i toward the vertex y_j ; $i,j = 1, 2, \ldots, n$ – Fig. 1. The calculations on the base of a CMG are connected with the following definitions (Chan & Mai, 1967, Donevsky & Nenov, 1979):

i.   By removing all outgoing from the vertex x_i edges and by adding the edges with transmission coefficients y_j from the vertex x_i directed toward the vertices y_j , $j=1, 2, \ldots, n$ one obtains the *graph* CMG,i;

ii. A *separation (S)* contains all vertices of CMG and a part of edges so that every vertex is incident to only one incoming and one only outgoing edge. The product of the transmission coefficients of all edges in a separations represents the corresponding *separation product (SP)*;

iii. Two edges with transmission coefficients a_{ij} and a_{ji} form a *symmetrical pair*.

iv. An edge which does not belong to a symmetrical pair is an *asymmetrical edge*.

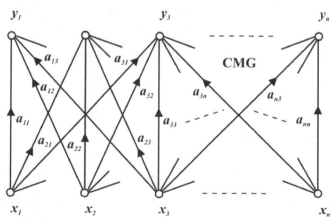

Fig. 1. Chan-Mai Signal-Flow Graph

An arbitrary unknown quantity x_i in $\mathbf{X}$ can be evaluated according to the expression

$$x_i = \frac{\sum\limits_{q=1}^{m}(signSP_q)SP_q(CMG,i)}{\sum\limits_{k=1}^{r}(signSP_k)SP_k(CMG,_k)} , \tag{4}$$

where

$$signSP_l = \begin{cases} (-1)^{N_{s,l}+N_{a,l}-1} & \text{for } N_{a,l} \neq 0 \\ (-1)^{N_{s,l}} & \text{for } N_{a,l} = 0 \end{cases} \tag{5}$$

$$l = q \text{ or } k$$

In (4) and (5) r is the number of the separations in CMG, m is the number of the separations in CMG,i , $N_{a,k}$ is the number of all asymmetrical edges in k-th separation of CMG, $N_{s,k}$ is the number of all symmetrical pairs in k-th separation of CMG, N_{aq} is the number of the asymmetrical edges in q-th separation of CMG,i, $N_{s,q}$ is the number of the symmetrical pairs in q-th separation of CMG,i, whereas $SP_q(CMG,i)$ and $SP_k(CMG)$ are the separation products of q-th separation of CMG,i and the separation products of k-th separation of CMG, respectively.

3. Nullor network Chan-Mai signal-flow graph

Suppose that an equivalent nullor network N with $m+1$ nodes, r passive branches and g nullors is given and the nodal equation of its passive part N_p (the part of N which is obtained by removing all nullors) is

$$\mathbf{Y}_p\mathbf{V}_p = \mathbf{I}_p \tag{6}$$

where

$$\mathbf{Y} = \begin{bmatrix} Y_{p,11} & \cdots & Y_{p,1m} \\ . & . & . \\ Y_{p,m1} & \cdots & Y_{p,mm} \end{bmatrix} \tag{7}$$

is the nodal matrix of N_p and

$$\left.\begin{aligned} \mathbf{V}_p &= \begin{bmatrix} V_{p,1} & V_{p,2} & \cdots & V_{p,m} \end{bmatrix}_t; \\ \mathbf{I}_p &= \begin{bmatrix} I_{p,1} & I_{p,2} & \cdots & I_{p,m} \end{bmatrix}_t \end{aligned}\right\} \tag{8}$$

are the nodal voltage and the nodal current vectors of N_p, respectively. Additionally we assume that between the nodes of all node pairs in N only one element or more than one but parallel connected elements exist.

The equation (1) can be represented graphically by using a CMG G_p (Chan & Mai, 1967). Further, taking into account the peculiarities of the nullators and the norators (Davies, 1966) the graph G_p can be transformed into the graph G of the actual network N according to the following

Rule 1:

i.   When a nullator is connected between the node k in N and the ground node $m+1$ one removes all vertices going out from the vertex V_k of G_p;
ii. When a norator is connected between the node k in N and the ground node $m+1$ one removes all vertices coming into the vertex I_k of G_p;
iii. When a nullator is connected between the nodes k and l in N one unites the vertices V_k and V_l in G_p;
iv. When a norator is connected between the nodes k and l in N one unites the vertices I_k and I_l in G_p.

The so obtained graph CMG G corresponds to the matrix equation

$$\mathbf{YV} = \mathbf{I} \tag{9}$$

where $\mathbf{Y}$ is an $(n \times n)$ nodal admittance matrix of N, $\mathbf{V}$ is the nodal voltage vector of N and $\mathbf{I}$ is the nodal current vector of for $n=m-g$.

4. Jacobian matrix determination

The matrices in (9) have the form:

$$\mathbf{Y} = \begin{bmatrix} Y_{11} & .. & Y_{1i} & .. & Y_{1n} \\ . & . & . & . & . \\ Y_{j1} & .. & Y_{ji} & .. & Y_{jn} \\ . & . & . & . & . \\ Y_{n1} & .. & Y_{nj} & .. & Y_{nn} \end{bmatrix}; \ \mathbf{V} = \begin{bmatrix} V_1 & V_2 & ... & V_n \end{bmatrix}_t; \ \mathbf{I} = \begin{bmatrix} I_1 & I_2 & ... & I_n \end{bmatrix}_t \tag{10}$$

In the common case every element Y_{ji} in (7) is an algebraic admittance sum

$$Y_{ji} = \sum_s y_s; \ j,i \in \{1,2,...,n\}; s \in \{1,2,...,r\}, \tag{11}$$

where y_s is the admittance of s-th branch of the network N_p.

The vectors $\mathbf{V}$ and $\mathbf{I}$ correspond to the unknown (dependent) variables and to independent variables of N, respectively and consequently

$$\mathbf{V} = \mathbf{Y}^{-1}\mathbf{I} \ . \tag{12}$$

Let us suppose that the admittance y_s changes its value to

$$y_s' = y_s + dy_s \ . \tag{13}$$

Symbolic Determination of Jacobian and Hessian Matrices and Sensitivities of Active Linear Networks
by Using Chan-Mai Signal-Flow Graphs

249

Usually the admittance y_s takes part in several (but no more then four) elements of (7) and then all these elements change their values (Nenov, 2004)

$$\left. \begin{array}{l} Y'_{ji} = Y_{ji} + dY_{ji} = Y_{ji} + \dfrac{\partial Y_{ji}}{\partial y_s} dy_s; \\[2mm] j,i \in \{1,2,...,n\}; s \in \{1,2,...,r\} \end{array} \right\} \tag{14}$$

and

$$\mathbf{Y'} = \mathbf{Y} + d\mathbf{Y}. \tag{15}$$

In a common case the admittance y_s influences the admittances Y_{ji}, Y_{jl}, Y_{ki} and Y_{kl}; $i,j,k,l \in \{1, 2, ..., n\}$. Then one obtains

$$\left. \begin{array}{l} d\mathbf{Y} = dy_s \mathbf{K}_s; \\[2mm] \mathbf{K}_s = \begin{bmatrix} 0 & . & 0 & . & 0 & . & 0 \\ . & . & . & & . & . & . \\ 0 & . & \dfrac{\partial Y_{ji}}{\partial y_s} & . & \dfrac{\partial Y_{jl}}{\partial y_s} & . & 0 \\ . & . & . & & . & . & . \\ 0 & . & \dfrac{\partial Y_{ki}}{\partial y_s} & . & \dfrac{\partial Y_{kl}}{\partial y_s} & . & 0 \\ . & . & . & & . & . & . \\ 0 & . & 0 & . & 0 & . & 0 \end{bmatrix}; \; i,j,k,l \in \{1,2,...,n\} \end{array} \right\} . \tag{16}$$

Note that the values of the derivatives in (16) are 1 or –1 because every admittance y_s takes part in (11) only once. Hence

$$\mathbf{V'} = \mathbf{V} + d\mathbf{V} \tag{17}$$

for

$$d\mathbf{V} = \left(\dfrac{\partial \mathbf{V}}{\partial Y_{ji}} \dfrac{\partial Y_{ji}}{\partial y_s} + \dfrac{\partial \mathbf{V}}{\partial Y_{jl}} \dfrac{\partial Y_{jl}}{\partial y_s} + \dfrac{\partial \mathbf{V}}{\partial Y_{ki}} \dfrac{\partial Y_{ki}}{\partial y_s} + \dfrac{\partial \mathbf{V}}{\partial Y_{kl}} \dfrac{\partial Y_{kl}}{\partial y_s} \right) dy_s, \tag{18}$$

or:

$$\left. \begin{array}{l} d\mathbf{V} = dy_s \sum\limits_{pq} \dfrac{\partial \mathbf{V}}{\partial Y_{pq}} \dfrac{\partial Y_{pq}}{\partial y_s}; \\[2mm] p,q \in \{1,2,...,n\}. \end{array} \right\} \tag{19}$$

By substituting $\mathbf{Y'}$ and $\mathbf{V'}$ in (9) instead $\mathbf{Y}$ and $\mathbf{V}$, respectively, it follows

$$[\mathbf{Y} + d\mathbf{Y}].[\mathbf{V} + d\mathbf{V}] = \mathbf{I}. \tag{20}$$

Having in mind that

$$dYdV \to 0 \tag{21}$$

the equation (20) yields

$$YdV = -dYV \tag{22}$$

or

$$dV = -Y^{-1}dYV . \tag{23}$$

Then we obtain

$$\frac{\partial V}{\partial y_s} dy_s = -dy_s . Y^{-1}K_s V \tag{24}$$

and the Jacobian matrix (Korn & Korn, 1968). for the change of the admittance y_s is

$$J = \begin{bmatrix} \dfrac{\partial V_1}{\partial y_1} & \dfrac{\partial V_1}{\partial y_2} & .. & \dfrac{\partial V_1}{\partial y_s} & .. & \dfrac{\partial V_1}{\partial y_r} \\[2mm] \dfrac{\partial V_2}{\partial y_1} & \dfrac{\partial V_2}{\partial y_2} & .. & \dfrac{\partial V_2}{\partial y_s} & .. & \dfrac{\partial V_2}{\partial y_r} \\[2mm] . & . & .. & . & .. & . \\[2mm] \dfrac{\partial V_n}{\partial y_1} & \dfrac{\partial V_n}{\partial y_2} & .. & \dfrac{\partial V_n}{\partial y_s} & .. & \dfrac{\partial V_n}{\partial y_r} \end{bmatrix} = \begin{bmatrix} J_1 & J_2 & .. & J_s & .. & J_r \end{bmatrix}, \tag{25}$$

where

$$J_s = \begin{bmatrix} \dfrac{\partial V_1}{\partial y_s} & \dfrac{\partial V_2}{\partial y_s} & .. & \dfrac{\partial V_n}{\partial y_s} \end{bmatrix}_t . \tag{26}$$

Taking into account (24) and (25) one obtains

$$\left. \begin{array}{l} J_s = -Y^{-1}K_s Y^{-1}I = -Y^{-1}K_s V = -Y^{-1}V_s; \\ V_s = K_s V. \end{array} \right\} \tag{27}$$

and according to (20) ÷ (22)

$$J = -Y^{-1}\begin{bmatrix} K_1 & .. & K_s & .. & K_r \end{bmatrix}V . \tag{28}$$

The expressions (22) show that in order to find the vector J_s it is necessary to follow the

Rule 2:

i.   Find the vector V by using the CMG G;
ii. Evaluate the vector V_s;
iii. Draw a new CMG G_s where the source vertices are the elements of the vector J_s and the sink vertices are the elements of the vector V_s;
iv. Find the source vertex variables in G_s.

Symbolic Determination of Jacobian and Hessian Matrices and Sensitivities of Active Linear Networks
by Using Chan-Mai Signal-Flow Graphs

251

Example A

The network N in Fig. 2 is given, where $m=6$; $r=9$; $g=2$. Here obviously $V_2=V_3=V_{23}$; $V_6=0$ and we wish to find the vector

$$J_3 = \left[\begin{array}{cccc} \dfrac{\partial V_1}{\partial(sC_3)} & \dfrac{\partial V_{23}}{\partial(sC_3)} & \dfrac{\partial V_4}{\partial(sC_3)} & \dfrac{\partial V_5}{\partial(sC_3)} \end{array}\right]_t . \tag{29}$$

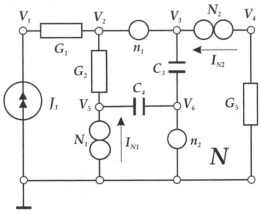

Fig. 2. Nullor Network N

In Fig. 3 the CMG G_p of the passive part of N is drawn (Nenov, 2004). Further following the *Rule* 2 we reach to the graph G in Fig. 4 for

$$\mathbf{Y} = \begin{bmatrix} G_1 & -G_1 & 0 & 0 \\ -G_1 & G_1+G_2 & 0 & -G_2 \\ 0 & sC_3 & G_5 & 0 \\ 0 & -sC_3 & 0 & -sC_4 \end{bmatrix};$$

$$\mathbf{V} = \begin{bmatrix} V_1 & V_2=V_3=V_{23} & V_4 & V_5 \end{bmatrix}_t; \mathbf{I} = \begin{bmatrix} J_1 & 0 & 0 & 0 \end{bmatrix}_t. \tag{30}$$

Because $Y_{32}=sC_3$; $Y_{42}=-sC_3$ and

$$\frac{\partial Y_{32}}{\partial(sC_3)} = 1; \quad \frac{\partial Y_{32}}{\partial(sC_3)} = -1. \tag{31}$$

from (16) and (31) we have

$$\mathbf{K}_3 = \begin{bmatrix} 0 & 0 & 0 & 0 \\ 0 & 0 & 0 & 0 \\ 0 & 1 & 0 & 0 \\ 0 & -1 & 0 & 0 \end{bmatrix};$$

$$\mathbf{V}_3 = \mathbf{K}_3\mathbf{V} = \begin{bmatrix} 0 & 0 & V_{23} & -V_{23} \end{bmatrix}_t. \tag{32}$$

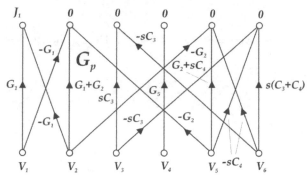

Fig. 3. CM Signal-Flow Graph G_p

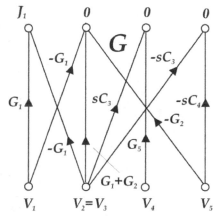

Fig. 4. CM Signal-Flow Graph G

Obviously, in the case we have to find the voltage V_{23} only. For this purpose a CM graph G_{23} is drawn (Fig. 5).

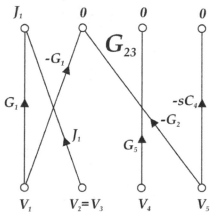

Fig. 5. CM Signal-Flow Graph G_{23}

Symbolic Determination of Jacobian and Hessian Matrices and Sensitivities of Active Linear Networks
by Using Chan-Mai Signal-Flow Graphs

253

According to (4) and (5) for the separations of the graph G (Fig. 6) we obtain

$$
\left.\begin{aligned}
SP_1 &= -G_1 s C_4 G_5 (G_1 + G_2); N_{a,1} = 4; N_{s,1} = 0; \\
SP_2 &= -G_1^2 s C_4 G_5; N_{a,2} = 2; N_{s,2} = 1; \\
SP_3 &= G_2 s C_3 G_5 : N_{a,3} = 2; N_{s,3} = 1
\end{aligned}\right\}
\tag{33}
$$

and for the unique separation of the graph G_{23} (Fig. 7):

$$
SP_{23,1} = J_1 G_1 s C_4 G_5; N_{a,23,1} = 2; N_{s,23,1}
\tag{34}
$$

Then the formulae (4) and (5) yield

$$
V_{23} = \frac{J_1 C_4}{G_2 (C_3 + C_4)}.
\tag{35}
$$

Having in mind (26) ÷ (29) we have

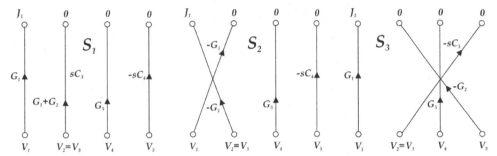

Fig. 6. Separations S_1, S_2 and S_3 of CM Graph G

$$
\mathbf{J}_3 = -\mathbf{Y}^{-1} \mathbf{V}_3 = \mathbf{Y}^{-1}(-\mathbf{V}_3)
\tag{36}
$$

and following **Rule 2** one draws the CM graph G_{J3} (Fig. 8). Obviously, the graphs G and G_{J3} have one and the same structure and consequently the expressions (33) hold for the source vertex quantities in (29) also. But for the nominator polynomials in (4) we have to draw according the **Rule 2** four new CM graphs – $G_{J3,1}$, $G_{J3,23}$, $G_{J3,4}$ and $G_{J3,5}$ – Fig. 9.

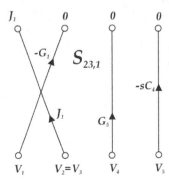

Fig. 7. Separation $S_{23,1}$ of CM Graph G_{23}

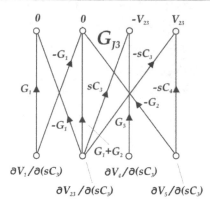

Fig. 8. CM Graph G_{J3}

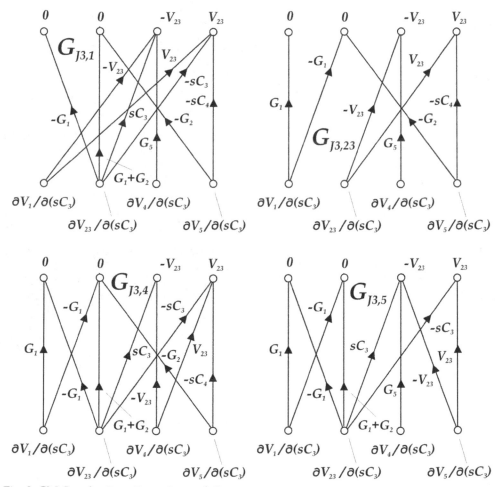

Fig. 9. CM Graphs $G_{J3,1}$, $G_{J3,23}$, $G_{J3,4}$ and $G_{J3,5}$

Symbolic Determination of Jacobian and Hessian Matrices and Sensitivities of Active Linear Networks
by Using Chan-Mai Signal-Flow Graphs

255

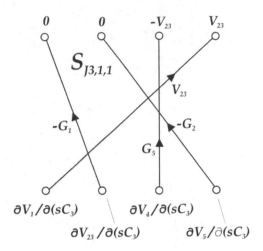

Fig. 10. Separation $S_{J3,1,1}$ of CM Graph $G_{J3,1}$

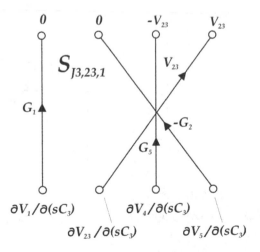

Fig. 11. Separation $S_{J3,23,1}$ of CM Graph $G_{J3,23}$

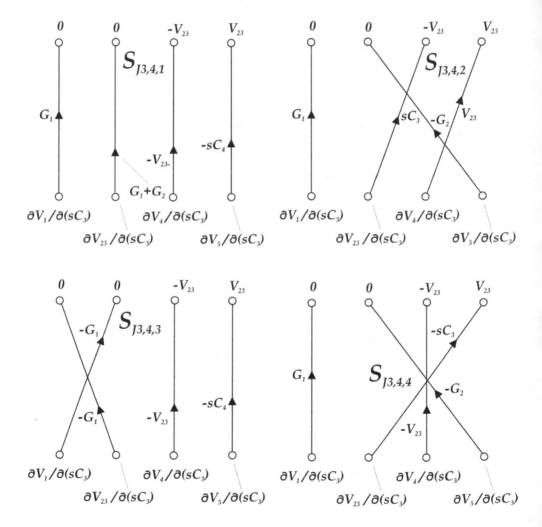

Fig. 12. Separations $S_{J3,4,1}$, $S_{J3,4,2}$, $S_{J3,4,3}$ and $S_{J3,4,4}$ of CM Graph $G_{J3,4}$

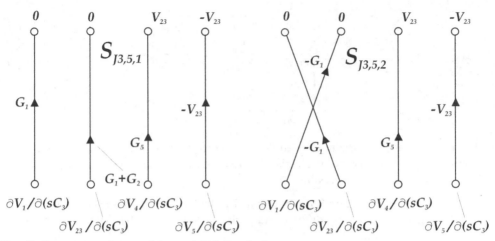

Fig. 13. Separations $S_{J3,5,1}$ and $S_{J3,5,2}$ of CM Graph $G_{J3,5}$

From Fig. 9 ÷ Fig. 13 it follows

$$\left.\begin{array}{l}
SP_{J3,1,1} = G_1 G_2 G_5 V_{23}; \ N_{a,J3,1,1} = 4; \ Ns_{,J3,1,1} = 0; \\
SP_{J3,23,1} = -G_1 G_2 G_5 V_{23}; \ N_{a,J3,23,1} = 4; \ Ns_{,J3,23,1} = 1; \\
SP_{J3,4,1} = G_1(G_1 + G_2)sC_4 V_{23}; \ N_{a,J3,4,1} = 4; \ Ns_{,J3,4,1} = 0; \\
SP_{J3,4,2} = -G_1 G_2 sC_3 V_{23}; \ N_{a,J3,4,2} = 4; \ N_{s,J3,4,2} = 0; \\
SP_{J3,4,3} = G_1^2 sC_4 V_{23}; \ N_{a,J3,4,3} = 2; \ N_{s,J3,4,3} = 1; \\
SP_{J3,4,4} = -G_1 G_2 sC_3 V_{23}; \ N_{a,J3,4,4} = 2; \ N_{s,J3,4,4} = 1; \\
SP_{J3,5,1} = G_1(G_1 + G_2)C_5 V_{23}; \ N_{a,J3,5,1} = 4; \ N_{s,J3,5,1} = 0; \\
SP_{J3,5,2} = G_1^2 G_5 V_{23}; \ N_{a,J3,5,2} = 2; \ N_{s,J3,5,2} = 1;
\end{array}\right\} \qquad (37)$$

Than by substituting (35) in (36) and by taking into consideration (33) from (4) and (5) we obtain the vector $\mathbf{J}_3$:

$$\mathbf{J}_3 = \left[\frac{J_1 C_4}{G_2 s(C_3 + C_4)^2} \quad \frac{J_1 C_4}{G_2 s(C_3 + C_4)^2} \quad \frac{J_1 C_4^2}{G_2 G_5 (C_3 + C_4)^2} \quad \frac{J_1 C_4}{G_2 s(C_3 + C_4)^2} \right]_t \qquad (38)$$

5. Hessian matrix determination

In many practical cases it is necessary and useful to find not only the first-order derivatives of a network function or variable (for example voltage V_w) among n variables with respect to some parameter (for example y_s) but their second-order derivatives with respect to the same or to an other parameter (for example y_t), too.

The matrix formed from all possible second-order derivatives of V_w with respect to the simultaneous changes of two parameters

$$
\mathbf{H}_w = \begin{bmatrix}
\dfrac{\partial^2 V_w}{\partial y_1^2} & \dfrac{\partial^2 V_w}{\partial y_1 \partial y_2} & \cdots & \dfrac{\partial^2 V_w}{\partial y_1 \partial y_n} \\[2ex]
\dfrac{\partial^2 V_w}{\partial y_2 \partial y_1} & \dfrac{\partial^2 V_w}{\partial y_2^2} & \cdots & \dfrac{\partial^2 V_w}{\partial y_2 \partial y_n} \\[1ex]
\cdot & \cdot & \cdot & \cdot \\
\dfrac{\partial^2 V_w}{\partial y_n \partial y_1} & \dfrac{\partial^2 V_w}{\partial y_n \partial y_2} & \cdots & \dfrac{\partial^2 V_w}{\partial y_n^2}
\end{bmatrix}
\tag{39}
$$

is the Hessian matrix or briefly Hessian (Korn & Korn, 1968, Wilde, 1978). Obviously for a network one exists a variety of Hessian matrices – every one matrix corresponds to a definite network function or variable.

The results obtained in section 3. can be applied to the derivation of a Hessian matrix as it will be explained below. By differentiating the vector $\mathbf{J}_s$ in (27) with respect to the admittance y_t one obtains

$$
\left.
\begin{aligned}
&\frac{\partial \mathbf{J}_s}{\partial y_t} = \frac{\partial^2 \mathbf{V}}{\partial y_s \partial y_t} = -\left[\frac{\partial \mathbf{Y}^{-1}}{\partial y_t}\mathbf{K}_s \mathbf{Y}^{-1} + \mathbf{Y}^{-1}\frac{\partial}{\partial y_t}\left(\mathbf{K}_s \mathbf{Y}^{-1}\right)\right]\mathbf{I}; \\
&s,t \in \{1,2,\ldots,n\}.
\end{aligned}
\right\}
\tag{40}
$$

Because the elements in $\mathbf{Y}$ depend linearly on the network element admittances and their derivatives with respect to the parameter y_s equal 1, -1 or 0 it holds

$$
\frac{\partial \mathbf{K}_s}{\partial y_t} = 0; \ \forall s; \ \frac{\partial \mathbf{Y}}{\partial y_t} = \mathbf{K}_t
\tag{41}
$$

and from (40) it follows

$$
\left.
\begin{aligned}
\frac{\partial^2 \mathbf{V}}{\partial y_s \partial y_t} &= \mathbf{Y}^{-1}\left(\mathbf{K}_t \mathbf{Y}^{-1}\mathbf{K}_s + \mathbf{K}_s \mathbf{Y}^{-1}\mathbf{K}_t\right)\mathbf{Y}^{-1}\mathbf{I} = \\
&= \mathbf{Y}^{-1}\mathbf{K}_{st}\mathbf{V} = \mathbf{Y}^{-1}\mathbf{V}_{st}; \\
\mathbf{K}_{st} &= \mathbf{K}_t \mathbf{Y}^{-1}\mathbf{K}_s + \mathbf{K}_s \mathbf{Y}^{-1}\mathbf{K}_t; \\
\mathbf{V}_{st} &= \mathbf{K}_{st}\mathbf{V}.
\end{aligned}
\right\}
\tag{42}
$$

The last result compared with the formulae (27) and (28) shows that we can find the vector $\partial^2 \mathbf{V}/\partial y_s \partial y_t$ in principle by using the same approach as for $\partial \mathbf{V}/\partial y_s$ in section 3.

However here we must pay attention to the obtaining of the matrix $\mathbf{K}_{st}$: In the common case the matrices $\mathbf{K}_s$ and $\mathbf{K}_t$ contain more than one nonzero element (1 or –1). Hence we can expressed each of them as a sum of no more then four addends

$$
\mathbf{K}_s = \sum_a \mathbf{K}_{s,a}; \ \mathbf{K}_t = \sum_b \mathbf{K}_{t,b}; \ a,b \le 4 ,
\tag{43}
$$

Symbolic Determination of Jacobian and Hessian Matrices and Sensitivities of Active Linear Networks
by Using Chan-Mai Signal-Flow Graphs

259

where each of the matrices $\mathbf{K}_{s,a}$ and $\mathbf{K}_{t,b}$ has only one nonzero element. Then as a result the expression of $\mathbf{K}_{st}$ in (42) is a sum of products of the kind

$$\mathbf{K}_{s,a}\mathbf{Y}^{-1}\mathbf{K}_{t,b} \text{ and } \mathbf{K}_{t,b}\mathbf{Y}^{-1}\mathbf{K}_{s,a}; \ \forall a, b. \cdot \tag{44}$$

The products in (44) are square matrices with only one nonzero element which is a definite element of $\mathbf{Y}^{-1}$. Let, for example, the nonzero element for the left-side matrix in (44) is on i-th row and on j-th column and the similar element for the right-side matrix is on k-th row and on l-th column. Then it is easy to see that the corresponding product in (44) contains the element $\pm z_{u,v}$; $u, v \in \{1, 2, ..., n\}$ on i-th row and on l-th column, where $z_{u,v}$ is an element of $\mathbf{Y}^{-1}$. The upper (lower) sign of this element holds for equal (non equal) signs of nonzero elements of $\mathbf{K}_{s,a}$ and $\mathbf{K}_{t,b}$ in (44), respectively.

The matrix $\mathbf{Y}^{-1}$ can be evaluated by using an auxiliary CM graph G_0 too. For this purpose let us consider the equation

$$\mathbf{X} = \mathbf{Y}^{-1}\mathbf{E}, \tag{45}$$

where

$$\mathbf{Y}^{-1} = \begin{bmatrix} z_{11} & z_{12} & \cdot & z_{1n} \\ z_{21} & z_{22} & \cdot & z_{2n} \\ \cdot & \cdot & \cdot & \cdot \\ z_{n1} & z_{n2} & \cdot & z_{nn} \end{bmatrix}; \ \mathbf{X} = \begin{bmatrix} x_1 & x_2 & \cdot & x_n \end{bmatrix}_t; \ \mathbf{E} = \begin{bmatrix} e_1 & e_2 & \cdot & e_n \end{bmatrix}_t \cdot \tag{46}$$

After multiplying in (45) for $\mathbf{X}$ one follows

$$\mathbf{X} = \begin{bmatrix} z_{11}e_1 + z_{12}e_2 + ... + z_{1n}e_n \\ z_{21}e_1 + z_{22}e_2 + ... + z_{2n}e_n \\ \cdot \\ z_{n1}e_1 + z_{n2}e_2 + ... + z_{nn}e_n \end{bmatrix}. \tag{47}$$

This means that if the CM graph G_0 corresponds to (45) the multipliers of $e_1, e_2, ..., e_n$ for every element of $\mathbf{X}$ are elements of $\mathbf{Y}^{-1}$. Note that in real cases a limited number of the elements of $\mathbf{Y}^{-1}$ are necessary only. Hence for determination of an element of the Hessian matrix $\mathbf{H}_{st}$ we can form the following:

Rule 3:

i.   Draw the CM graph G_p of the nullor network under consideration;
ii. Transform the graph G_p into the graph G, according to the *Rule 1* in section 2. and compose the vectors $\mathbf{V}$ and $\mathbf{I}$;
iii. Determine the vector $\mathbf{V}$ from G;
iv. Write the matrices $\mathbf{K}_s$ and $\mathbf{K}_t$;
v.   Determine the matrix $\mathbf{Y}^{-1}$ by using the auxiliary CM graph G_0;
vi. Determine the matrix $\mathbf{K}_{st}$;
vii. Determine the matrix $\mathbf{V}_{st}$;
viii. Draw a CM graph G_{st} in accordance with $\mathbf{V}_{st}$;
ix. Determine the elements of the vector $\partial^2 \mathbf{V}/\partial y_s \partial y_t$ from G_{st}.

Note that by following the above sequence we obtain $2n$ elements of n Hessian matrices simultaneously, *because* $\partial^2 V/\partial y_s \partial y_t = \partial^2 V/\partial y_t \partial y_s$ – Fig. 14.

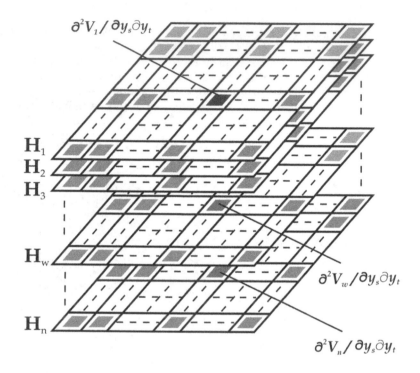

Fig. 14. A Set of n Hessian Matrices

Example B

Suppose that we want to determine the vector $\partial^2 V/\partial(sC_3)\partial(sC_4)$ for the network N in Fig.2. Because the items **i, ii** and **iii** of the *Rule 3* were fulfilled in the **Example A** we have to continue further: Here the matrices K_3 and K_4 are

$$K_{31} = \begin{bmatrix} 0 & 0 & 0 & 0 \\ 0 & 0 & 0 & 0 \\ 0 & 1 & 0 & 0 \\ 0 & 0 & 0 & 0 \end{bmatrix}; \; K_{32} = \begin{bmatrix} 0 & 0 & 0 & 0 \\ 0 & 0 & 0 & 0 \\ 0 & 0 & 0 & 0 \\ 0 & -1 & 0 & 0 \end{bmatrix};$$

$$K_3 = K_{31} + K_{32}; \; K_4 = \begin{bmatrix} 0 & 0 & 0 & 0 \\ 0 & 0 & 0 & 0 \\ 0 & 0 & 0 & 0 \\ 0 & 0 & 0 & -1 \end{bmatrix}$$

(48)

and from (42) ÷ (46) it follows

Symbolic Determination of Jacobian and Hessian Matrices and Sensitivities of Active Linear Networks
by Using Chan-Mai Signal-Flow Graphs

261

$$\mathbf{K}_{34} = \begin{bmatrix} 0 & 0 & 0 & 0 \\ 0 & 0 & 0 & 0 \\ 0 & 0 & 0 & -z_{24} \\ 0 & -z_{43} + z_{44} & 0 & z_{24} \end{bmatrix}. \tag{49}$$

We can find the nonzero elements of $\mathbf{K}_{34}$ by using the auxiliary CM graph G_0 drawn in Fig. 15. By comparing (47) with (49) one settles we need only these addends of elements x_2 and x_4 in (47) that content the quantities e_4 and e_3, e_4, respectively. According to the Chan-Mai procedure we draw the graphs $G_{0,2}$ and $G_{0,4}$ - Fig. 16 and Fig. 17.

$$z_{24} = z_{44} = -\frac{1}{s(C_3 + C_4)}; z_{43} = 0. \tag{50}$$

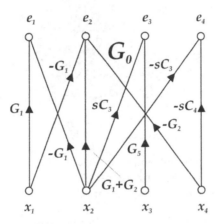

Fig. 15. The auxiliary CM graph G_0

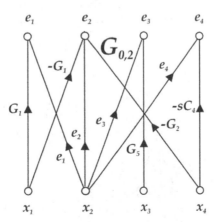

Fig. 16. The CM graph $G_{0,2}$

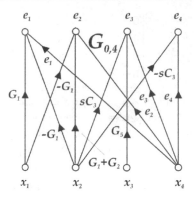

Fig. 17. The CM graph $G_{0,4}$

From Fig. 18 ÷ Fig. 19 we obtain the products

$$
\left.\begin{aligned}
SP_{0,2,1} &= -G_1 s C_4 G_5 e_2; \; N_{a,0,2,1} = 4; N_{s,0,2,1} = 0; \\
SP_{0,2,2} &= G_1 s C_4 G_5 e_1; \; N_{a,0,2,2} = 2; N_{s,0,2,2} = 1; \\
SP_{0,2,3} &= -G_1 G_2 G_5 e_4; \; N_{a,0,2,3} = 2; N_{s,0,2,3} = 1; \\
SP_{0,4,1} &= G_1(G_1 + G_2) G_5 e_4; \; N_{a,0,4,1} = 4; N_{s,0,4,1} = 0; \\
SP_{0,4,2} &= G_1^2 G_5 e_4; \; N_{a,0,4,2} = 2; N_{s,0,4,2} = 1; \\
SP_{0,4,3} &= -G_1 s C_3 G_5 e_2; \; N_{a,0,4,3} = 2; N_{s,0,4,3} = 1;
\end{aligned}\right\} \tag{51}
$$

Note that with the exception of the sink and source quantities the graph G_0 is isomorphic to the graph G in Fig. 4. That is why the expressions (33) remain valid for the denominator in (4) also. Than for the elements of the vector (47) from (51) and (33) it follows:

$$
\left.\begin{aligned}
x_2 &= \frac{G_1 s C_4 G_5 e_1 + G_1 s C_4 G_5 e_2 - G_1 G_2 G_5 e_4}{G_1 G_2 G_5 s(C_3 + C_4)}; \\
x_4 &= \frac{-G_1 s C_3 G_5 e_2 - G_1 G_2 G_5 e_4}{G_1 G_2 G_5 s(C_3 + C_4)}
\end{aligned}\right\} \tag{52}
$$

or taking into consideration (46) and (47)

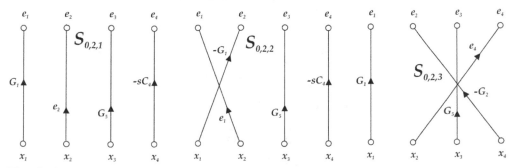

Fig. 18. Separations $S_{0,2,1}$, $S_{0,2,2}$ and $S_{0,2,3}$ of CM graph $G_{0,2}$

Symbolic Determination of Jacobian and Hessian Matrices and Sensitivities of Active Linear Networks
by Using Chan-Mai Signal-Flow Graphs

263

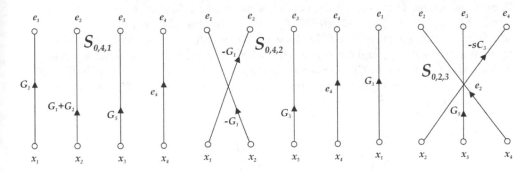

Fig. 19. Separations $S_{0,4,1}$, $S_{0,4,2}$ and $S_{0,4,3}$ of CM graph $G_{0,4}$

$$z_{24} = z_{44} = -\frac{1}{s(C_3 + C_4)}; \ z_{43} = 0 \Big\}. \tag{53}$$

Now we return to (42) and (49) and obtain

$$\mathbf{K}_{34} = \begin{bmatrix} 0 & 0 & 0 & 0 \\ 0 & 0 & 0 & 0 \\ 0 & 0 & 0 & \dfrac{1}{s(C_3 + C_4)} \\ 0 & \dfrac{1}{s(C_3 + C_4)} & 0 & -\dfrac{1}{s(C_3 + C_4)} \end{bmatrix}; \ \mathbf{V}_{34} = \mathbf{K}_{34}\mathbf{V} = \mathbf{K}_{34}\begin{bmatrix} V_1 \\ V_{23} \\ V_4 \\ V_5 \end{bmatrix} = \begin{bmatrix} 0 \\ 0 \\ \dfrac{V_5}{s(C_3 + C_4)} \\ -\dfrac{V_{23} + V_5}{s(C_3 + C_4)} \end{bmatrix} \Big\}. \tag{54}$$

$$\left.\begin{aligned}
SP_{H,34,1,1} &= -G_1 G_2 G_5 \frac{V_{23} + V_5}{s(C_3 + C_4)}; \ N_{a,J3,1,1} = 4; \ Ns_{,J3,1,1} = 0; \\[6pt]
SP_{H,34,23,1} &= G_1 G_2 G_5 \frac{V_{23} + V_5}{s(C_3 + C_4)}; \ N_{a,J3,23,1} = 4; \ Ns_{,J3,23,1} = 1; \\[6pt]
SP_{H,34,4,1} &= -G_1(G_1 + G_2)sC_4 \frac{V_5}{s(C_3 + C_4)}; \ N_{a,J3,4,1} = 4; \ Ns_{,J3,4,1} = 0; \\[6pt]
SP_{H,34,4,2} &= G_1 G_2 sC_3 \frac{V_{23} + V_5}{s(C_3 + C_4)}; \ N_{a,J3,4,2} = 4; \ N_{s,J3,4,2} = 0; \\[6pt]
SP_{H,34,4,3} &= -G_1^2 sC_4 \frac{V_5}{s(C_3 + C_4)}; \ N_{a,J3,4,3} = 2; \ N_{s,J3,4,3} = 1; \\[6pt]
SP_{H,34,4,4} &= G_1 G_2 sC_3 \frac{V_5}{s(C_3 + C_4)}; \ N_{a,J3,4,4} = 2; \ N_{s,J3,4,4} = 1; \\[6pt]
SP_{H,34,5,1} &= -G_1(G_1 + G_2)C_5 \frac{V_{23} + V_5}{s(C_3 + C_4)}; \ N_{a,J3,5,1} = 4; \ N_{s,J3,5,1} = 0; \\[6pt]
SP_{H,34,5,2} &= -G_1^2 G_5 \frac{V_{23} + V_5}{s(C_3 + C_4)}; \ N_{a,J3,5,2} = 2; \ N_{s,J3,5,2} = 1;
\end{aligned}\right\}. \tag{55}$$

In order to determine the second derivatives of the vector $\partial^2 \mathbf{V}/\partial (sC_3)\partial (sC_4)$ and having in mind (42) one draws the CM graph $G_{H,34}$ shown in Fig. 20. In the case we have a simplification of the analysis on the base of the graph $G_{H,34}$ because the substantial difference between $G_{H,34}$ and G_{J3} consists in the sink and source vertex signal expressions – instead of $-V_{23}$ and V_{23} in G_{J3} the corresponding signals in $G_{H,34}$ are $V_5 /s(C_3+C_4)$ and $-(V_{23}+V_5) /s(C_3+C_4)$. Owing to this peculiarity further we use directly (37) after substituting sink vertex signals, namely:

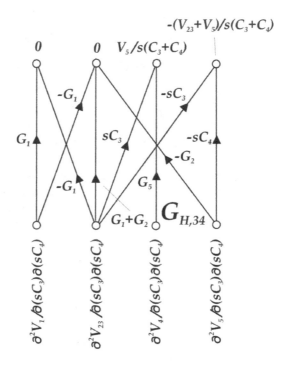

Fig. 20. CM Graph $G_{H,34}$

The voltage V_5 can be find similarly to V_{23} from CM graph G and it is:

$$V_5 = -\frac{C_3 J_1}{s(C_3 + C_4)}. \tag{56}$$

Then by using (33), (35), (55) and (56) one obtains the vector

$$\frac{\partial^2 \mathbf{V}}{\partial (sC_3)\partial (sC_4)} = \left[\frac{J_1(C_4 - C_3)}{G_2 s^2 (C_3 + C_4)^3} \quad \frac{J_1(C_4 - C_3)}{G_2 s^2 (C_3 + C_4)^3} \quad -2\frac{J_1 C_3 C_4}{s G_2 G_5 (C_3 + C_4)^3} \quad \frac{J_1(C_4 - C_3)}{G_2 s^2 (C_3 + C_4)^3} \right]_t, \tag{57}$$

Its elements are a part of elements in the Hessian matrices $\mathbf{H}_1$, $\mathbf{H}_{23}$, $\mathbf{H}_4$ and $\mathbf{H}_5$ with respect to the admittances sC_3 and sC_4.

6. First and second-order quadratic sensitivity sums

The sensitivity is an important parameter for the evaluation of practical suitability of electrical networks. For this purpose usually one uses the first-order sensitivity and the second-order sensitivity, defined by the well known formulae (Cederbaum, 1984; Chua & Lin, 1975)

$$S_x^F = \frac{\partial F}{\partial x} \cdot \frac{x}{F} ; \tag{58}$$

and

$$S_{x,y}^F = \frac{\partial^2 F}{\partial x \partial y} \cdot \frac{xy}{F} , \tag{59}$$

respectively and where F is a network function or variable and x, y are changeable network element parameters.

Obviously, the derivatives in these expressions can be determined according to the above described method based on Chan-Mai signal-flow graphs. Besides very often we are interested in a global index as a quadratic sum of sensitivities (first- or second-order):

$$\sum_i (S_x^{F_i})^2 = \sum_i \left(\frac{\partial F_i}{\partial x} \cdot \frac{x}{F_i} \right)^2 \tag{60}$$

and

$$\sum_i (S_{x,y}^{F_i})^2 = \sum_i \left(\frac{\partial^2 F_i}{\partial x \partial y} \cdot \frac{xy}{F_i} \right)^2 , \tag{61}$$

where $i \in \{1, 2, ..., n\}$.

Without loss of generality further we assume that the functions F_i are the elements of the voltage vector $\mathbf{V}$. Then the sum (60) can be derived with the help of the expressions of the corresponding Jacobian matrix subvectors $\mathbf{J}_i$ and of the voltage vector $\mathbf{V}$:

$$\left. \begin{array}{l} \sum_i (S_x^{V_i})^2 = x^2 \mathbf{J}_{i,t} (\mathbf{M}^{-1})^2 \mathbf{J}_i ; \\ \mathbf{M} = diag\{V_1, V_2, ..., V_i, ..., V_n\} \end{array} \right\}. \tag{62}$$

If from the elements of the Hessian matrices $\mathbf{H}_i$ one forms the vector

$$\mathbf{h}_{xy} = \begin{bmatrix} h_{1,xy} & h_{2,xy} & \cdots & h_{i,xy} & \cdots & h_{n,xy} \end{bmatrix}_t ; \, h_{i,xy} = \frac{\partial^2 V_i}{\partial x \partial y} \tag{63}$$

the sum (61) can be rewritten as

$$\sum_i (S_{xy}^{V_i})^2 = x^2 y^2 \mathbf{h}_{i,t} (\mathbf{M}^{-1})^2 \mathbf{h}_i \,.$$ (64)

7. Conclusions

A topological method for obtaining the Jacobian and Hessian matrices and their use for quadratic first- or second-order sensitivity sums calculation of active networks is presented. It is based on the replacement of the investigated network N by using a nullor equivalent circuit and on the representation of the circuit passive part N_p by a Chan-Mai signal-flow graph G_p. The Jacobian and the Hessian matrix elements of the nullor network can be obtained by means of the some dependent variables of some Chan-Mai graphs derived from G. The substantial advantage of the method consists in the use mainly of isomorphic graphs. Two examples illustrate the proposed method.

8. Acknowledgement

The author would like to thank Higher School of Transport "Todor Kableshkov", Sofia, Bulgaria for the financial support for the publishing this work.

9. References

Cederbaum, I. (1984). "Some Applications of Graph Theory to Network Analysis and Synthesis", *IEEE Tr. on Circuits and Systems,* vol. CAS-31, 1, 1984, pp. 64-68

Chan, S.P., Mai, H.N. (1967). "A Flow-Graph Method for the Analysis of Linear Systems", *IEEE Tr. on Circuit Theory*, 9, 1967, pp. 350-354

Chua, L.O., Lin, P.-M. (1975). *"Computer-Aided Analysis of Electronic Circuits"*, Prentice-Hall Inc. Englewood Clifs, New Jersey, 1975

Davies, A. C. (1966). "Matrix Analysis of Network Containing Nullators and Norators", *Electronics Letters*, vol. 2, ,2, 1966, pp. 48 -49

Donevsky, B.D., Nenov,G.A. (1979). *"Application of Graphs for the Analysis and Synthesis of Electronic Circuits"*, "Technica", Sofia, 1979 (in Bulgarian)

Korn, G., A., Korn, T.M. (1968). *"Mathematical Handbook"*, Mc Graw-Hill Co. New York. 1968

Nenov, G. A. (2004). "Evaluation of Nullor Nertwork Jacobian Matrix by using Chan-Mai Signal-Flow Graphs", *Proceedings of the SMACD'04*, Wroclaw, 2004, Poland, pp. 79-82

Wilde, D.J. (1978). *"Globally Optimal Design"*, John Wiley & Sons, New York, 1978

13

Research Progress of Complex Electric Power Systems: Graph Theory Approach

Yagang Zhang, Zengping Wang and Jinfang Zhang
*State Key Laboratory of Alternate Electrical Power System
with Renewable Energy Sources
(North China Electric Power University),
China*

1. Introduction

Electric power system is one of the most complex artificial systems in this world, which safe, steady, economical and reliable operation plays a very important part in guaranteeing socioeconomic development, even in safeguarding social stability. In early 2008, the infrequent disaster of snow and ice that occurred in the south of China had confirmed it again. The complexity of electric power system is determined by its characteristics about constitution, configuration, operation, organization, etc., which has caused many disastrous accidents, such as the large-scale blackout of America-Canada electric power system on August 14, 2003, the large-scale blackout of Italy electric power system on September 28, 2003. In order to resolve this complex and difficult problem, some methods and technologies that can reflect modern science and technology level have been introduced into this domain, such as computer and communication technology, control technology, superconduct and new materials technology and so on. Obviously, no matter what we adopt new analytical method or technical means, we must have a distinct recognition of electric power system itself and its complexity, and increase continuously analysis, operation and control level (Yuan, 2007; Ye, 2003; Xue, 2002).

A fault is defined as a departure from an acceptable range of an observed variable or calculated parameter associated with systems. It may arise in the basic technological components or in its measurement and control instruments, and may represent performance deterioration, partial malfunctions or total breakdowns. Fault analysis implies the capability of determining, either actively or passively, whether a system is functioning as intended or as modeled. The goal of fault analysis is to ensure the success of the planned operations by recognizing anomalies of system behavior. A system with faults does not necessarily imply that the system is not functioning. Detecting a fault involves identifying a characteristic of the system, which when a fault occurs, can be distinguished from other characteristics of the system. According to nonlinear complex systems, we have carried out large numbers of basic researches (Zhang et al., 2010; Zhang et al., 2006; Zhang et al., 2007; Zhang & Wang, 2008). In this chapter, basing on graph theory and multivariate statistical analysis theory, we will discuss the complexity in electric power system.

The fault in electric power system can not be completely avoided. When electric power system operates from normal state to failure or abnormal operates, its electric quantities (current, voltage and their angles, etc.) may change significantly. In our researches, after some accidents, utilizing real-time measurements of phasor measurement unit (PMU) (Phadke & Thorp, 2008; Wang et al., 2007; Rakpenthai et al., 2005; Peng et al., 2006), basing on graph theory and multivariate statistical analysis theory, we are using mainly Breadth-first search (BFS), Depth-first search (DFS) and cluster analysis technology (Arifin & Asano, 2006; Otazu & Pujol, 2006; Park & Baik, 2006; Tola et al., 2008; Zhao et al., 2008; Templ et al., 2008), and seeking after for the uniform laws of electrical quantities' marked changes. Then we can carry out fast and exact analysis of fault component. Finally we can accomplish fault isolation.

2. Electric circuit theory

Let's consider a circuit with resistors(R), inductors (L), and capacitors(C), which has one element of each connected in a loop (Robinson, 2004). The part of the circuit containing one element is a branch. The points where the branches connect are nodes. There are three branches and nodes in Figure. 1.

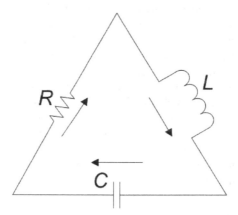

Fig. 1. RLC electric circuit

Let i_R, i_L and i_C be the current in the resistor, inductor and capacitor respectively, and let v_R, v_L and v_C be the voltage drop across the three branches of the circuit. Kirchhoff's voltage law states that the sum of the voltage drops around any loop is zero, that is, $v_R + v_L + v_C = 0$. Kirchhoff's current law states that the total current flowing into a node must equal the current flowing out of that node, namely, $|i_R| = |i_L| = |i_C|$ with the correct choice of signs.

A resistor is determined by a relationship between the current i_R and voltage v_R. Here, a linear resistor given by $v_R = R\,i_R$, where $R > 0$ is a constant, which is determined by Ohm's law. An inductor is characterized by giving the time derivative of the current $\dfrac{di_L}{dt}$, Faraday's law has proved that,

$$L\frac{di_L}{dt} = v_L \tag{1}$$

where the constant $L > 0$ is called the inductance. A capacitor is characterized by giving the time derivative of the voltage $\frac{dv_C}{dt}$, in terms of the current i_C,

$$C\frac{dv_C}{dt} = i_C \tag{2}$$

where the constant $C > 0$ is called the capacitance.

Furthermore, let's define two variables $x = i_R = i_L = i_C$, $y = v_C$, and $v_R = R\,x$, $v_L = -v_C - v_R = -y - R\,x$. So, this system can be expressed as,

$$\frac{d}{dt}\begin{pmatrix} x \\ y \end{pmatrix} = \begin{pmatrix} -\dfrac{R}{L} & -\dfrac{1}{L} \\ \dfrac{1}{C} & 0 \end{pmatrix}\begin{pmatrix} x \\ y \end{pmatrix} \tag{3}$$

Its characteristic equation is,

$$\lambda^2 + \frac{R}{L}\lambda + \frac{1}{LC} = 0 \tag{4}$$

which has two roots:

$$\lambda = -\frac{R}{2L} \pm \sqrt{\frac{R^2C - 4L}{4L^2C}} \tag{5}$$

Actually this system always has eigenvalues with negative real parts. If $R^2 \geq \frac{4L}{C}$, then the eigenvalues are real; else $R^2 < \frac{4L}{C}$, they are complex.

3. Search principles in graph theory

Many real world situations can conveniently be described by means of a diagram consisting of a set of points together with lines joining certain pairs of these points. In mathematics and computer science, graph theory is the study of graphs: mathematical structures used to model conjugated relations between objects from a certain collection. A graph is an abstract notion of a set of nodes and connection relations between them, that is, a collection of vertices or nodes and a collection of edges that connect pairs of vertices. A graph may be undirected, meaning that there is no distinction between the two vertices associated with each edge, or its edges may be directed from one vertex to another.

Applications of graph theory are primarily, but not exclusively, concerned with labeled graphs and various specializations of these. Structures that can be represented as graphs are ubiquitous, and many problems of practical interest can be represented by graphs. For example, in electric circuit theory, the Kirchhoff's voltage law and Kirchhoff's current law are only concerned with the structures and properties of the electric circuit. Then, any concrete electric circuit can be abstracted as a graph (Bondy & Murth, 1976). Here, let's give a simple electric circuit (See Figure 2), and its structure can be expressed as a graph (See Figure 3).

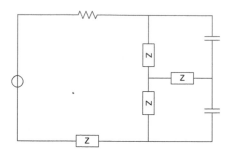

Fig. 2. A simple electric circuit

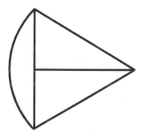

Fig. 3. A graph based on the simple electric circuit

Graph theory can be used to model many different physical and abstract systems such as transportation and communication networks, models for business administration, political science, and psychology and so on. Efficient storage and algorithm design techniques based on the graph representation make it particularly useful for utilizing computer. There are many algorithms that can be applied to resolve different kinds of problems, such as Breadth-first search, Depth-first search, Bellman-Ford algorithm, Dijkstra's algorithm, Ford-Fulkerson algorithm, Kruskal's algorithm, Nearest neighbor algorithm, Prim's algorithm, etc. Hereinto, Breadth-first search (BFS) is a graph search algorithm that begins at the root node and explores all the neighboring nodes. Then for each of those nearest nodes, it explores their unexplored neighbor nodes, and so on, until it finds the goal.

BFS is an uninformed search method that aims to expand and examine all nodes of a graph or combinations of sequence by systematically searching through every solution. In other words, it exhaustively searches the entire graph or sequence without considering the goal until it finds it. From the standpoint of the algorithm, all child nodes obtained by expanding a node are added to a first-in, first-out (FIFO) queue. In typical implementations, nodes that have not yet been examined for their neighbors are placed in some container (such as a queue or linked list) called "open" and then once examined are placed in the container "closed" (Knuth, 1997).

BFS can be used to solve many problems in graph theory, for example:

- Testing whether graph is connected, and finding all connected components in a graph;
- Computing a spanning forest of graph;
- Computing, for every vertex in graph, a path with the minimum number of edges between start vertex and current vertex or reporting that no such path exists;
- Computing a cycle in graph or reporting that no such cycle exists.

The Depth-first search (DFS) is an algorithm for traversing or searching a tree, tree structure, or graph. One starts at the root and explores as far as possible along each branch before backtracking (Thomas et al., 2001).

In formal way, DFS is an uninformed search that progresses by expanding the first child node of the search tree that appears and going deeper and deeper until a goal node is found, or until it reaches a node which has no child node. Then the search backtracks, and it will return to the most recent node that it has not finished exploring. The space complexity of DFS is much lower than BFS. It also lends itself much better to heuristic methods of choosing a likely-looking branch. Time complexity of both algorithms is proportional to the number of vertices plus the number of edges in the graphs they traverse.

4. Cluster analysis

Theories of classification come from philosophy, mathematics, statistics, psychology, computer science, linguistics, biology, medicine, and other areas. Cluster analysis can also be named classification, which is concerned with researching the relationships within a group of objects in order to establish whether or not the data can be summarized validly by a small number of clusters of similar objects. That is, cluster analysis encompasses the methods used to:

- Identify the clusters in the original data;
- Determine the number of clusters in the original data;
- Validate the clusters found in the original data.

Cluster analysis is commonly applied for statistical analyses of large amounts of experimental data exhibiting some kind of redundancy, which allows for compression of data to amount feasible for further exploration. This permits further mining of each cluster independently or, alternatively, constructing a high level view of the data set by replacing each cluster with its best single representative. Cluster analysis has great strength in data analysis and has been applied successfully to the researches of various fields. The effectiveness of a cluster approach depends on many choices. These include the choice of a cluster algorithm, an appropriate feature subspace, and a similarity metric defined over this subspace. In addition, cluster algorithms typically have a set of tunable parameters inherent to them that can heavily influence their performance. For example, many algorithms require the number of clusters desired, the maximum number of iterations, learning rate, its change schedule, etc. While some of these choices are obvious for simple artificial datasets. The most common clustering algorithm choices are hierarchical cluster analysis.

5. Fault analysis based on BFS and DFS

Now let us consider IEEE9-Bus system, Figure 4 is presents the IEEE 9-Bus system electric diagram. In the structure of electric power system, Bus1 appears single-phase to ground fault. Through simulation experiments, using these actual measurement data of corresponding variables, we can carry through fault analysis of fault component and non-fault component.

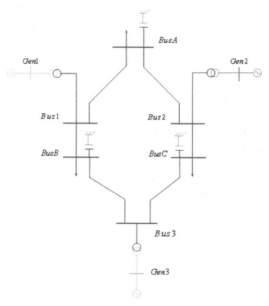

Fig. 4. Electric diagram of IEEE 9-Bus system

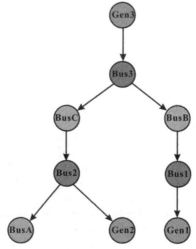

Fig. 5. BFS diagram of IEEE 9-Bus system

The adjacency matrix of IEEE9-Bus system can be expressed as follows,

$$
\begin{array}{c c}
 & \begin{array}{ccccccccc} \text{Bus1} & \text{Bus2} & \text{Bus3} & \text{BusA} & \text{BusB} & \text{BusC} & \text{Gen1} & \text{Gen2} & \text{Gen3} \end{array} \\
\begin{array}{c} \text{Bus1} \\ \text{Bus2} \\ \text{Bus3} \\ \text{BusA} \\ \text{BusB} \\ \text{BusC} \\ \text{Gen1} \\ \text{Gen2} \\ \text{Gen3} \end{array} &
\left(\begin{array}{ccccccccc}
0 & 0 & 0 & 1 & 1 & 0 & 1 & 0 & 0 \\
0 & 0 & 0 & 1 & 0 & 1 & 0 & 1 & 0 \\
0 & 0 & 0 & 0 & 1 & 1 & 0 & 0 & 1 \\
1 & 1 & 0 & 0 & 0 & 0 & 0 & 0 & 0 \\
1 & 0 & 1 & 0 & 0 & 0 & 0 & 0 & 0 \\
0 & 1 & 1 & 0 & 0 & 0 & 0 & 0 & 0 \\
1 & 0 & 0 & 0 & 0 & 0 & 0 & 0 & 0 \\
0 & 1 & 0 & 0 & 0 & 0 & 0 & 0 & 0 \\
0 & 0 & 1 & 0 & 0 & 0 & 0 & 0 & 0
\end{array}\right)
\end{array}
$$

By the simulation experiments, we can get node phase voltage at T_{-1}, T_0 (Fault), T_1, T_2 and T_3 five times, see Table 1.

Bus	T_{-1}	T_0(Fault)	T_1	T_2	T_3
Gen1	1.0100	0.7275	0.6924	0.6814	0.6747
Gen2	1.0100	0.8762	0.8476	0.8327	0.8134
Gen3	1.0100	0.8449	0.8071	0.7909	0.7710
Bus1	1.0388	0	0	0	0
Bus2	1.0430	0.7622	0.7350	0.7217	0.7049
Bus3	1.0534	0.7600	0.7275	0.7134	0.6960
BusA	1.0319	0.7540	0.7248	0.7114	0.6944
BusB	1.0222	0.2512	0.2404	0.2356	0.2294
BusC	1.0061	0.2470	0.2381	0.2336	0.2276

Table 1. The Node Phase Voltage At T_{-1}, T_0 (Fault), T_1, T_2 And T_3 Five Times

Figure.5 is the BFS process of IEEE9-Bus system. In this diagram, Gen1 is the first generator node, it is also one of the terminals of BFS, and Bus1 is just the only node that connects with it. Combined the information characters of electrical measurements that have marked changes, the difference of Bus1 and other Buses is distinct. At the beginning, Bus1 has just been set as single-phase to ground, which is a typical bus-bar fault. In the final analysis, both of these two aspects are consistent, and we can identify effectively fault location based on BFS.

Figure.6 is the DFS process of IEEE9-Bus system. In this diagram, the difference of Bus1 and other Buses is more distinct. Gen1 is the terminal of DFS, and Bus1 is just the only node that connects with it. In the beginning, we have set the Bus1 as single-phase to ground fault. Both of these two aspects are consistent. So, we can also identify effectively fault location based on DFS.

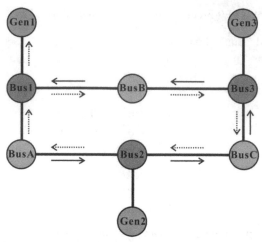

Fig. 6. DFS diagram of IEEE 9-Bus system

6. Fault analysis based on hierarchical cluster analysis

The hierarchical cluster analysis does not require us to specify the desired number of clusters K, instead affording a cluster dendrogram. In practice, the choice can be based on some domain specific and often have subjective components. There are three steps to hierarchical cluster analysis. First, we must identify an appropriate proximity measure. Second, we need to identify the appropriate cluster method for the data. Finally, an appropriate stopping criterion is needed to identify the number of clusters in the hierarchy. The distance or similarity metric used in cluster is crucial for the success of the cluster method. Euclidean distance and Pearson correlation are among the most frequently used.

Now let us continue to consider IEEE9-Bus system. According to the results of the simulation experiments (Table 1), basing on node phase voltage, we carry out hierarchical cluster analysis. Figure.7 is the dendrogram of hierarchical cluster analysis based on node phase voltage.

Let us explain the entire process of cluster analysis in detail. The entire cluster analysis process is carried out according to the principle of similarity from high to low (distance from near to far), the order is,

Steps 1: BusC is combined with BusB to form the new BusB;
Steps 2: Bus3 is combined with Bus2 to form the new Bus2;
Steps 3: BusA is combined with Bus2 to form the new Bus2;
Steps 4: Bus2 is combined with Gen1 to form the new Gen1;
Steps 5: Gen3 is combined with Gen2 to form the new Gen3;
Steps 6: Gen2 is combined with Gen1 to form the new Gen1;
Steps 7: BusB is combined with Bus1 to form the new Bus1;
Steps 8: Gen1 is combined with Bus1 to form the new Bus1.

From the entire hierarchical cluster process analysis, Bus1 has the lowest similarity to other nodes (the farthest distance to other nodes). It can also be found easily out from Figure.7

that Bus1 has remarkable difference with other buses, and the fault characteristic is obvious. These results are entirely identical to the fault location set in advance, so we can also confirm exactly fault location by the hierarchical cluster analysis.

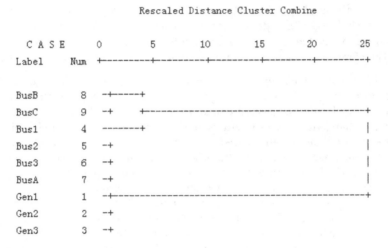

```
          Dendrogram using Average Linkage (Between Groups)


                      Rescaled Distance Cluster Combine


          C A S E        0       5       10      15      20      25
          Label    Num   +---------+---------+---------+---------+---------+


          BusB      8     -+-----+
          BusC      9     -+      +------------------------------------------+
          Bus1      4     --------+                                          |
          Bus2      5     -+                                                 |
          Bus3      6     -+                                                 |
          BusA      7     -+                                                 |
          Gen1      1     -+------------------------------------------------+
          Gen2      2     -+
          Gen3      3     -+
```

Fig. 7. The dendrogram of hierarchical cluster analysis based on node phase voltage

7. New approach for finding connected routes of power network

The connectivity analysis is the essence of the power system topology analysis, which is also playing the basic function in many kinds of advanced application software for power system analysis and calculation (Monticelli, 1999). The actual power network structures are shown by the result of the power system topology analysis processor, as the bus/branch model. Hence, the aim of topology analysis is to map the bus-section/switching-device model in the physical equipments level into the bus/branch model for a number of advanced functions in Energy Management System (EMS) or Distribution Management System (DMS) (Monticelli, 1999), such as power flow calculation, state estimation, dispatcher training simulator (DTS) and so on.

The main methods used in connectivity analysis of power system are based on the Graph theory, including tree-search based method and matrix-based method (Zhu et al., 2002). The former one can also be classified into two algorithms according to the different search patterns, named as Depth First Search (DFS) and Breadth First Search (BFS). By now, this topology analysis method has a widely application in power system analysis software (e.g. EMS or DMS). However, if the loop structure exists in the current network, the efficiency of tree-search based method will be low (Zhu et al., 2002). Matrix-based method is a systematic method based on adjacent matrix (Goderya et al., 1980), which can clearly depict the connected relationship between the two nodes belonged to the same branch. As the aim of connectivity analysis, whether any two nodes is connected or not and how many connected pieces all nodes could be mapped into will be the ultimate target of topology analysis. Therefore, the complete connected matrix will be needed to gain a global connectivity information among nodes, which can be obtained by the self-multiplying of the above mentioned adjacent matrix with the number of operations no more than $n-1$ (where, the n is the total number of the nodes). Obviously, the calculation burden of the matrix-based method will increase sharply as the expanding of the network scale. So in the substation or plants of power system where the number of nodes is not large, the matrix-based one also can play the role in grouping the connected physical nodes and then mapping the connected pieces into buses. Substation configuration has taken the most part of the total time needed in topology analysis (Zhang et al., 2010). Therefore, the substation configuration will be paid more attention in this paper.

The basic object of the above two methods is the vertexes/edges model mapped from the physical connections, in which the edges are corresponding to the switch devices, and the vertex will be the electrical connected points or physical buses. Therefore, once the state of the switches is changed, the connectivity analysis process will be restarted. Neither in tree-search based method nor in matrix-based method, the repeat search and calculation can not be avoided, which seriously effects the efficiency of the tracking of the status change happened in switch devices. In order to reduce the on-line topology analysis burden, the reference (Zhang et al., 2010) has established one method based on the graphic characteristic of the main connections, in which the each element of the complete connected matrix is represented by a set of connected routes with the open-close state of edges as variables. If the status of the each edge is determined, the value of these connected routes could be gained and the relationship between the mentioned nodes is confirmed. Hence, the large amount of repeat calculation is avoided. However, the connected routes finding algorithm is based on the type of electrical main connections used as the rule-based method, which is not systematic method and not suit for the irregular connections.

In this section, a new connected routes finding algorithm is proposed based on the adjoint matrix of the symbolic adjacent matrix, and a simplifying method is also applied to extract the connected routes information readily. Compare to the graphic feature based method, the new finding algorithm is an systematic one, which is suitable for various network connection structures.

8. The new connected routes finding algorithm

According to the Graph Theory (Diestel, 2000), the matrix could be used as the representation of one special graph, which is easily be analyzed and calculated by the computer. The two kinds of matrix usually used in analysis are incident matrix A_{inc} and adjacent matrix A_{adj} respectively. The A_{inc} depicts the connected relationship between the nodes and edges, and the A_{adj} gives a description of the relationship between two nodes. The aim of connectivity analysis is to group the nodes into different connected pieces, A_{adj} could meet this requirement and usually be used to do connectivity analysis.

For a special simple network structure as shown in Figure.8, the vertex set is $V(G) = [v1, v2, v3, v4, v5, v6]$, and the edge set is $E(G) = [e1, e2, e3, e4, e5, e6]$. The mapping relationship set between the vertexes and edges is $\varphi(G): E \rightarrow V \times V$, such as $\varphi(e1) = \{v1, v2\}$, $\varphi(e2) = \{v2, v3\}$, $\cdots$, $\varphi(e6) = \{v5, v6\}$.

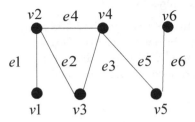

Fig. 8. A simple graph G with six vertexes and six edges

The adjacent matrix A_{adj} of graph G in Figure.8 is following:

$$A_{adj} = \begin{bmatrix} 1 & 1 & 0 & 0 & 0 & 0 \\ 1 & 1 & 1 & 1 & 0 & 0 \\ 0 & 1 & 1 & 1 & 0 & 0 \\ 0 & 1 & 1 & 1 & 1 & 0 \\ 0 & 0 & 0 & 1 & 1 & 1 \\ 0 & 0 & 0 & 0 & 1 & 1 \end{bmatrix} \qquad (6)$$

Where, if there is a edge between vertex i and vertex j, the value of $A_{adj}(i,j)$ is equal to 1; otherwise, $A_{adj}(i,j) = 0$. Especially, the diagonal element of A_{adj} is set to 1. Obviously, this matrix A_{adj} can not describe the relationship between any two nodes, which are not incident to the same edge. Therefore, the completed connected matrix A is needed, which can depict the connected relationship between any two nodes in one graph. Corresponding to connected graph G in Figure. 8, the elements of matrix A are equal to 1, as following:

$$A = \begin{bmatrix} 1 & 1 & 1 & 1 & 1 & 1 \\ 1 & 1 & 1 & 1 & 1 & 1 \\ 1 & 1 & 1 & 1 & 1 & 1 \\ 1 & 1 & 1 & 1 & 1 & 1 \\ 1 & 1 & 1 & 1 & 1 & 1 \\ 1 & 1 & 1 & 1 & 1 & 1 \end{bmatrix} \qquad (7)$$

In a summary, the matrix-based method for the connectivity analysis is to find the completed connected matrix A from original adjacent matrix A_{adj}. According to the Graph Theory, the matrix A is obtained from the formulation (8):

$$A = A_{adj}^{n-1} \qquad (8)$$

where, n is the number of vertexes in the graph G. In this process, the computation burden centers on the square of the matrix. It is worth to point out that the repeat calculation has taken the most part of the total time for the determination process of connected relationship among nodes.

8.1 The basic principle of new algorithm

Different from the traditional matrix-based method, the edges in the graph G are treated as the independent variables with the open-close two statuses; that is, if the edge is opened, the value of this corresponding variable is set to 0, otherwise, the value is equal to 1. Therefore, the adjacent matrix A_{adj} is the function of the current edge status, as following based on the Figure. 8.

$$A_{adj}(E) = \begin{bmatrix} 1 & e1 & 0 & 0 & 0 & 0 \\ e1 & 1 & e2 & e4 & 0 & 0 \\ 0 & e2 & 1 & e3 & 0 & 0 \\ 0 & e4 & e3 & 1 & e5 & 0 \\ 0 & 0 & 0 & e5 & 1 & e6 \\ 0 & 0 & 0 & 0 & e6 & 1 \end{bmatrix} \tag{9}$$

where, E is organized as a vector, such as $[e1,e2,e3,e4,e5,e6]$. Once the state of the each edge is determined, the vector E will be established and the matrix $A_{adj}(E)$ is also formed by substituting the current edge status. Take Figure.8 as an example, all the edges are closed and the current edge status vector E is equal to $[1,1,1,1,1,1]$.

The adjoint matrix of A_{adj} is defined as A_{adj}^*, and corresponding to formulation (9), the detailed element representations in the first row of A_{adj}^* are extracted for the in-depth analysis, which are as following:

$$\begin{aligned} A_{adj}^*(1,1) &= e2^2 e5^2 + e2^2 e6^2 - e2^2 - 2e2e3e6^2 e4 \\ &+ 2e4e3e2 + e3^2 e6^2 - e6^2 + e6^2 e4^2 \\ &- e3^2 - e5^2 + 1 - e4^2 \\ A_{adj}^*(1,2) &= e1e3^2 - e1e3^2 e6^2 + e1e5^2 + e1e6^2 - e1 \\ A_{adj}^*(1,3) &= -e1e2e5^2 - e1e2e6^2 + e1e2 - e1e4e3 \\ &+ e1e3e6^2 e4 \\ A_{adj}^*(1,4) &= e1e6^2 e2e3 - e1e2e3 - e1e6^2 e4 + e1e4 \\ A_{adj}^*(1,5) &= e1e5e2e3 - e1e5e4 \\ A_{adj}^*(1,6) &= -e1e6e5e2e3 + e1e6e5e4 \end{aligned} \tag{10}$$

The detailed analysis of the above representations in (10) is:

As to non-diagonal element, the representation has included all the routes between the two nodes indexed by the column and row number. For instance, in Figure.8, there is only one route between node $v1$ and node $v2$, which is $e1$. Comparing to the $A_{adj}^*(1,2)$, the route $e1$ is included in this element's representation as one individual term stamped by one rectangle,

$$A_{adj}^*(1,2) = e1e3^2 - e1e3^2e6^2 + e1e5^2 + e1e6^2 - \boxed{e1} \tag{11}$$

Similarly, there are two connected routes between node $v1$ and $v4$, respectively which are $e1e4$ and $e1e2e3$. As a result in the representation of $A_{adj}^*(1,4)$, these two routes are playing the roles as two terms signed by two rectangles,

$$A_{adj}^*(1,4) = e1e6^2e2e3 - \boxed{e1e2e3} - e1e6^2e4 + \boxed{e1e4} \tag{12}$$

As to diagonal element, there is not route between the node and itself. As shown in $A_{adj}^*(1,1)$, the representation is very complicated. However, in the practical analysis, the value of the diagonal element is set to 1. Obviously, this term is also existing in the $A_{adj}^*(1,1)$ as

$$\begin{aligned} A_{adj}^*(1,1) = &\, e2^2e5^2 + e2^2e6^2 - e2^2 - 2e2e3e6^2e4 \\ &+ 2e4e3e2 + e3^2e6^2 - e6^2 + e6^2e4^2 \\ &- e3^2 - e5^2 + \boxed{1} - e4^2 \end{aligned} \tag{13}$$

In a summary, the non-diagonal element of the adjoint matrix A_{adj} has contained all the routes between the relevant two nodes. However, it is very low efficient to finding all the connected routes between the nodes just by the observation method. Therefore, in the next subsection, one simplifying method is proposed to make a rapidly and accurately finding.

8.2 Simplifying method

Having taken many different network structures cases into consideration, the needed terms which represent the connected routes between the mentioned two nodes can be extracted by the following three steps:

As to one special representation of the element in adjoint matrix A_{adj}^*,

Step1: Firstly, if the highest power of one term in the polynomial is not less than 2, this term should be removed;
Step2: Secondly, if the absolute value of coefficient of one term is not equal to 1, this term also should be removed;
Step3: Thirdly, the coefficient of all the left terms is set to its absolute value, which is 1 in fact.

After the above mentioned three steps, the representation of element in adjoint matrix A_{adj} can be simplified into the needed connected routes formation. In order to make a detailed depiction, the $A_{adj}^*(1,2)$ is picked and taken as example.

For $A_{adj}^*(1,2) = e1e3^2 - e1e3^2e6^2 + e1e5^2 + e1e6^2 - e1$:

These terms such as $e1e3^2$, $-e1e3^2e6^2$, $e1e5^2$, $e1e6^2$ will be removed in the first step, because the highest power of them are equal to 2 respectively; and the simplified result $A_{adj}^*(1,2)\_s1$ in this step is as following:

$$A_{adj}^*(1,2)\_s1 = s_1(A_{adj}^*(1,2)) = -e1 \tag{14}$$

In the second step, no term will be removed, because the left term $-e1$ does not meet the condition that absolute value of coefficient is not equal to 1. Hence, there is:

$$A_{adj}^*(1,2)\_s2 = s_2(A_{adj}^*(1,2)\_s1) = -e1 \tag{15}$$

In the final step, the coefficients of terms in $A_{adj}^*(1,2)\_s2$ are substituted by their absolute values, and ultimate organization formation of connected routes is obtained as following:

$$A_{adj}^*(1,2)\_s3 = s_3(A_{adj}^*(1,2)\_s2) = e1 \tag{16}$$

Similarly, the first row in the adjoint matrix A_{adj} can be transformed into the connected routes formation as following

$$
\begin{aligned}
A_{adj}^*(1,1)\_s3 &= 1 \\
A_{adj}^*(1,2)\_s3 &= e1 \\
A_{adj}^*(1,3)\_s3 &= e1e2 + e1e4e3 \\
A_{adj}^*(1,4)\_s3 &= e1e2e3 + e1e4 \\
A_{adj}^*(1,5)\_s3 &= e1e5e2e3 + e1e5e4 \\
A_{adj}^*(1,6)\_s3 &= e1e6e5e2e3 + e1e6e5e4
\end{aligned}
\tag{17}
$$

According to the above simplifying method, the elements in the other rows of the adjoint matrix A_{adj} also can be simplified and organized as the connected routes form. When the matrix A_{adj}_s3 is formed as the ultimate result, the current edge statuses can be substituted as individual variables and the completed connected matrix A also can be obtained equally. Obviously, the following formulation will be established:

$$A = A_{adj}\_s3(E) \tag{18}$$

9. The application for substation configuration of power system topology analysis

In substation configuration, the network structure is more complicated than the network connection among substations or plants. The number of nodes in one substation or plants is not large, so the matrix-based method can be adopted for the operation of low dimension matrix. Generally, one electrical connection can be mapped into a graph, in which the switch device is corresponding to the edge, and the physical bus and electrical connected point are transformed into vertexes.

In this part, the more complicated electrical main connection—angle connection with four angles is taken as an example. One typical connection of this type is show in Figure. 9.

The corresponding adjacent matrix A_{adj} is:

$$A_{adj} = \begin{bmatrix} 1 & e1 & 0 & e2 \\ e1 & 1 & e3 & 0 \\ 0 & e3 & 1 & e4 \\ e2 & 0 & e4 & 1 \end{bmatrix} \tag{19}$$

As the adjoint matrix of the symmetrical matrix A_{adj}, the upper triangular of A_{adj}^{*} is:

$$A_{adj}^{*} = \begin{bmatrix} -e4^2+1-e3^2 & e1e4^2-e1-e4e3e2 & e1e3+e2e4 & -e1e3e4-e2+e3^2e2 \\ & -e4^2+1-e2^2 & -e3-e2e1e4+e2^2e3 & e4e3+e2e1 \\ & & -e2^2+1-e1^2 & -e4+e1^2e4-e1e2e3 \\ & & & -e3^2+1-e1^2 \end{bmatrix} \tag{20}$$

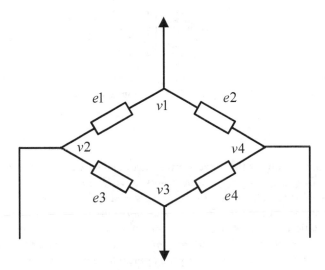

Fig. 9. A typical angle connection with four angles

After the treatment of the simplifying method proposed in Section II, the final connected routes matrix is represented by the upper triangular as following:

$$A_{adj}^{*}\_s3 = \begin{bmatrix} 1 & e1+e4e3e2 & e1e3+e2e4 & e1e3e4+e2 \\ & 1 & e3+e2e1e4 & e4e3+e2e1 \\ & & 1 & e4+e1e2e3 \\ & & & 1 \end{bmatrix} \tag{21}$$

Hence, the completed connected matrix A can be determined according to matrix $A_{adj}^{*}\_s3$ with the current edges status vector E.

If the $e1$ and $e4$ are opened with the other two edges closed, the current edge status vector $E = [1,0,0,1]$, then

$$A = A^*_{adj} \_s3(E) = \begin{bmatrix} 1 & 1 & 0 & 0 \\ 1 & 1 & 0 & 0 \\ 0 & 0 & 1 & 1 \\ 0 & 0 & 1 & 1 \end{bmatrix} \tag{22}$$

Therefore, the vertex $v1$ and $v2$ are grouped into one connected pieces, and the left two vertexes are mapped into the second bus.

Sometimes, the edge $e4$ is opened from closed, and the edge $e2$ and $e3$ are closed, the current edge status vector has become as $E = [1,1,1,0]$; therefore, the complete connected matrix A is equal to:

$$A = A^*_{adj} \_s3(E) = \begin{bmatrix} 1 & 1 & 1 & 1 \\ 1 & 1 & 1 & 1 \\ 1 & 1 & 1 & 1 \\ 1 & 1 & 1 & 1 \end{bmatrix} \tag{23}$$

In other words, all the four vertexes are connected with each, and only one connected piece is formed with the current edge statuses. In this way, this connection is mapped into one bus.

10. Conclusion

Electric power system is one of the most complex artificial systems in this world, which safe, steady, economical and reliable operation plays a very important part in guaranteeing socioeconomic development, even in safeguarding social stability. The complexity of electric power system is determined by its characteristics about constitution, configuration, operation, organization, etc. However, no matter what we adopt new analytical method or technical means, we must have a distinct recognition of electric power system itself and its complexity, and increase continuously analysis, operation and control level.

The characteristic of the adjoint matrix has been given an in-depth analysis, which is deduced from the adjacent matrix represented as the function of current edge status in the network. By the use of the simplifying method, the needed connected routes information could be extracted readily and accurately. Combination the original analysis of the adjoint marix and the connected routes extracting process, a novel connected routes finding algorithm is established. In this way, the complete connected matrix is finally formed as the function of the edge state, and each element could be represented by a set of connected routes provided by the proposed new algorithm. Once the status of each edge is determined, the connectivity between any two nodes is easily to be confirmed by substituting the current statuses into the corresponding connected routes set. These connected routes set could be found in the off-line way based on the novel routes finding algorithm, which will be able to save the on-line analysis time, especially for the topology analysis of power system. Finally, the case study on the angle connection in substation has validated the efficiency of the established novel routes finding algorithm.

The fault in electric power system can not be completely avoided. When electric power system operates from normal state to failure or abnormal operates, its electric quantities (current, voltage and their angles, etc.) may change significantly. In our researches, utilizing real-time measurements of PMU, we are using mainly graph theory and multivariate statistical analysis theory to seek after for the uniform laws of electrical quantities' marked changes. Then we can carry out fast and exact analysis of fault component. Finally we can accomplish fault isolation.

These researches have proven that the complexity of electric power system can be explored successfully by analysis and calculation based on graph theory and multivariate statistical analysis theory.

11. Acknowledgements

This research was supported partly by the National Natural Science Foundation of China (50837002), the Fundamental Research Funds for the Central Universities (11MG37), the Natural Science Foundation of Hebei Province and the Training Plan of Top-notch Doctoral Candidates in "211 Project" of NCEPU.

12. References

Arifin, A.Z. & Asano, A. (2006). Image segmentation by histogram thresholding using hierarchical cluster analysis. *Pattern Recognition Letters*, Vol. 27, 1515-1521, ISSN: 0167-8655

Bondy, J.A. & Murth, U.S.R. (1976). *Graph Theory with Applications*, Elsevier Science Publishing Co., Inc., ISBN: 978-0-444-19451-0, New York

Diestel, R. (2000). *Graph Theory*, Springer-Verlag, ISBN: 978-3-642-14278-9, NewYork

Goderya, F., Metwally, A.A & Mansour, O. (1980). Fast detection and identification of islands in power networks. *IEEE Transactions on Power Apparatus and Systems*, Vol. PAS-99(1), 217-221, ISSN: 0018-9510

Knuth, D.E. (1997). *The Art Of Computer Programming*, Third Edition, Addison-Wesley, ISBN: 978-0-201-89684-8, Boston

Monticelli, A. (1999). *State Estimation in Electric Power Systems: A Generalized Approach*, Kluwer Academic Publisher, ISBN: 978-0-792-38519-6, Massachusetts

Otazu, X. & Pujol, O. (2006). Wavelet based approach to cluster analysis. Application on low dimensional data sets. *Pattern Recognition Letters*, Vol. 27, 1590-1605, ISSN: 0167-8655

Park, H.S. & Baik, D.K. (2006). A study for control of client value using cluster analysis. *Journal of Network and Computer Applications*, Vol.29, 262-276, ISSN: 1084-8045

Peng, J.N., Sun, Y.Z. & Wang, H.F. (2006). Optimal PMU placement for full network observability using Tabu search algorithm. *International Journal of Electrical Power & Energy Systems*, Vol. 28, 223-231, ISSN: 0142-0615

Phadke, A.G. & Thorp, J.S. (2008). *Synchronized Phasor Measurements and Their Applications*, Springer verlag, ISBN: 978-0-387-76535-8

Rakpenthai, C., Premrudeepreechacharn, S., Uatrongjit, S. & Watson, N.R. (2005). Measurement placement for power system state estimation using decomposition technique. *Electric Power Systems Research*, Vol. 75, 41-49, ISSN: 0378-7796

Robinson, R.C. (2004). *An Introduction to Dynamical Systems: Continuous and Discrete*, Pearson Education, ISBN: 978-0-131-43140-9, New Jersey

Templ, M., Filzmoser, P. & Reimann, C. (2008). Cluster analysis applied to regional geochemical data: Problems and possibilities. *Applied Geochemistry*, Vol. 23, 2198-2213, ISSN: 0883-2927

Thomas, H.C., Charles, E.L., Ronald, L.R. & Clifford, S. (2001). *Introduction to Algorithms*, Second Edition, MIT Press and McGraw-Hill, ISBN: 978-0-262-03293-3, Cambridge

Tola, V., Lillo, F., Gallegati, M. & Mantegna, R.N. (2008). Cluster analysis for portfolio optimization. *Journal of Economic Dynamics and Control*, Vol.32, 235-258, ISSN: 0165-1889

Wang, C., Dou, C.X., Li, X.B. & Jia, Q.Q. (2007). A WAMS/PMU-based fault location technique. *Electric Power Systems Research*, Vol. 77, 936-945, ISSN: 0378-7796

Xue, Y.S. (2002). Interactions between power market stability and power system stability. *Automation of Electric Power Systems*, Vol. 26, 1-6, ISSN: 1000-1026.

Ye, L. (2003). Study on sustainable development strategy of electric power in China in 2020. *Electric Power*, Vol. 36, 1-7, ISSN: 1004-9649

Yuan, J.X. (2007). *Wide Area Protection and Emergency Control to Prevent Large Scale Blackout*, China Electric Power Press, ISBN : 978-7-508-35182-7, Beijing

Zhang, J.F., Wang, Z.P. & Zhang, Y.G. (2010). A new substation configuration algorithm based on the graphic characteristic of the main electrical connection. *Proc. 2010 5th International Conference on Critial Infrastructure(CRIS 2010)*

Zhang, J.F., Wang, Z.P., Zhang, Y.G. & Ma, J. (2010). A novel method of substation configuration based on the virtual impedance. *Proc. 2010 Asia-Pacific Power and Energy Engineering Conf. (APPEEC 2010)*

Zhao, W.X., Hopke, P.K. & Prather, K.A. (2008). Comparison of two cluster analysis methods using single particle mass spectra. *Atmospheric Environment*, Vol. 42, 881-892, ISSN: 1352-2310

Zhang, Y.G. & Wang, C.J. (2007). Multiformity of inherent randomicity and visitation density in n-symbolic dynamics. *Chaos, Solitons and Fractals*, Vol. 33, 685-694, ISSN: 0960-0779

Zhang, Y.G., Wang, C.J. & Zhou, Z. (2006). Inherent randomicity in 4-symbolic dynamics. *Chaos, Solitons and Fractals*, Vol. 28, 236-243, ISSN: 0960-0779

Zhang, Y.G. & Wang, Z.P. (2008). Knot theory based on the minimal braid in Lorenz system. *International Journal of Theoretical Physics*, Vol. 47, 873-880, ISSN: 0020-7748

Zhang, Y.G., Xu, Y. & Wang, Z.P. (2010). Dynamical randomicity and predictive analysis in cubic chaotic system. *Nonlinear Dynamics*, Vol.61, 241-249, ISSN: 0924-090X

Zhu, Y.L., Sidhu, T.S., Yang, M.Y. & Huo, L.M. (2002). An AI-based automatic power network topology processor. *Electric Power Systems Research*, Vol. 61, 57-65, ISSN: 0378-7796

Permissions

The contributors of this book come from diverse backgrounds, making this book a truly international effort. This book will bring forth new frontiers with its revolutionizing research information and detailed analysis of the nascent developments around the world.

We would like to thank Yagang Zhang, for lending his expertise to make the book truly unique. He has played a crucial role in the development of this book. Without his invaluable contribution this book wouldn't have been possible. He has made vital efforts to compile up to date information on the varied aspects of this subject to make this book a valuable addition to the collection of many professionals and students.

This book was conceptualized with the vision of imparting up-to-date information and advanced data in this field. To ensure the same, a matchless editorial board was set up. Every individual on the board went through rigorous rounds of assessment to prove their worth. After which they invested a large part of their time researching and compiling the most relevant data for our readers. Conferences and sessions were held from time to time between the editorial board and the contributing authors to present the data in the most comprehensible form. The editorial team has worked tirelessly to provide valuable and valid information to help people across the globe.

Every chapter published in this book has been scrutinized by our experts. Their significance has been extensively debated. The topics covered herein carry significant findings which will fuel the growth of the discipline. They may even be implemented as practical applications or may be referred to as a beginning point for another development. Chapters in this book were first published by InTech; hereby published with permission under the Creative Commons Attribution License or equivalent.

The editorial board has been involved in producing this book since its inception. They have spent rigorous hours researching and exploring the diverse topics which have resulted in the successful publishing of this book. They have passed on their knowledge of decades through this book. To expedite this challenging task, the publisher supported the team at every step. A small team of assistant editors was also appointed to further simplify the editing procedure and attain best results for the readers.

Our editorial team has been hand-picked from every corner of the world. Their multi-ethnicity adds dynamic inputs to the discussions which result in innovative outcomes. These outcomes are then further discussed with the researchers and contributors who give their valuable feedback and opinion regarding the same. The feedback is then collaborated with the researches and they are edited in a comprehensive manner to aid the understanding of the subject.

Apart from the editorial board, the designing team has also invested a significant amount of their time in understanding the subject and creating the most relevant covers. They scrutinized every image to scout for the most suitable representation of the subject and create an appropriate cover for the book.

The publishing team has been involved in this book since its early stages. They were actively engaged in every process, be it collecting the data, connecting with the contributors or procuring relevant information. The team has been an ardent support to the editorial, designing and production team. Their endless efforts to recruit the best for this project, has resulted in the accomplishment of this book. They are a veteran in the field of academics and their pool of knowledge is as vast as their experience in printing. Their expertise and guidance has proved useful at every step. Their uncompromising quality standards have made this book an exceptional effort. Their encouragement from time to time has been an inspiration for everyone.

The publisher and the editorial board hope that this book will prove to be a valuable piece of knowledge for researchers, students, practitioners and scholars across the globe.

List of Contributors

Khmaies Ouahada and Hendrik C. Ferreira
Department of Electrical and Electronic Engineering Science, University of Johannesburg, Auckland Park, 2006, South Africa

Bozydar Dubalski, Slawomir Bujnowski, Damian Ledzinski, Antoni Zabludowski and Piotr Kiedrowski
University of Technology and Life Sciences, Bydgoszcz, Poland

Yulia Kempner
Holon Institute of Technology, Israel

Vadim E. Levit
Ariel University Center of Samaria, Israel

Zh.G. Nikoghosyan
Institute for Informatics and Automation Problems, National Academy of Sciences, Armenia

Tetsuya Yoshida
Graduate School of Information Science and Technology, Hokkaido University, Japan

Angel M. Nuñez, Lucas Lacasa, Jose Patricio Gomez and Bartolo Luque
Universidad Politécnica de Madrid, Spain

Qi Xuan and Li Yu
Zhejiang University of Technology, China

Fang Du
Johns Hopkins University, USA

Tie-Jun Wu
Zhejiang University, China

Ali Hamlili
ENSIAS, Mohamed V-Souissi University, Rabat, Morocco

Ping Li
Department of Mathematics, West Virginia University, Morgantown, WV, USA

Guizhen Liu
School of Mathematics and System Science, Shandong University, Jinan, China

Csaba Szabó and Branislav Sobota
Technical University of Košice, Slovak Republic

G. Georgiadis, M. Papatriantafilou, L. A. Tuan and L. Bertling
Chalmers University of Technology, Sweden

P. H. Nguyen and W. L. Kling
Eindhoven University of Technology, The Netherlands

Georgi A. Nenov
Higher School of Transport "T. Kableshkov", Sofia, Bulgaria

Yagang Zhang, Zengping Wang and Jinfang Zhang
State Key Laboratory of Alternate Electrical Power System with Renewable Energy Sources (North China Electric Power University), China